Modern Chemistry: A Comprehensive Approach

Modern Chemistry: A Comprehensive Approach

Edited by
Bruce Horak

Willford Press
www.willfordpress.com

Published by Willford Press,
118-35 Queens Blvd., Suite 400,
Forest Hills, NY 11375, USA

ISBN: 978-1-68285-452-5

Cataloging-in-Publication Data

Modern chemistry : a comprehensive approach / edited by Bruce Horak.
p. cm.
Includes bibliographical references and index.
ISBN 978-1-68285-452-5
1. Chemistry. I. Horak, Bruce.
QD31.3 .M63 2018
540--dc23

For information on all Willford Press publications
visit our website at www.willfordpress.com

WILLFORD PRESS

Printed in China.

Contents

Preface

This book was inspired by the evolution of our times; to answer the curiosity of inquisitive minds. Many developments have occurred across the globe in the recent past which has transformed the progress in the field.

Chemistry is the science which studies the properties and composition of organic and inorganic matter. The subject can be divided into several branches like analytical chemistry, biochemistry, neurochemistry, theoretical chemistry and nuclear chemistry. This book is a compilation of chapters that discuss the most vital concepts and emerging trends in the field of modern chemistry. Coherent flow of topics, student-friendly language and extensive use of examples make this book an invaluable source of knowledge.

This book was developed from a mere concept to drafts to chapters and finally compiled together as a complete text to benefit the readers across all nations. To ensure the quality of the content we instilled two significant steps in our procedure. The first was to appoint an editorial team that would verify the data and statistics provided in the book and also select the most appropriate and valuable contributions from the plentiful contributions we received from authors worldwide. The next step was to appoint an expert of the topic as the Editor-in-Chief, who would head the project and finally make the necessary amendments and modifications to make the text reader-friendly. I was then commissioned to examine all the material to present the topics in the most comprehensible and productive format.

I would like to take this opportunity to thank all the contributing authors who were supportive enough to contribute their time and knowledge to this project. I also wish to convey my regards to my family who have been extremely supportive during the entire project.

Editor

1

The Categorization and Structural Prediction of Transition Metal Carbonyl Clusters Using 14n Series Numerical Matrix

Enos Masheija Rwantale Kiremire

Correspondence: Enos Masheija Rwantale Kiremire, Department of Chemistry and Biochemistry,University of Namibia, Private Bag 13301, Windhoek. E-mail: kiremire15@yahoo.com

Abstract

A matrix table of valence electron content of carbonyl clusters has been created using the 14n-based series. The numbers so generated form an array of series which conform precisely with valence electron contents of carbonyl clusters. The renowned 18 electron rule is a special case of 14n+4 series. Similarly, the 16 electron rule is another special case of the 14n+2 series. Categorization of the carbonyl clusters using the matrix table of series has been demonstrated. The table is so organized that clusters numerically represented can easily be compared and analyzed. The numbers that are diagonally arranged from right to left represent capping series. The row from right to left represents a decrease in valence electron content with increase in cluster linkages. The variation of cluster shapes of constant number of skeletal elements especially four or more may be monitored or compared with the variation with the valence electron content.

Keywords: valence electron content, carbonyl clusters, isolobal series, 18 and 16 electron rule, valence electron matrix table, shapes of carbonyl clusters, capping and decapping series, diagonal relationship, fragments

1. Introduction

The borane and transition metal carbonyl clusters have been of great interest to chemists for considerable long time due to the special geometries, unusual bonding framework and promising wide range of industrial applications (Stock, 1933; Kobayashi, 2007). The Wade-Mingos rules have been extremely helpful in deducing the cluster geometries (Mingos, 1984;Wade, 1976; Jemmis, 1984, 2001,2002,2003,2005,2006,2008;King, 2002; Welch, 2013). Furthermore, it has been found that carbonyl clusters and some main group clusters can be categorized using 4n and 14n series (Kiremire, 2014, 2015). Since the 14n series (14n±q) run parallel to 4n series (4n±q), the 4n series can be used instead of the 14n series. This is because it is easier to use 4n rather than 14n series. We have also demonstrated that the 14n and 4n series can be used to categorize and predict shapes of clusters of carbonyls, boranes and heteroboranes (Kiremire, 2014,2015). In this paper, a numerical sequences based on 14n series that give precise valence contents of clusters and their arrangements into series is presented. In this way, using the valence content of a cluster formula, the cluster can readily be categorized and its possible geometry predicted.

2. Results and Discussion

2.1 Creation of a Valence Electron Content Table for Carbonyl Clusters

A matrix of numbers has been created using the 14n-based series. The series have been chosen from the range S = 14n-12 to 14n+12 and for n = 1 to 30. Consider the first row of the matrix, when S = 14n-12, and n = 1, S = 14(1) -12 = 2. For S = 14n-10, when n =1, S = 14(1)-10 = 4, and S = 14n-8 =14(1) -8 = 6. The rest of S values are, 8, 10, 12,14,16,18, 20,22, 24 and 26. The S values are similarly obtained for the second row when n =2 and for other rows. These results are shown in Table 1.

2.2 The Extraction of Carbonyl Clusters from the Table

We can use Table 1 to extract carbonyl clusters as needed. Furthermore, it is quite interesting to discover that the table gives precise valence content of the cluster skeletal fragments including the stable mono-skeletal carbonyl complexes that obey the famous 18-electron rule. For instance, when n = 1, and S = 14n+0 = 14. This means the valence content is 14 for a single skeletal element. Suitable transition metal fragments are [$Os(CO)_3$], [$ReH(CO)_3$] and [$RhH(CO)_2$]. For n = 1 and V = 18, suitable fragments could be $Ni(CO)_4$, $Fe(CO)_5$, $Cr(CO)_6$, $HCo(CO)_4$, $HMn(CO)_5$, and $H_2Fe(CO)_4$. Clearly, the widely used and long-standing 18-electron rule of mono-skeletal element clusters is part and parcel of the entire cluster series universe of the transition metal carbonyl complexes. Let us take another example of n = 6 and S =

86. This refers to a valence electron content of 86 for 6 skeletal elements. When we look at the table, this corresponds to series S = 14n+2. The cluster formula corresponding to this will be given by F = 14n+2 = $[Os(CO)_3](6)$+ CO = $Os_6(CO)_{18}$+CO = $Os_6(CO)_{19}$. However, we know that the cluster exists as $Os_6(CO)_{18}^{2-}$ and has an octahedral shape (O_h). The special number [14] represents a vital building block for carbonyl series. Other transition metal carbonyl fragments of one skeletal element with 14 valence electron content include $RhH(CO)_2$, and $ReH(CO)_3$. We can construct carbonyl clusters of these fragments that correspond to S = V = 86 and n = 6 where S means series and V valence electron content. Let us proceed with rhodium, F = 14n+2 = $[Rh(H)(CO)_2](6)$+CO = $Rh_6(H)_6(CO)_{12}$+CO = $Rh_6(CO)_3(CO)_{12}$+CO = $Rh_6(CO)_{16}$. This rhodium cluster is well known and is the first one to be found to have an octahedral symmetry (Cotton and Wilkinson, 1980). In these simple operations in terms of valence electrons, [14] in the formula represents the series building block for transition metal carbonyl clusters and for the main group elements the building block is [4]. But if we use the 4n series for transition metal carbonyl clusters, [4] should be interpreted as [14] in order to generate cluster formulas. It can be seen that the [2H] = [CO] in terms of valence electron contribution to a cluster fragment or formula. Also [2] in F =14n+2 can be replaced by [CO] or a negative charge of (-2) in terms of valence electron content. That is why the osmium cluster $Os_6(CO)_{19}$ may be written as $Os_6(CO)_{18}^{2-}$ in terms of the series. The rhenium cluster can be derived in the same manner, F = 14n+2 = [Re(H)(CO)3](6)+CO = $Re_6(H)_6(CO)_{18}$+CO = $Re_6(CO)_3(CO)_{19}$ = $Re_6(CO)_{22}$. The following known octahedral rhenium carbonyls , $Re_6(C)(CO)_{19}^{2-}$, $Re_6(C)(CO)_{19}(H)^-$, $Re_6(C)(CO)_{18}(H)^{3-}$, $Re_6(C)(CO)_{18}(H)_2^{2-}$,and $Re_6(H)_7(CO)_{18}^-$, all are electronically equivalent in terms of valence electron content of 86. Let us determine an osmium cluster of the series S = 14n+4 when n = 1. For n = 1 means we will be focusing on one skeletal atom of a fragment. Such a fragment will have S = 14(1)+4 = 18 valence electrons surrounding it. Suppose we assume such fragment has an osmium atom, then the [14] building block will be [$Os(CO)_3$, V = 8+3x2 = 14]. Therefore the formula fragment of the series S = 14n+4 when n =1 will be given by F = 14n+4 = $[Os(CO)_3](1)$+2CO = $Os(CO)_5$. Consider another element say, chromium. The building block of [14] will have the corresponding fragment of [$Cr(CO)_4$, V = 6+4x2 = 14]. Hence the formula of the carbonyl complex corresponding to 14n+4 for one skeletal atom, n =1 will be given by F = $[Cr(CO)_4](1)$+2CO = $Cr(CO)_4$+2CO = $Cr(CO)_6$. Thus, a carbonyl cluster of a number of given fragments of a known transition metal can be derived in the same manner.

2.3 Categorization of carbonyl Clusters Using the 14n Series Matrix Table

The use of Table 1 to categorize a given cluster is straight forward. First of all, the valence content(S or V) of the cluster is calculated. Then the number (n) of the skeletal elements is noted. Finally you identify on the matrix where the value V corresponding to a given n is. The corresponding cluster series is read off on the top of the matrix table. A few examples will be used to illustrate this. Take $Os_3(CO)_{12}$ cluster, n = 3, V = 48. Moving along n = 3 axis and reaching 48, the vertical movement arrives at the series S = 14n+6. This means the cluster belongs to ARACHNO series. The clusters,$Os_4(CO)_{14}$; n = 4, V = 60,S = 14n+4, NIDO; $Os_5(CO)_{16}$, n = 5, V = 72, S = 14n+2, CLOSO;$Os_6(CO)_{18}$; n = 6, V = 84, S = 14n+0, mono-capped, C^1C[M-5] on trigonal bipyramid cluster;$Os_6(CO)_{18}^{2-}$ cluster, n = 6, V = 86, S = 14n+2, CLOSO;$Os_8(CO)_{22}^{2-}$, n = 8, V = 110, S = 14n-2, bi-capped cluster, C^2C[M-6] on an octahedral geometry;$Os_{10}(CO)_{26}^{2-}$, n = 10, V= 134, S = 14n-6, C^4C[M-6] is a tetra-capped octahedral cluster; and $Pd_{23}(CO)_{20}L_{10}$, n = 23, S = 14n-32, $C^{17}C$[M-6]-has 17 capped atoms and 6 nuclear cluster atoms possibly forming an octahedral shape. In the case of this palladium cluster, its valence content of 290 for n = 23 does not appear in Table 1. However its cluster series can be obtained by simple extrapolation. For n = 23, V= 310 when S = 14n – 12. Now when n = 23 and V = 290 along the n = 23 row, V undergoes a decrease of 310-290 = 20. Therefore S value has to decrease by the same amount. This means the required S = 14n-12-20 = S = 14n-32, and Cp = $C^{17}C$[M-6]. This means that the cluster has a closo octahedral complex surrounded by 17 atoms. A closo cluster refers to S = 14n+2 series. For [M-6] closo nuclear cluster, n = 6 and V = 86 as shown in Table 1. A collection of known carbonyl clusters categorized in this manner are given in Table 2. The matrix table was also utilized to categorize a number of ruthenium carbonyl clusters generated using UV Laser Desorption Mass Spectroscopy (Critchey, et al 1999). The results are shown in Tables 3 and 4.

2.4 Vertical Movement Along the Series Matrix

2.4.1 The Islands of Chemical Stability

The 18-Electron Rule-14n+4

The matrix table contains some important information. Each number in the matrix (see Table 1) represents one or more fragments some of which may be molecules or clusters. Let us consider, (1,18) set. This means one transition metal skeletal element surrounded by 18 electrons. This type of notation will be used in this paper. There are many transition metal elements which satisfy this condition. Among others, these include $Fe(CO)_5$(1,18), $Ni(CO)_4$(1,18), $Cr(C_6H_6)_2$(1,18), $Fe(C_5H_5)_2$(1,18), $Cr(CO)_6$(1,18), $Mo(CO)_6$(1,18),$W(CO)_6$(1,18), $Mn(CO)_6^+$(1,18),$V(CO)_6^-$(1,18), $(C_5H_5)_2FeH^-$(1,18),$(C_4H_4)Fe(CO)_3$(1,18),$(C_5H_5)Fe(CO)_2(Cl)$(1,18) and$(C_5H_5)Ni(NO)$(1,18). For one skeletal element surrounded by 18 electrons belongs to the first member of the S = 14n+4 series as shown in the matrix Table 1. Clearly this is the basis of the well known 18-electron rule which has been widely used in chemistry for a long time (Langmuir, 1921; Tolman, 1972; Miessler, et al, 2014).

Adding 14 to 18 produces the second element 32 written as (2,32). This means two skeletal elements surrounded by 32 valence electrons. This set also belongs to S = 14n+4series. Its value is given by S = 14(2)+4 = 32. Some of the fragments which fulfill this condition are $(C_5H_5)_2Rh(CO)_2$(2,32), $Re_2(C_5H_5)_2(CO)_4$(2,32), $Re_2H_2(CO)_8$(2,32) and $Os_2(CO)_8$(2,32). The other sets of numbers are produced in the same way by adding 14 to the previous number of the valence content.

The next set of cluster set for the S = 14n+4 series is (3, 46). The cluster examples include $Os_3H_2(CO)_{12}$, $Re_3H_3(CO)_{10}^{2-}$, and $Ru_3(CO)_{10}^{2-}$. Other sets of S = 14n+2 are (4, 60); $Os_4(CO)_{14}$, $Os_4H_4(CO)_{12}$, $Co_4(CO)_{12}$, $Rh_4(CO)_{12}$, $Ir_4(CO)_{12}$, $Re_4H_6(CO)_{12}^{2-}$,(5,74); $Os_5H_2(CO)_{16}$, $Os_5(C)(CO)_{14}^{2-}$, $Ru_5(C)(CO)_{15}$, and $Ru_5(N)(CO)_{14}^{-}$

2.4.2 Cluster linkage of k = 2n-½q for S = 14n+q

The cluster number k can be calculated from the series formula. For S = 14n+4, k =2n-2. Therefore, when n = 1, k = 2(1)-2 = 0; n= 2, k = 2(2)-2 = 2; n = 3, k = 2(3)-2 = 4; n = 4, k =2(4)-2 = 6; n = 5, k = 2(5)-2 = 10-2 = 8. The metal-metal skeletal linkages is related to geometrical skeletal structures for clusters of low nuclearity. For the 14n+4 series k = 0 for n = 0 and k = 2 for n = 2. In the case of k =2 for n =2 two metal atoms joined together by two linkages. Thus, there is a metal-metal bonding of bond order 2. That is, M=M, and hence the following complexes$(C_5H_5)_2Rh_2(CO)_2$, $Re_2(C_5H_5)_2(CO)_4$, $Re_2H_2(CO)_8$ and $Os_2(CO)_8$ are expected to have metal-metal double bond. For n =3, k = 4 implies that there 4 linkages within the 3 metal atoms. The possible linkages are of the skeletal elements M-2 to M-5 for 14n+4 series are shown in Figure 1.

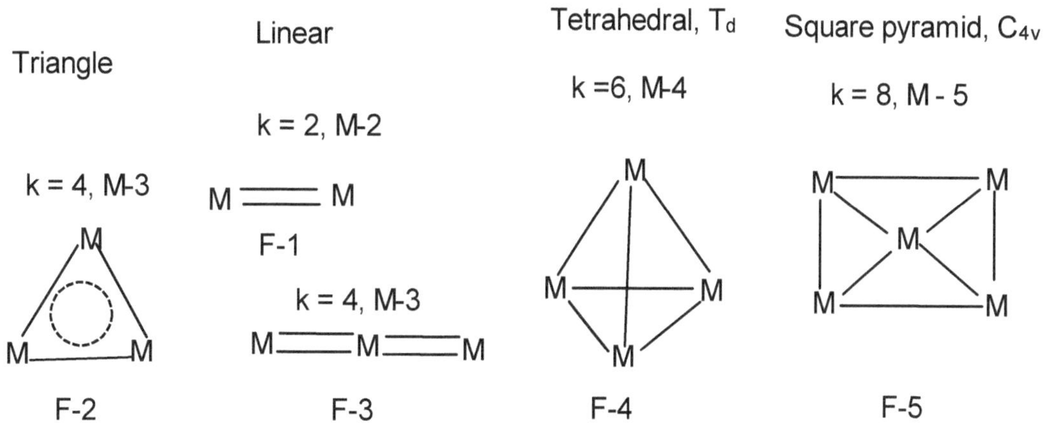

Figure 1. Sketches of Possible Linkages for M-2 to M-5 for S = 14n+4.

2.4.3 The 16-Electron Rule-14n+2

There is another interesting set of numbers,(1,16). This means a single skeletal element surrounded by 16 electrons. The osmium fragment $Os(CO)_4$ fulfills this condition. Some metal complexes include, $RhCl(PPh_3)_3$, $(C_5H_5)_2TiCl_2$, $IrCl(L)_2(CO)$ where, L = 2-elctron donor, $Ir(L)_2(CO)_2^{-}$, $(C_5H_5)_2Fe_2(CO)_4$, and $HPtCl(L)_3$. The (1,16) set of numbers in the matrix Table 1 is the first member of S = 14n+2 series. A number of (1,16) complexes play vital roles in in catalysis(Tolman, 1972). They include, $Ni(CO)_3$(1,16) derived from $Ni(CO)_4$(1,18), $RhCl(PPh_3)_3$(1,16), and $HCo(CO)_3$(1,16) derived from HCo(CO)4(1,18). Could this be the basis of the 16-electron rule? Many more mono-skeletal clusters that fulfill this condition could be found. These are some of the fragments or complexes for mono-skeletal that obey the 16 electron rule. Let us consider 2 skeletal elements of 14n+2 series. Some examples of the complexes that are members of 14n+2 represented by (2, 30) include $Mo_2(C_5H_5)_2(CO)_4$, and (3,44);$Os_3(CO)_{10}$, (4,58); $Os_4(CO)_{13}$, $Re_4H_5(CO)_{12}^{-}$;(5,72); $Os_5(CO)_{16}$, $Os_5(CO)_{15}^{2-}$, $PtRh_4(CO)_{12}^{2-}$,(6,86); $Os_6(CO)_{18}^{2-}$,$Co_6(CO)_{16}$, $Rh_6(CO)_{16}$, $Ir_6(CO)_{16}$, $Rh_6(C)(CO)_{15}^{2-}$,$Re_6(C)(CO)_{19}^{2-}$, (7,100); $Os_7H_2(CO)_{21}$, $Rh_7(N)(CO)_{15}^{2-}$ and $Co_7(N)(CO)_{15}^{2-}$.

2.4.4 Cluster linkage of k = 2n-½q for S = 14n+q

The cluster number k for the series 14n+2 are given below. For 14n+2, k = 2n-1 and hence for n = 1, k = 2(1)-1 = 1; and n= 2, k = 2(2)-1 = 3; n = 3, k = 2(3)-1 = 5; n = 4, k = 2(4)-1 = 7; n = 5, k = 2(5)-1 = 9, n = 6, k = 2(6)-1= 11 and n = 7, k = 2(7)-1 = 13. Let us take n = 2 and k = 3, n = 5, k = 9 and n = 6, k = 11 and sketch possible geometries for them. These are shown in Figure 2.

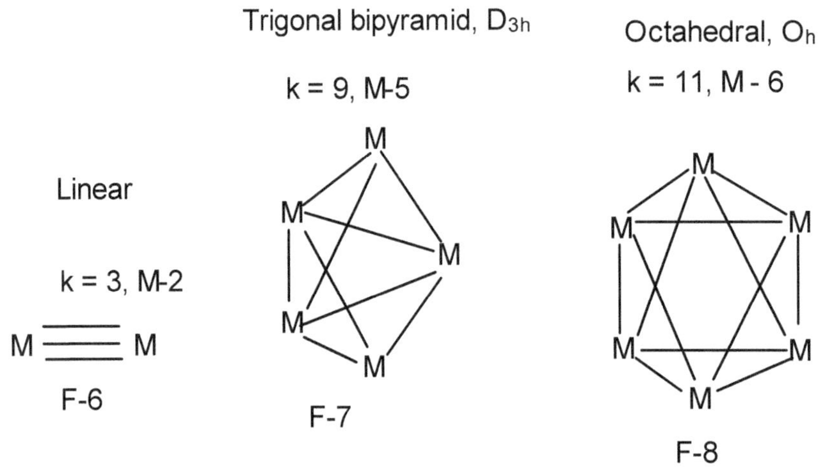

Figure 2. Sketches of Possible Linkages for M-2 , M-5, M-6 for S = 14n+2.

Let us consider a few examples from the arachno family S = 14n+6, k = 2n-3. The set (2,34) in which 2 skeletal atoms are surrounded by 34 valence electrons. Examples include $Mn_2(CO)_{10}$, and $Co_2(CO)_8$. For the case of the set (3,48); $M_3(CO)_{12}$, M = Fe, Ru and Os; (6, 90); $Co_6(C)(CO)_{15}^{2-}$. For n = 2, k = 2(2)-3 = 1; n = 3, k = 2(3)-3 = 3; and n = 6, k =2(6)-3 = 9. The shapes are for M-2 for k =1, is a single metal-metal bond and for M-3, k = 3 is a triangle and M-6, k = 9, the shape normally observed is trigonal prism. The sketches of the structures are shown in Figure 3.

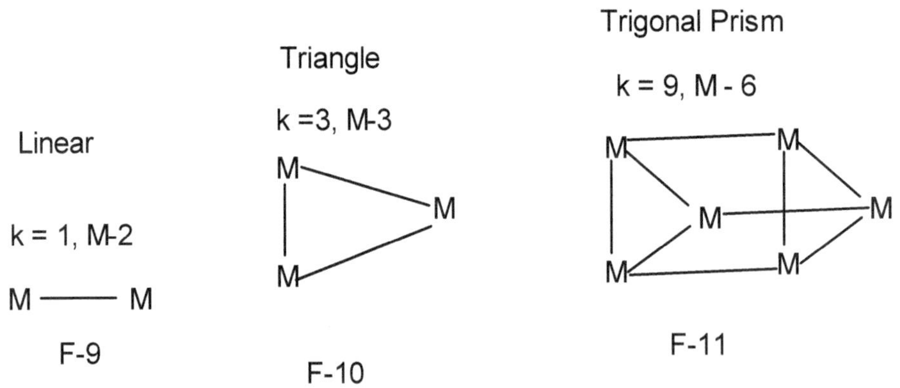

Figure 3. Sketches of Possible Linkages for M-2 , M-3, M-6 for S = 14n+6.

2.5 Correlation between Transition Metal Carbonyl Clusters with Main Group Clusters

Let us consider S = 14n+6 for n =2, and k =2n-3 = 1 The value of S = 14(2)+6 = 34. This in agreement with the valence content of $Mn_2(CO)_{10}$ and the presence of single metal-metal bond in the complexes. It has been shown that 14n-based series of transition metal carbonyl clusters and fragments run parallel to the 4n-based series of main group clusters and fragments. Hence, 14n +6 will correspond to 4n+6 of the main group elements. While the [14] electron content is utilized to form a backbone of transition metal carbonyl clusters, the [4] electron content can be used to construct the backbone of main group clusters. A natural suitable fragment is carbon [C] or boron hydrogen [BH]. In order to derive the formula main group cluster, corresponding to n = 2, we can apply F = 4n+6 = [C](2)+6H = C_2H_6. This means that $Mn_2(CO)_{10}$ and C_2H_6 are isolobal or in simple terms equivalent. Furthermore, this implies that the fragment CH_3 ⟺ $Mn(CO)_5$. Since $Co_2(CO)_8$ is also of the set (2,34), and belonging to S =14n+6, it is also isolobal to C_2H_6 and the fragment $Co(CO)_4$ ⟺ CH_3. Let us consider another case of 14n+2, k = 2n-1 for n =2, k = 3. The 14n+2 runs side by side with 4n+2. That is, 14n+2 ⟺ 4n+2. In order to find its equivalent hydrocarbon we must derive it from the series F = 4n+2 for n = 2, this means F = [C](2)+2H = C_2H_2. As we know well,C_2H_2 has a carbon-carbon triple bond, C≡C. Similarly, $Mo_2(C_5H_5)_2(CO)_4$ is expected to have a triple bond around which the supporting ligands congregate. Its possible structure is shown in Figure 4 and that of C_2H_2 is inserted in the background for comparison. It is readily deduced that CH fragment is isolobal to $(C_5H_5)Mo(CO)_2$ fragment.

The complex $(C_5H_5)_2Rh_2(CO)_2$ belongs to 14n+4 series with k = 2. The series 14n+4$\Leftrightarrow$ 4n+4. The hydrocarbon corresponding to this for n =2 is given by F = 4n+4 = [C](2)+4H = C_2H_4. For k = 2 for n = 2, means that both the rhodium complex and C_2H_4 have double bonds. Their shapes are sketched in Figure 5. The structure of related diborane is added comparison.

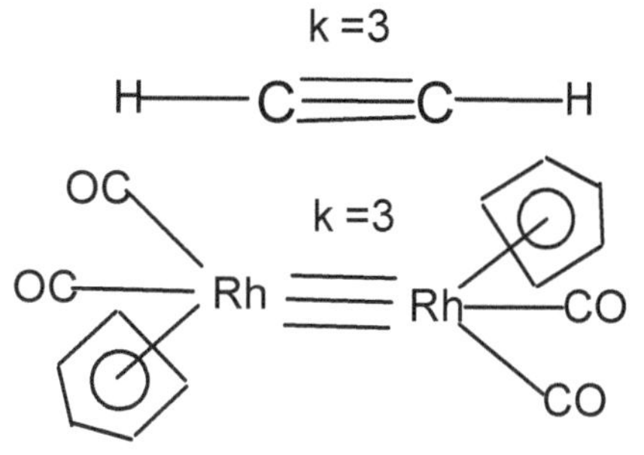

Figure 4. Possible shape of $Rh_2(C_5H_2)_2(CO)_4$

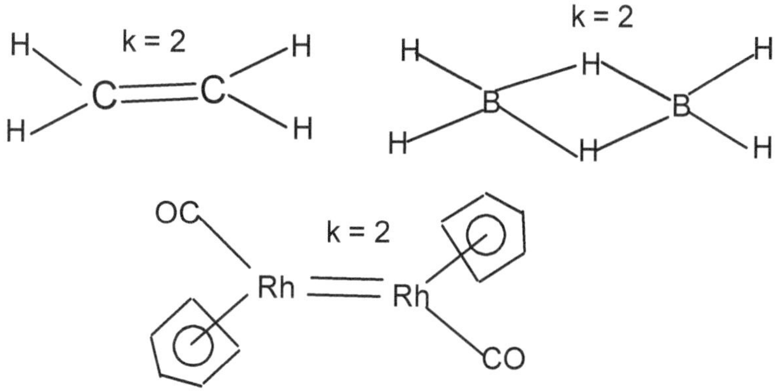

Figure 5. Possible shape of $Rh_2(C_5H_2)_2(CO)_2$

We can also derive a borane equivalent compound in the same way we did for C_2H_4 molecule. Thus, F = 4n+4 for n = 2, F = [BH](2)+ 4H = B_2H_2+4H = B_2H_6.

Let us consider the complex $Co_6(C)(CO)_{15}^{2-}$. We can code the complex as (6, 90). This means that the 6 skeletal elements are surrounded by 90 valence electrons. It also belongs to 14n+6 series with k = 2n-3 = 2(6)-3 = 9. From the series, we can deduce that the parallel main group series is S = 4n+6. For n = 6, we can also derive the corresponding hydrocarbon formula as F = 4n+6 = [C](6)+6H = C_6H_6. The carbon atom with a valence content of 4 can be used as a building block replacing [4] in the formula and replacing n by 6 and the last 6 in the formula substituting it with 6 H atoms which are assumed to supply six electrons. If we were considering carbonyl clusters, 6 could be replaced by 3 CO molecules. In summary $Co_6(C)(CO)_{15}^{2-}$ $\Leftrightarrow$$C_6H_6$. For both clusters, k = 2n-3 = 9. The shape of $Co_6(C)(CO)_{15}^{2-}$ is trgonal prism (see F-9 Figure 3). One of the isomers of C_6H_6 is sketched in Figure 6, F-12 for benzene and F-13 for prismane. The structure of F-13 is similar to that of F-11. However, in case of C_6H_6, many more isomers can be generated which can satisfy the requirement of k =9. Some selected known fragments and complexes for CLOSO, NIDO and ARACHNO series respectively are indicated in trees T-1, T-2 and T-3.

2.6 Diagonal Movement along the Cluster Matrix

When Table 1 is examined carefully, some interesting information is discovered along the diagonal. Take the case of (6,86) cluster. This set means that 6 skeletal elements are surrounded by 86 valence electrons. This electron environment usually allows the cluster to adopt an octahedral geometry (O_h). Furthermore, the cluster is a member of 14n+2 CLOSO series. A horizontal movement to the left gives (6,84). This corresponds to a loss of a CO ligand or 2 electrons from (6,86). The set (6,84) is a beginning of capping series for CLOSO 14n+2 of (5,72). However, a diagonal movement by one box in Table 1 gives (7,98). This is the first member of the capping series of (6,86)-an octahedral closo cluster, which is a member 14n+2 series. THIS MEANS THAT THE MOVEMENT ACROSS THE DIAGONAL GIVES CLOSO CAPPING SERIES. Hence, the capping sets for octahedral (6,86) are (7,98),(8,110), (9,122), (10,134), (11,146), (12,158) and (13,170). Selected carbonyl clusters that are in agreement with these numbers are given in capping tree T-4. The set (7,98) represents the first capping of the octahedral unit with a symbol, $C^1C[M-6]$ of the series

S = 14n+0. The next set (8,110) represents the second capping with symbol C^2C[M-6] of the series S = 14n-2. The other sets represent higher series. The capping series for trigonal bipyramid of closo series 14n+2 represented by (5,72) are (6,84), (7,76), (8,108), (9,120), ((10, 132), 11, 144)and (12, 156). Some selected clusters that are in agreement with this are shown in the tree T-5. Sketches illustrating the capping on an octahedral closo cluster are shown in Figure 6. Let us take a last example on capping series. Starting with (1,18) which represents the 18 electron rule, we move down the diagonal from right to left. The next set of numbers is (2,30). This means 2 skeletal atoms surrounded by 30 valence electrons for transition metal carbonyl clusters. This set belongs to 14n+2 closo series. The corresponding main group series will be 4n+2 series. A hydrocarbon of two skeletal atoms that belongs to the series is given by F = 4n+2 = [C](2)+2H = C_2H_2. This molecule has a triple bond(C≡C). Similarly any 2 skeletal atoms of transition metals of the set(2,30) are expected to have a metal-metal triple bond, M≡M. Since (2,30) belongs to 14n+2 series, the next capping set will belong to 14n+0. The capping involves adding [12] electrons which correspond to [12] = [14-2] = 14n-2 series and k = 2n-(-1)= 2n+1. Since n = 1, then k = 2(1)+1 = 3. Thus, capping involves a step-wise increase of k value by 3. The entire capping set in Table 1 is (1,18)→(2,30) →(3,42) →(4,52) →(5,66) →(6,78) →(7,90) →(8,102) →(9,114). A complex that is one of the representatives of the closo series 14n+2 is $Mo_2(Cp)_2(CO)_4$. It has a triple bond Mo≡Mo. The last capping of (2,30) from Table 1 is (9, 114). This set belongs to S = 14n-12 series. Its value V = 14(9)-12 = 126-12 =114. Thus, the series formulation is correct. Let us generate a palladium carbonyl cluster that corresponds to (9,114). This is readily obtained from F = 14n-12 = $[Pd(CO)_2](9)$-6CO = $Pd_9(CO)_{18}$-6CO =$Pd_9(CO)_{12}$. The capping symbol for this is Cp = $C^1+C^6 = C^7C$[M-2]. The symbol makes sense as [M-2] represents (2,30) which we have already established as being a closo system of 14n+2 series. Selected capping clusters based on trigonal bipyramid and octahedral closo clusters are shown in trees T-4 and T-5.

2.7 The Movement Across the Row in the Matrix Table

Let us consider the first row of numbers. The set (1,18) means having one skeletal element surrounded by 18 valence electrons. This is the basis of the famous 18-electron rule. There many complexes that obey this rule. These include, $Ni(CO)_4$, $Fe(CO)_5$, $Cr(CO)_6$, $Fe(C_5H_5)_2$, and $(C_4H_4)Fe(CO)_3$. These complexes obey the S = 14n+4 series. These belong to the first member of the series S = 14n+4 (1,18).The next first set of the series S = 14n+2 is (1,16). There are also complexes that obey the 16 electron rule. They include complexes such as $RhClL_3$, L = PPh_3, (1,16)Wilkinson catalyst, $PtCl_3(C_2H_4)^-$(1,16), $(Cp)_2TiCl_2$ (1,16),Cp = C_5H_5,$HRuClP_4$ (1,16), P =$P(OPh)_3$ and $IrClL_2(CO)$(1,16). Although such complexes were and are considered to be some of the exceptions to the 18-electron rule, it appears that are in fact the mono-skeletal complexes of the S = 14n +2 series. The complexes that satisfy (1,16) condition may be considered to obey the 16-Electron rule. The complexes $(Cp)_2Zr(R)^+$,(1,14),$(Cp)_2Zr(R)(C_2H_4)^+$(1,14), HgR_2(1,14), and $MeAu(PPh_3)$(1,14) could be regarded as the complexes that begin the S= 14n+0 series in the same way as the fragment $Os(CO)_3$. Let us focus on the cluster members of higher nuclearity n= 6. Consider the set (6,90) of 14n+6 and k = 9. This type of carbonyl cluster has 6 skeletal elements and surrounded by 90 valence electrons. The carbonyl cluster complexes that fulfill this condition take the shape of a trigonal prism (see Figure 3, F-11). When the valence electron content decreases to 86 keeping the same number of skeletal elements, the shape is transformed from trigonal prism to octahedral(6,86) shape (O_h),(see Figure, F-8). The set (6,86) is diagonal to the set (5,74). Hence, (6,86) may be considered to be a capping set of (5,74), S = 14n+4, NIDO and k = 8. The set (5,74) tends to adopt a square pyramid shape while the (6,86) of 14n+2, CLOSO, k =11 normally takes up an octahedral shape. The set (6,84) of S = 14n+0 lies on the diagonal of (5,72) a CLOSO system of S = 14n+2, k = 9 with a trigonal bipyramid shape(see Figure 2, F-7). The (6,84) is a mono-capped cluster based on trigonal bipyramid(see Figure 7). Although it is regarded as a mono-capped trigonal bipyramid, it lies on the diagonal of very important clusters: (3,48), (4,60), (5,72), (6,84), (7,96), (8, 108) and so on. If the sets (4,60), (5,72) and (6,84) are examined closely, the set (4,60) represents a tetrahedral geometry. According to the series, the capping begins with 14n+0 which is in line with (6,84). This means that one skeletal atom is considered to be capping on 5 closo atoms [M-5] which is a trigonal bipyramid (5,72). But (5,72) can be regarded as mono-capped tetrahedral (4,60) although (4,60) is not a member of 14+0 series.The capping atom is associated with 12 capping valence electrons. In other words, (6,84) may be viewed as mono-capped trigonal bipyramid or bicapped tetrahedral or tricapped (3,48)-trigonal planar. The numbers after (5,72) should be regarded as capping series of trigonal bipyramid. Similarly the numbers after (6,90) trigonal prism diagonal such as (7,102), (8, 114), (9, 126), (10,138), (11, 150), and (12, 162) may be regarded as capping series of the trigonal prism. Let us consider the following structural changes (6,90)→(6,86)→(5,74)→(5,72). The set (6,90) belongs to 14n+6 arachno series and k = 2n-3 = 2(6)-3 = 9. But the set (6, 86) belongs to 14+2 closo series, k = 2n-1 = 2(6)-1 = 11, and (5,74) belongs to 14n+4 with k = 2n-2 = 2(5)-2 = 8. The last set (5,72) belongs to 14n+2 closo series with k value of 2n-1 = 2(5)-1 = 9. The (6,90) has a trigonal prism or a raft shape. The (6, 86) closo cluster has an octahedral shape. Hence, the transformation (6,90)→(6,86) is structural change from trigonal prism or raft to an octahedral geometry. This involves the decrease of valence electron content by 4 while the number of skeletal atoms remains constant. The change from(6,86)→(5,74) involves the change of geometry from octahedral to square pyramid while the change from (5,74)→(5,72) involves transformation of geometry from square pyramid to trigonal bipyramid. These geometrical transformations are sketched in Figure 8. Correlation of valence electron content with shapes of clusters has been of intense research by chemists for some time (Hoffmann, et all, 1986; Bruce, 1986). Let us look at the simple parallel series of main group elements. We can start with (2,34) →(2,32) →(2,30) →2,28) →(2,26) →(2,24). Since 14n±q⇔4n±q, all we need to do is to remove 10n, in this case (2x10 =20) from the valence content. If we do that the parallel 4n series will be (2,14) →(2,12) →(2,10) →(2,8) →(2,6) →(2,4). The set (2,14) belongs to 4n+6 with k = 2n-3. A hydrocarbon the matches this is F = 4n+6 = [C](2)+6H = C_2H_6. The fluorine molecule F_2 (F—F) also satisfies this requirement. This molecule has k value = 2(2)-3 = 1. This means it has a single carbon- carbon bond $H_3C—CH_3$. The set (2,12) corresponds to 4n+4 with a hydrocarbon F = [C](2) +4H = C_2H_4. The molecule, C_2H_4 has a double bond (C=C). The oxygen molecule has a double bond (O=O) and 12 valence electrons. The set (2,10) corresponds to S = 4n+2 with k = 2n-1. A suitable hydrocarbon is given by F= 4n+2 = [C](2)+2H = C_2H_2. This has a triple bond (HC≡CH) with 10 valence electrons. Other diatomic species that fulfill this condition include C≡O, $N≡O^+$, and $C≡N^-$. The set (2, 8) belongs to S = 4n+0 series with k = 2n = 2(2) = 4. The chemical species that satisfy this condition are C_2, CN^+, BN, and CB^-. These species have recently (Shaik, et al, 2012) been shown to have a quadruple bond M≣M. It is quite clear that keeping the number of skeletal elements constant and decreasing the valence electron content results in the increase of the number of skeletal linkages. In the above examples we moved through the following phases, M-M→M=M→M≡M→M≣M by simply successively decreasing the valence electron content by 2.

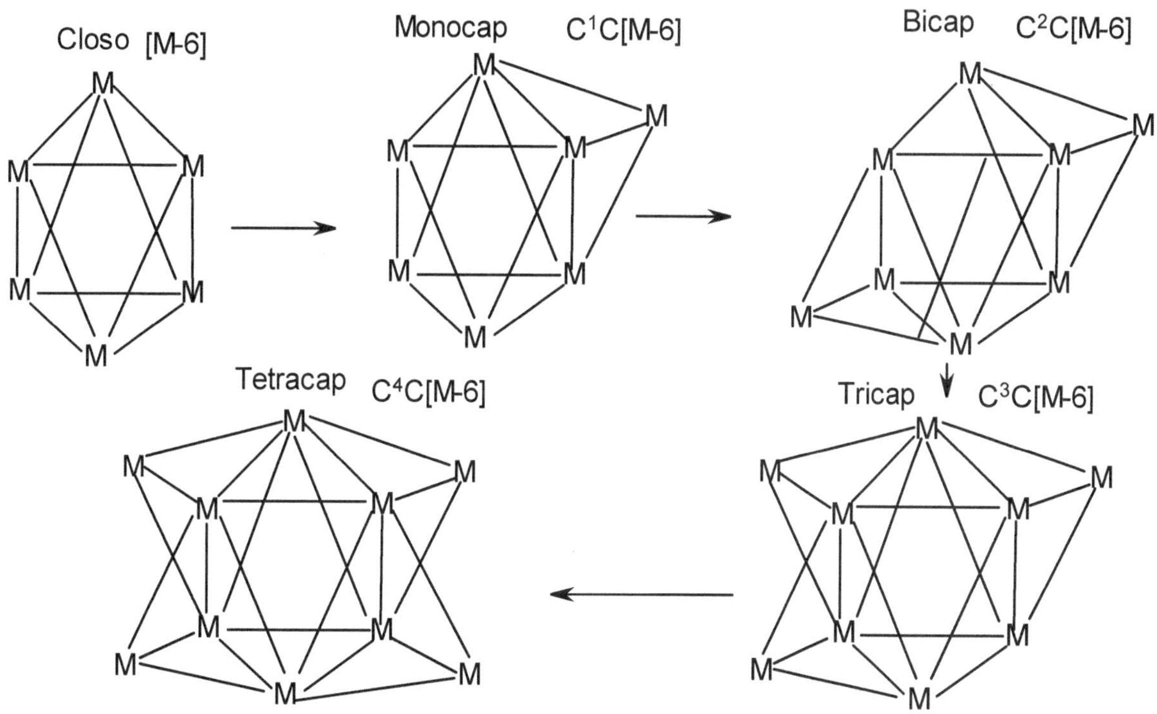

Figure 6. Sketches of some of the capping series based on Octahedral cluster

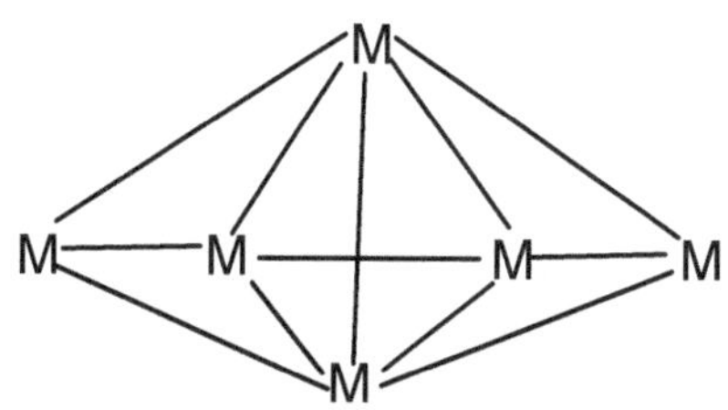

Figure 7. Sketch of skeletal elements of (6,84)

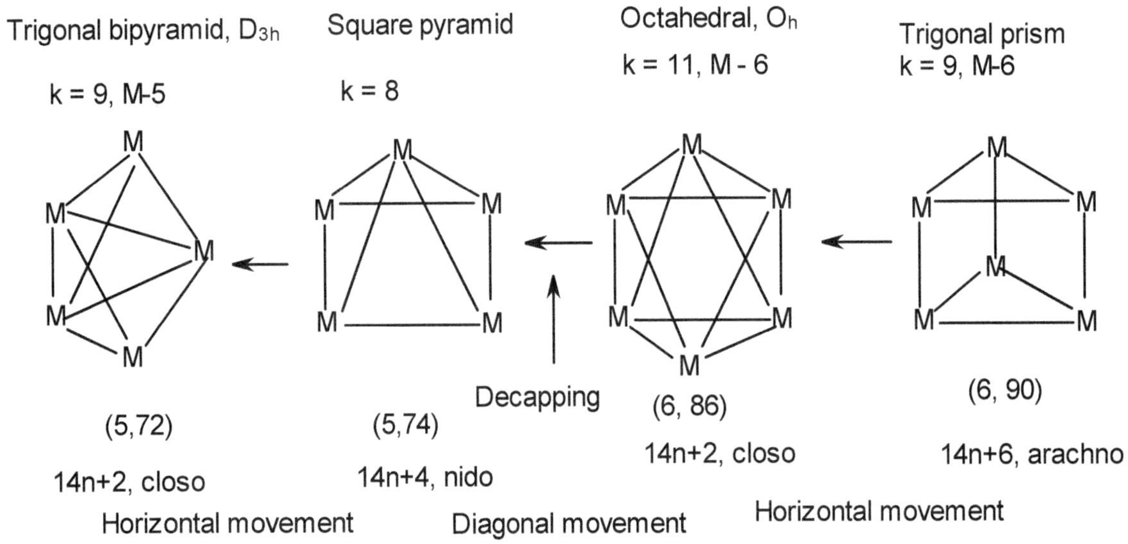

Figure 8. Sketches showing an example of geometrical changes along the row (horizontal), and along the diagonal

Table 1. The Valence Content of Carbonyl Clusters in Numerical Matrix

14n-12	14n-10	14n-8	14n-6	14n-4	14n-2	14n+0	14n+2	14n+4	14n+6	14n+8	14n+10	14n+12	n
2	4	6	8	10	12	14	16	18	20	22	24	26	1
16	18	20	22	24	26	28	30	32	34	36	38	40	2
30	32	34	36	38	40	42	44	46	48	50	52	54	3
44	46	48	50	52	54	56	58	60	62	64	66	68	4
58	60	62	64	66	68	70	72	74	76	78	80	82	5
72	74	76	78	80	82	84	86	88	90	92	94	96	6
86	88	90	92	94	96	98	100	102	104	106	108	110	7
100	102	104	106	108	110	112	114	116	118	120	122	124	8
114	116	118	120	122	124	126	128	130	132	134	136	138	9
128	130	132	134	136	138	140	142	144	146	148	150	152	10
142	144	146	148	150	152	154	156	158	160	162	164	166	11
156	158	160	162	164	166	168	170	172	174	176	178	180	12
170	172	174	176	178	180	182	184	186	188	190	192	194	13
184	186	188	190	192	194	196	198	200	202	204	206	208	14
198	200	202	204	206	208	210	212	214	216	218	220	222	15
212	214	216	218	220	222	224	226	228	230	232	234	236	16
226	228	230	232	234	236	238	240	242	244	246	248	250	17
240	242	244	246	248	250	252	254	256	258	260	262	264	18
254	256	258	260	262	264	266	268	270	272	274	276	278	19
268	270	272	274	276	278	280	282	284	286	288	290	292	20
282	284	286	288	290	292	294	296	298	300	302	304	306	21
296	298	300	302	304	306	308	310	312	314	316	318	320	22
310	312	314	316	318	320	322	324	326	328	330	332	334	23
324	326	328	330	332	334	336	338	340	342	344	346	348	24
338	340	342	344	346	348	350	352	354	356	358	360	362	25
352	354	356	358	360	362	364	366	368	370	372	374	376	26
366	368	370	372	374	376	378	380	382	384	386	388	390	27
380	382	384	386	388	390	392	394	396	398	400	402	404	28
394	396	398	400	402	404	406	408	410	412	414	416	418	29
408	410	412	414	416	418	420	422	424	426	428	430	432	30

Table 2. Known Carbonyl Clusters Categorized Using the Cluster Matrix Table

CLUSTER	VALENCE CONTENT	n	CODE	SERIES	CAPPING SYMBOL
$Re_2(CO)_{10}$	34	2	(2,34)	4n+6	ARACHNO
$Re_3H_4(CO)_{12}$	46	3	(3,46)	4n+4	NIDO
$Os_3(CO)_{12}$	48	3	(3,48)	4n+6	ARACHNO
$Ir_4(CO)_{11}L$	60	4	(4,60)	4n+4	NIDO
$Re_4H_6(CO)_{12}^{2-}$	60	4	(4,60)	4n+4	NIDO
$Rh_4(CO)_{12}$	60	4	(4,60)	4n+4	NIDO
$Re_5(C)(H)(CO)_{16}^{2-}$	74	5	(5,74)	4n+4	NIDO
$Rh_5(CO)_{15}^{-}$	76	5	(5,76)	4n+6	ARACHNO
$Os_5(CO)_{19}$	78	5	(5,78)	4n+8	HYPHO
$Rh_6(CO)_{16}$	86	6	(6,86)	4n+2	CLOSO
$Os_6(CO)_{18}H_2$	86	6	(6,86)	4n+2	CLOSO
$Re_6H_7(CO)_{18}^{-}$	86	6	(6,86)	4n+2	CLOSO
$Re_6(C)(CO)_{19}^{2-}$	86	6	(6,86)	4n+2	CLOSO
$Re_6H_8(CO)_{18}^{2-}$	88	6	(6,88)	4n+4	NIDO
$Os_6(CO)_{21}$	90	6	(6,90)	4n+6	ARACHNO
$Os_7(CO)_{21}$	98	7	(7,98)	4n+0	$C^{1}C$[M-6]
$Re_7H(C)(CO)_{21}^{3-}$	98	7	(7,,98)	4n+0	$C^{1}C$[M-6]
$Rh_7(CO)_{16}^{3-}$	98	7	(7,98)	4n+0	$C^{1}C$[M-6]
$Os_6Pt(CO)_{18}H_8$	106	8	(8,106)	4n-6	$C^{4}C$[M-4]
$Os_8(CO)_{18}^{2-}$	102	8	(8,102)	4n-10	$C^{6}C$[M-2]
$Os_6Pt_2(CO)_{16}(COD)_2$	110	8	(8,110)	4n-2	$C^{2}C$[M-6]
$Os_8(CO)_{22}^{2-}$	110	8	(8,110)	4n-2	$C^{2}C$[M-6]
$Re_8(C)(CO)_{24}^{2-}$	110	8	(8,110)	4n-2	$C^{2}C$[M-6]
$Rh_8(C)(CO)_{19}$	114	8	(8,114)	4n+2	CLOSO
$Os_8(CO)_{24}^{2-}$	114	8	(8,114)	4n+2	CLOSO
$Ir_9(CO)_{19}(H)^{2-}$	120	9	(9,120)	4n-6	$C^{4}C$[M-5]
$Rh_9(CO)_{19}^{3-}$	122	9	(9,122)	4n-4	$C^{3}C$[M-6]
$Os_9(CO)_{24}^{2-}$	122	9	(9,122)	4n-4	$C^{3}C$[M-6]
$Os_6Pt_4(CO)_{22}(COD)$	124	10	(10,124)	4n-16	$C^{9}C$[M-1]
$Rh_{10}(CO)_{21}^{2-}$	134	10	(10,134)	4n-6	$C^{4}C$[M-6]
$Os_{10}(CO)_{26}^{2-}$	134	10	(10,134)	4n-6	$C^{4}C$[M-6]
$Rh_{11}(CO)_{22}^{3-}$	148	11	(11,148)	4n-6	$C^{4}C$[M-7]
$Rh_{12}(C)_2(CO)_{25}$	166	12	(12,166)	4n-2	$C^{2}C$[M-10]
$Rh_{13}(H)_2(CO)_{24}^{3-}$	170	13	(13,170)	4n-12	$C^{7}C$[M-6]
$Rh_{14}(CO)_{25}^{4-}$	180	14	(14,180)	4n-16	$C^{9}C$[M-5]
$Rh_{15}(CO)_{27}^{3-}$	192	15	(15,192)	4n-18	$C^{10}C$[M-5]
$Rh_{17}(CO)_{30}^{3-}$	216	17	(17,216)	4n-22	$C^{12}C$[M-5]
$Rh_{23}(N)_4(CO)_{38}^{-}$	304	22	(22,304)	4n-4	$C^{3}C$[M-19]

Table 3. Carbonyl Clusters from Research Paper Categorized Using the Cluster Matrix Table

CLUSTER	VALENCE CONTENT	n	CODE	SERIES	CAPPING SYMBOL
$Ru_9(CO)_{23}^{2-}$	120	9	(9,120)	4n-6	$C^4C[M-5]$
$Ru_9(CO)_{22}^{2-}$	118	9	(9,118)	4n-8	$C^5C[M-4]$
$Ru_8(CO)_{21}^{2-}$	108	8	(8,108)	4n-4	$C^3C[M-5]$
$Ru_8(CO)_{20}^{2-}$	106	8	(8,106)	4n-6	$C^4C[M-4]$
$Ru_7(CO)_{20}^{2-}$	98	7	(7,98)	4n+0	$C^1C[M-6]$
$Ru_7(CO)_{19}^{2-}$	96	7	(7,96)	4n+0	$C^1C[M-6]$
$Ru_7(CO)_{18}^{2-}$	94	7	(7,94)	4n+0	$C^1C[M-6]$
$Ru_6(CO)_{21}^{2-}$	92	6	(6,92)	4n+8	HYPHO
$Ru_6(CO)_{20}^{2-}$	90	6	(6,90)	4n+6	ARACHNO
$Ru_6(CO)_{18}^{2-}$	86	6	(6,86)	4n+2	CLOSO
$Ru_6(CO)_{17}^{2-}$	84	6	(6,84)	4n+0	MONOCP= $C^1C[M-5]$
$Ru_6(CO)_{16}^{2-}$	82	6	(6,82)	4n-2	$C^2C[M-4]$
$Ru_6(CO)_{15}^{2-}$	80	6	(6,80)	4n-4	$C^3C[M-3]$
$Ru_6(CO)_{14}^{2-}$	78	6	(6,78)	4n-6	$C^4C[M-2]$
$Ru_5(CO)_{15}^{2-}$	72	5	(5,72)	4n+2	CLOSO
$Ru_5(CO)_{14}^{2-}$	70	5	(7,70)	4n+0	MONOCP= $C^1C[M-4]$
$Ru_5(CO)_{13}^{2-}$	68	5	(5,68)	4n-2	$C^2C[M-3]$
$Ru_4(CO)_{13}^{2-}$	60	4	(4,60)	4n+4	NIDO
$Ru_4(CO)_{12}^{2-}$	58	4	(4,58)	4n+2	CLOSO
$Ru_3(CO)_{11}^{2-}$	48	3	(3,48)	4n+6	ARACHNO
$Ru_3(CO)_{10}^{2-}$	46	3	(3,46)	4n+2	CLOSO
$Ru_3(CO)_9^{2-}$	44	3	(3,44)	4n+2	CLOSO
$HOs_5(CO)_{15}^{-}$	72	5	(5,72)	4n+2	CLOSO
$PtRu_5(C)(CO)_{15}^{2-}$	86	6	(6,86)	4n+2	CLOSO
$Os_{10}(C)(CO)_{24}^{2-}$	134	10	(10,134)	4n-6	$C^4C[M-6]$
$Os_{17}(CO)_{36}^{2-}$	210	17	(17,210)	4n-28	$C^{15}C[M-2]$
$Os_{20}(CO)_{40}^{2-}$	242	20	(20,242)	4n-38	$C^{20}C[M-0]$
$Pt_3\,Ru_{10}(C)_2(CO)_{32}^{2-}$	184	13	(13,184)	4n+2	CLOSO
$Pd_6Ru_6(CO)_{24}^{2-}$	158	12	(12,158)	4n-10	$C^6C[M-6]$
$Pd_4(CO)_5L_4$	58	4	(4,58)	4n+2	CLOSO
$Pd_{10}(CO)_{12}L_6$	136	10	(10,136)	4n-4	$C^3C[M-7]$
$Pd_{12}(CO)_{12}L_6$	156	12	(12,156)	4n-12	$C^7C[M-5]$
$Pd_{16}(CO)_{13}L_9$	204	16	(16,204)	4n-20	$C^{11}C[M-5]$
$Pd_{23}(CO)_{20}L_{10}$	290	23	(23,290)	4n-32	$C^{17}C[M-6]$

Table 4. Further Categorization of Carbonyl Clusters from Research Paper Using the Cluster Matrix Table

CLUSTER	VALENCE CONTENT	n	CODE	SERIES	CAPPING SYMBOL
$Ru_3(CO)_{12}$	48	3	(3,48)	4n+6	ARACHNO
$Ru_4(CO)_{14}$	60	4	(4,60)	4n+4	NIDO
$Ru_5(CO)_{16}$	72	5	(5,72)	4n+2	CLOSO = TBP
$Ru_6(CO)_{18}$	84	6	(6,84)	4n+0	MONOCP = $C^1C[M-5]$
$Ru_7(CO)_{20}$	96	7	(7,96)	4n-2	BICP = $C^2C[M-5]$
$Ru_8(CO)_{22}$	108	8	(8,108)	4n-4	TRICP = $C^3C[M-5]$
$Ru_9(CO)_{24}$	120	9	(9,120)	4n-6	TETRACP = $C^4C[M-5]$
$Ru_{10}(CO)_{26}$	132	10	(10,132)	4n-8	PENTACP = $C^5C[M-5]$
$Ru_{11}(CO)_{28}$	144	11	(11,144)	4n-10	HEXACP = $C^6C[M-5]$
$Ru_{12}(CO)_{30}$	156	12	(12,156)	4n-12	HEPTACP= $C^7C[M-5]$

14n+2, CLOSO

n		
1	16 →	$Os(CO)_4$, $Rh(Cl)(L)_3$, L = PPh_3, $HPt(Cl)(L_2)$, $PtCl(C_2H_4)^-$, $(C_5H_5)_2TiCl_2$, $Ir(Cl)(L)_2(CO)$
2	30 →	$Mo_2(Cp)_2(CO)_4$
3	44 →	$Os_3(CO)_{10}$,
4	58 →	$Os_4(CO)_{13}$
5	72 →	$Os_5(CO)_{16}$,
6	86 →	$Os_6(CO)_{18}^{2-}$, $Co_6(CO)_{16}$, $Rh_6(CO)_{16}$, $Ir_6(CO)_{16}$, $Rh_6(C)(CO)_{15}^{2-}$, $Re_6(C)(CO)_{19}^{2-}$
7	100 →	$Os_7(H)_2(CO)_{21}$, $Rh_7(N)(CO)_{15}^{2-}$, $Co_7(N)(CO)_{15}^{2-}$

T-1

14n+4, NIDO

n		
1	18 →	$Fe(CO)_5$, $Ru(CO)_5$,$Os(CO)_5$,$Ni(CO)_4$, $Cr(C_6H_6)_2$,$Fe(C_5H5)_2$, $Cr(CO)_6$, $Mo(CO)_6$, $W(CO)_6$, $(C_5H_5)_2Fe(H)^-$, $(C_4H_4)Fe(CO)_3$
2	32 →	$Rh_2(Cp)_2(CO)_2$, $Re_2(Cp)_2(CO)_4$,$Re_2(H)_2(CO)_8$
3	46 →	$Os_3(H)_2(CO)_{10}$,$Re_3(H)_3(CO)_{10}{}^{2-}$
4	60 →	$Os_4(CO)_{14}$, $Os_4(H)_4(CO)_{12}$, $Co_4(CO)_{12}$, $Rh_4(CO)_{12}$,$Ir_4(CO)_{12}$, $Re_4(H)_6(CO)_{12}{}^{2-}$
5	74 →	$Os_5(H)_2(CO)_{16}$,$Os_5(C)(CO)_{15}$,$Os_5(C)(CO)_{14}{}^{2-}$,$Fe_5(C)(CO)_{12}L_3$, L =PMe2Ph, $Ru_5(C)(CO)_{15}$, $Ru_5(N)(CO)_{14}{}^-$,$Re_5(C)(CO)_{16}(H)^{2-}$
6	88 →	
7	102 →	

T-2

14n+6, ARACHNO

n

1 20 → $[Os(CO)_6$

2 34 → $Mn_2(CO)_{10}$, $Tc_2(CO)_{10}$, $Re_2(CO)_{10}$, $Fe_2(CO)_9$, $Co_2(CO)_8$, $Os_2(CO)_8^{2-}$, $Mo_2(Cp)_2(CO)_6$, $Os_2(Cp)_2(CO)_4$, $Fe_2(Cp)_2(CO)_4$, $Cr_2(Cp)_2(CO)_6$, $Cr_2(Cp)_2(NO)_4$, $FeCo(Cp)(CO)_6$

3 48 → $Os_3(CO)_{12}$, $Ru_3(CO)_{12}$, $Fe_3(CO)_{12}$, $Re_3(H)_3(CO)_{12}$, $Os_3(H)_2(CO)_{11}$

4 62 → $Os_4(CO)_{15}$, $Re_4(CO)_{16}^{2-}$

5 76 → $Os_5(CO)_{18}$, $Rh_5(CO)_{15}^{-}$, $Ni_5(CO)_{12}^{2-}$, $Rh_5(CO)_{14}I^{2-}$, $Ni_3Cr_2(CO)_{16}^{2-}$, $Ni_3Mo_2(CO)_{16}^{2-}$, $Ni_3W_2(CO)_{16}^{2-}$, $RuIr_4(CO)_{15}^{2-}$, $PtRh_4(CO)_{14}^{2-}$

6 90 → $Co_6(C)(CO)_{15}^{2-}$, $Os_6(P)(CO)_{18}^{-}$, $Rh_6(C)(CO)_{15}^{2-}$, $Rh_6(N)(CO)_{15}^{-}$

7 104 →

T-3

CAPPING SERIES BASED ON TRIGONAL BIPYRAMID, [M-5]

n		
5	76 →	$Os_5(CO)_{16}$
6	84 →	$Os_6(CO)_{18}$
7	96 →	$Ru_7(CO)_{19}^{2-}$
	108 →	$Ru_8(CO)_{22}$
9	120 →	$Ru_9(CO)_{24}$
10	132 →	$Ru_{10}(CO)_{26}$
11	144 →	$Ru_{11}(CO)_{28}$
12	156 →	$Ru_{12}(CO)_{30}$

T-4

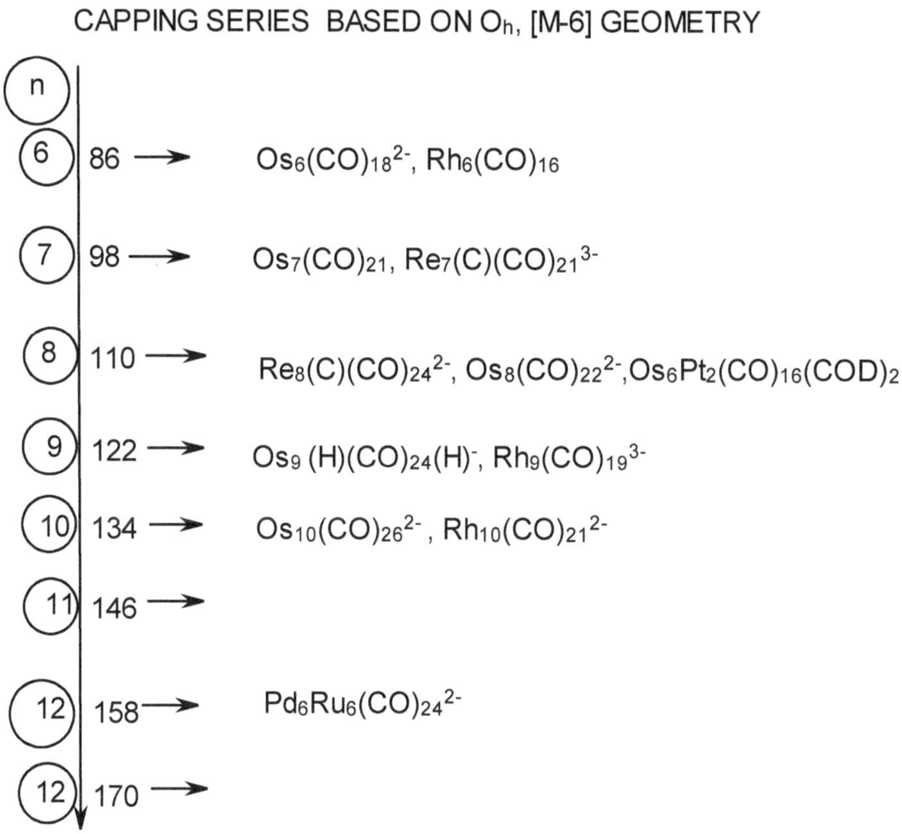

T-5

3. Conclusion

The array of matrix numbers in Table 1 were generated from the series S = 14n±q. It has been found that the electron valence content of carbonyl metal complexes conform to the numbers of the series matrix. The matrix clearly shows that the mono-skeletal carbonyl complexes are the first members of certain type of series which develop into multi-skeletal cluster complexes of the same series. For instance, the cluster $Fe(CO)_5$(1,18) is one of the first member of 14n+4 series. Other multi-nuclear complexes skeletal of the same series include $Rh_2(Cp)_2(CO)_2$(2,32),$Os_4(CO)_{14}$(4,60),and $Os_5(C)(CO)_{15}$(5,74). The other series such as 14n+2 beginning with (1,16) $IrClL_2(CO)$(1,16),L = is a phosphorous based ligand bolongs to other clan members of higher nuclearity clusters such $Mo_2(Cp)_2(CO)_4$(2,30),$Os_5(CO)_{16}$(5,72)and $Rh_6(CO)_{16}$(6,86). The first members of the 14n+6, 14n+4, 14n+2, 14n+0,14n-2, 14n-4, 14n-6 will have the valence electron content of 20, 18, 16, 14, 12, 10 and 8 respectively. It is proposed that the sets (1,18) and (1,16) form the basis of the well-known 18 and 16 electron rules which are prevalent in organometallic chemistry. We may probably regard complexes such as $Ni(OH_2)_6^{2+}$(1,20), $Ni(CN)_4^{2-}$(1,20), $(C_5H_5)_2TiCl_2$(16), $RhCl(PPh_3)_3$(1,16), MR_2(M = Zn, Cd, Hg) (1,14),$TiCl_4$(1,8) as first members of the respective series in Table 1. Clearly, the 18-electron rule of 14n+4 series that has a wide range of complexes has enjoyed the attention of science for a long time(Langmuir, 1921; Miessler,2014). The prevalence of 18-electron based complexes could imply that the 18-electron system (island of chemical stability)is more stable than 16, 14 or any other valence electron systems. The valence electron content around a cluster of skeletal elements has great influence on the shape of the skeletal elements regardless of their origin. That is, (6, 86) has a tendency to form an octahedral shape (O_h). For example, [M-6, 86, M= Os, Rh, Re) of S = 14n+2 closo series are all expected to have an octahedral geometry. The valence electron table as written (Table 1) presents capping series along the diagonal from right to left. However, the capping series as defined by 14n+0 column-begins with mono-capping [C^1C] series with the very series 14n+0 column. The horizontal movement from right to the left in Table 1 results in the successive increase of k value by 1. On the other hand the movement along the diagonal gives a stepwise change in cluster k value by 3. The movement along the column involves the successive variation of k value by 2. The series matrix(Table 1) implies that the clusters especially the transition metal carbonyls are all interrelated via series. Thus, the 18-electron rule complexes (1,18) of 14n+4 are interrelated with those of (2,32), (3,46), (4,60) and so on which belong to the S =14n+4 series. There is also an interesting correlation between 14n±q and 4n±q series.

Acknowledgement

The author wishes to acknowledge the University of Namibia for providing an enabling environment and financial support, NAMSOV, Namibia for generous financial support and my wife, Merab Kambamu Kiremire for her continued encouragement to write up this paper.

References

Bruce, R. B. (1986). Metal Cluster Topology: Osmium Carbonyl Clusters. *Inorg. Chim. Acta, 116*, 99-107. http://dx.doi.org/10.1016/S0020-1693(00)82162-3

Cotton, F. A., & Wilkinson, G. (1980). Advanced Inorganic Chemistry, 4th Edition, John Wiley & Sons, New York.

Critchey, G., Dyson, P. J., Johnson, B. F. G., McIndoe, J. S., Reilly, R. K., & Langridge-Smith, P. R. R. (1999). Reactivity and Characterization of Transition Metal Carbonyl Clusters Using UV Laser Desorption Mass Spectroscopy. *Organometallics, 18*, 4090-4097. http://dx.doi.org/10.1021/om990075f

Jemmis, E. D. (2005). Building relationships between polyhedral boranes and elemental boron. *Inorg. Chem. 18*, 620-628.

Jemmis, E. D., & Balakrishnarajan, M. M. (2001). Polyhedral boranes and elemental boron. Direct structural relations and diverse electronic requirements. *J. Am. Chem. Soc., 123*, 4324-4330. http://dx.doi.org/10.1021/ja0026962

Jemmis, E. D., & Jayasree, E. G. (2003). Analogies between boron and carbon. *Acc. Chem. Res., 36*, 816-824. http://dx.doi.org/10.1021/ar0300266

Jemmis, E. D., & Pavankumar, P. N. V. (1984). Stability of Polyhedral Borane anions and Carboranes. *Proc. Indian Acad. Sci., 93*(3), 479-489.

Jemmis, E. D., & Prasad, D. L. V. K. (2008). Unknowns in the chemistry of Boron. *Current Science, 95*(10), 1277-1283.

Jemmis, E. D., Balakrishnarajan, M. M., & Pancharatna, P. D. (2001). Unifying electron counting rule for Macropolyhedral Boranes, Metallaboranes, and Metallocenes. *J. Am. Chem. Soc., 123*(18), 4313-4323. http://dx.doi.org/10.1021/ja003233z

Jemmis, E. D., Balakrishnarajan, M. M., & Pancharatna, P. D. (2002). Electronic Requirements for Macropolyhedral Boranes. *Chem. Rev., 102*(1), 93-144. http://dx.doi.org/10.1021/cr990356x

Jemmis, E. D., Jayasree, E. G., & Parameswaran, P. (2006). Hypercarbons in polyhedral structures. *Chem. Soc. Rev., 35*, 157-168. http://dx.doi.org/10.1039/B310618G

King, R. B. (2002). Spherical Chemical Bonding Models for three-diamensional aromaticity in deltahedral boranes. PINSA, 68, A, 473-486.

Kiremire, E. M. (2014). Numerical Sequence of Borane Series. Orient. *J. Chem., 30*(3), 1055. http://dx.doi.org/10.13005/ojc/300317

Kiremire, E. M. (2015a). Categorization and Structural Determination of Simple and More Complex Carbonyl Clusters of Rhenium and Osmium Using k values and the Cluster Table. *Orient. J. Chem., 31*(1), 293-302. http://dx.doi.org/10.13005/ojc/310133

Kiremire, E. M. (2015b). Classification of Transition Metal Carbonyl Clusters Using the 14n Rule Derived from Number Theory. *Orient. J. Chem., 31*(2), 605-618. http://dx.doi.org/10.13005/ojc/310201

Kiremire, E. M. (2015d). Isolobal Series of Chemical Fragments. *Orient. J. Chem., 31*(Spl. Edn), 59-70.

Kiremire, E. M. R. (2015c). *Capping and Decapping Series. Int. J. Chem., 7*(2), 186-197. http://dx.doi.org/10.5539/ijc.v7n2p186

Kobayashi, T., & Xu, Q. (2007). *Journal of Power Sources, 168*, 167-171. http://dx.doi.org/10.1016/j.jpowsour.2007.03.009

Langmuir, I. (1921). Types of Valence. *Science, 54*, 59-67. http://dx.doi.org/10.1126/science.54.1386.59

Miessler, G, Fischer, P., & Tarr, D. (2014). Inorganic Chemistry, 5th Edition, Pearson Education, Inc., Upper Saddle River

Mingos, D. M. P. (1984). *Acc. Chem. Res., 17*(9), 311-319. http://dx.doi.org/10.1021/ar00105a003

Shaik, S., Danovich, D., Wu, W., Su, P., Rzepa, H. S., & Hiberty, P. C. (2012). Quadruple bonding in C2 and analogous eight valence electron species. *Nature Chemistry, 4*, 195-200. http://dx.doi.org/10.1038/nchem.1263

Stock, A. (1933). Hydrides of Boron and Silicon, Cornell University Press. Ithaca, New York.

Tolman, C, A. (1972). The 16 and 18 Electron Rule in Organometallic Chemistry and Homogeneous Catalysis. *Chemical Soc. Rev., 1*, 337-353. http://dx.doi.org/10.1039/cs9720100337

Wade, A. (1976). *Adv. Inorg. Chem. Radiochem., 18*, 1-66. http://dx.doi.org/10.1016/S0065-2792(08)60027-8

Wheeler, R. A., & Hoffmann, R. (1986). A New Magic Cluster Electron Count and Metal-Metal Multiple Bonding. *J. Am. Chem. Soc., 108*, 6605-6610. http://dx.doi.org/10.1021/ja00281a025

2

The Application of the 4n Series Method to Categorize Metalloboranes

Enos Masheija Rwantale Kiremire

Correspondence: Enos Masheija Rwantale Kiremire, Department of Chemistry and Biochemistry, University of Namibia, Private Bag 13301, Windhoek, Namibia. E-mail: kiremire15@yahoo.com

Abstract

Metalloboranes have been categorized by using 4n series method. More than one hundred metalloboranes have been analyzed and characterized by the method. The clusters have been found to be centered within the series range S = 4n+6 to 4n-8. According to the classification, the clusters of series S= 4n+0 are mono-capped, S = 4n-2, bi-capped, S = 4n-4, tri-capped, S = 4n-6, tetra-capped and S= 4n-8, penta-capped. Whereas the known stable boranes have almost non-existent capped boranes, such clusters are prevalent within metalloborane complexes. This implies that metal fragments have the potency of stabilizing the fragile capped boranes.

Keywords: 4n series, metalloboranes, capped-clusters, isolobal, numerical, valency

1. Introduction

The Wade-Mingos rules (Wade, 1971; Mingos, 1972, 1984, 1991; Cox, et al, 1981) have widely been utilized to categorize clusters. The 4n series method of categorizing clusters that has recently been developed (Kiremire, 2015a; 2015b) complements the PSEPT method. The method is a systematic numerical approach of electron-counting of valence electrons in atoms, fragments, molecules and clusters. The categorization of metalloboranes using the 4n series method is hereby presented in this paper.

2. Results and Discussion

2.1 Background to the Application of the 4n Series Method

The evolution of the 4n series method for analyzing and categorizing clusters has been discussed extensively (Kiremire, 2015a, 2015b). In this paper, the 4n series method is being extended to classification of metallaborane clusters. Adequate examples have been provided to demonstrate the flexibility and the power of the method so as to enable the readers to be more familiar with it. The cluster complexes used have been taken from a wide spectrum of sources but mostly from the work by Fehlner and Grimes (Fehlner, 2002; Callahan, et al, 1974; Grimes, 1974). In addition, the vital isolobal relationship between transition metals and main group elements will be utilized (Hoffmann, 1982; Kiremire, 2015a). The relationship is summarized below.

$$14n+q \underset{+10n}{\overset{-10n}{\rightleftharpoons}} 4n+q$$

This means that the valence electrons from the transition metal fragments have to be lowered to the level of the main group fragments by subtracting 10 unit sets of electrons. If it is one fragment, a numerical value of 10 is subtracted, for two, 2x10, three, 3x10 and so on. It has been found much easier to work with 4n series rather than 14n series. The cluster number k is defined as k = 2n-½ q. In terms of categorization of clusters, S =4n+2 represents the closo cluster, 4n+4(nido), 4n+6 (arachno), 4n+8 (hypho) and 4n+10 (klapo). The capping series start from S =4n+0(mono-capped), 4n-2(bi-capped), 4n-4(tri-capped), 4n-6(tetra-capped) and so on. The proposed capping symbol which was discussed in earlier work is Cp = C^1C (mono-capped), Cp = C^2C (bi-capped), Cp = C^3C (tri-capped) and Cp = C^4C (tetra-capped) (Kiremire, 2015b). The S = 4n+0 represents a mono-capped cluster and the symbol C^1 is used. For any additional +1(-2), that is, the series S = 4n-2, we get another capping and hence an additional symbol C^1 is added. Thus, the series S = 4n+1(-2), the net capping symbol becomes, Cp = $C^1+C^1 = C^2$, and for S = 4n+3(-2) = 4n-6, Cp = $C^1+C^3 = C^4$. Thus, S = 4n-6 series represents, a tetra-capped cluster. Let us consider the following examples (most of which are known from the application of Mingo-Wades rules) as illustrations of the application of the 4n series method of categorizing clusters.

2.2 B_5H_9

The molecular formula is decomposed into mono-fragments with electron valence content of four. Hence, B_5H_9 is split into five [BH] fragments. These fragments consist of 5 boron atoms and 5 hydrogen atoms. Since, B_5H_9 has 5 boron atoms and 9 hydrogen atoms, the remaining 4 hydrogen atoms will be added separately. This is summarized by the work out below.

$$5[BH] \rightarrow 5[3+1] = 5[4+0] \rightarrow 4n+0(n=5)$$

Balancing the hydrogen atoms: $[9-5](H) = 4(H) \rightarrow 0+4(n=0)$

$S = 4n+4(n=5)$; Nido cluster,

$$k = 2n-2 = 2(5)-2 = 8$$

The skeletal shape of B_5H_9 is shown in F-1 as a projection and F-2 as square pyramid. A diagram showing the transformation of series into an equivalent hydrocarbon is given in Scheme 1. The skeletal shape of the hydrocarbon analogue C_5H_4 as an isomer corresponding to B_5H_9 is shown in F-3. Most of the examples have been taken from an outstanding review by Fehlner (Felner, 2002).

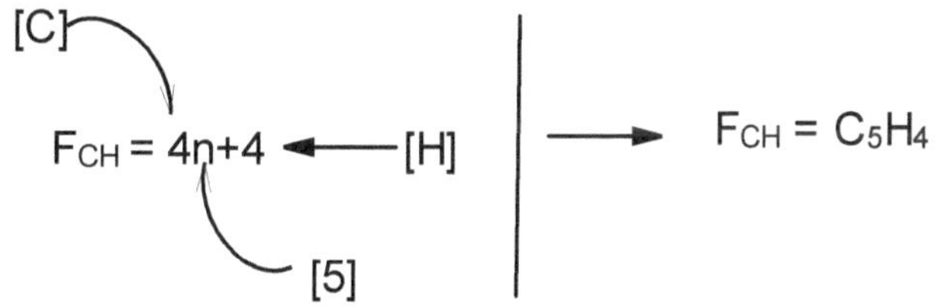

Scheme 1. Converting a series Formula into a corresponding Hydrocarbon molecular formula

This means that B_5H_9 is equivalent to C_5H_4.

The B_5H_9 cluster has 5 skeletal atoms M-5 and is a nido type of cluster. Therefore, it is derived from M-6 closo cluster which has an octahedral shape. Hence, the B_5H_9 cluster has a square pyramid pyramid shape with k value = 8.

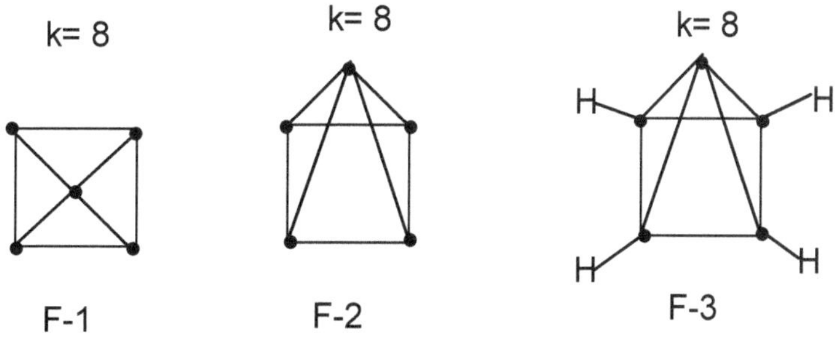

2.3 $C_2B_3H_7$

$$2[C] = 2[4+0] \rightarrow 4n+0(n=2)$$

$$3[BH] = 3[4+0] \rightarrow 4n+0(n=3)$$

$$[7-3](H) = 4(H) \rightarrow 0+4(n=0)$$

$$S = 4n+4(n=5);\ \text{Nido series},\ k = 8$$

The shape will be similar to that of B_5H_9 discussed above. Again, in this example the molecular formula has been decomposed into mono-fragments each with valence electron content of four in line with 4n series method.

2.4 $C_5H_3Me_2^+$

$$5[C] \rightarrow 4n+0(n=5)$$
$$3H \rightarrow 0+3(n=0)$$
$$2Me \rightarrow 0+2(n=0)$$

Charge, q→0-1(n=0), q = charge

S = 4n+4(n=5); Nido cluster, k = 8. Therefore, $C_5H_3Me_2^+$ cluster is also expected to have a shape similar to that of B_5H_9.

2.5 $(C_4H_4)Fe(CO)_3$

$$1[Fe(CO)_3] = 1[8+6] = 1[14+0] \rightarrow 1[14+0-10] = 1[4+0] \rightarrow 4n+0(n=1)$$
$$4[C] \rightarrow 4n+0(n=4)$$
$$4H \rightarrow 0+4(n=0)$$
$$S = 4n+4(n=5);\ \text{Nido series},\ k = 2n-2 = 2(5)-2 = 8$$

Hence, the skeletal shape is similar to that of B_5H_9.

2.6 $B_3H_7Fe_2(CO)_6$

$$2[Fe(CO)_3] \rightarrow 4n+0(n=2)$$
$$3[BH] \rightarrow 4n+0(n=3)$$
$$[7-3](H) = 4H \rightarrow 0+4(n=0)$$
$$S = 4n+4(n=5);$$

Nido cluster, k =8. Shape is similar to that of B_5H_9.

2.7 $Fe_5(C)(CO)_{15}$

$$5[Fe(CO)_3] \rightarrow 4n+0(n=5)$$
$$C \rightarrow 0+4(n=0)$$

S = 4n+4(n=5). The $Fe_5(C)(CO)_{15}$ cluster is member of Nido clan of clusters and its skeletal is similar to that of B_5H_9 cluster.

2.8 $(Cp^*Ir)_2B_4H_8$

$$2[Cp^*Ir] \rightarrow 2[5+9] = 2[14+0] \rightarrow 4n+0(n=2)$$
$$4[BH] \rightarrow 4[4+0] \rightarrow 4n+0(n=4)$$
$$4H \rightarrow 0+4(n=0)$$
$$S = 4n+4(n=6);\ \text{NIDO}$$

F_B = 4n+4(n=6)→[BH](6)+4(H) =B_6H_6+4H = B_6H_{10}. Since this is a nido cluster, its shape is related that of the nearest closo system $B_7H_7^{2-}$.This is in agreement to the Rudolph series (Rudolph, 1976). The cluster B_6H_{10} has a pentagonal pyramid shape (k = 10). The hydrocarbon equivalent is given by F_{CH} =[C](6)+4(H) = C_6H_4. The k value = 2n-2=2(6)-2 = 10. A sketch of the ideal skeletal shape of B_6H_{10} is given in F-4. This is the expected ideal shape for $(Cp^*Ir)_2B_4H_8$. This is in agreement with the observed shape (Fehlner, 2002).

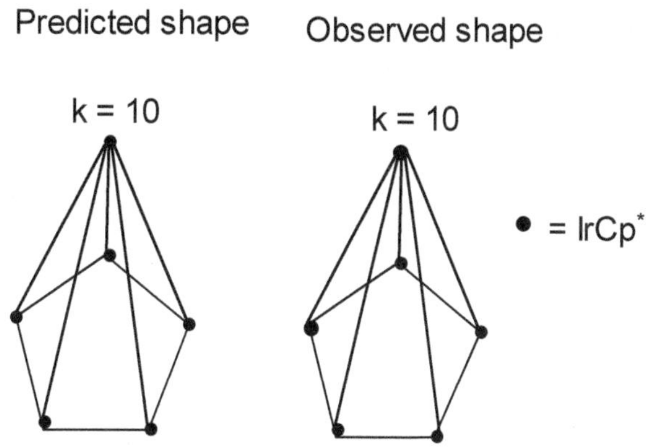

F-4 Ideal skeletal shape of $(Cp^*Ir)_2B_4H_8$

Two of the skeletal points (enlarged size) refer to the two IrCp* fragments.

2.9 $(Cp^*Ru)_2B_4H_8$

2[Cp*RuH]→4n+0(n=2). This step involves the lowering of [14] valence electron equivalent to [4] valence electron equivalent so as to use 4n series method of categorization.

4[BH]→4n+0(n=4)

(H)=2H→0+2

S = 4n+2(n=6); Closo series; k = 2n-1 = 2(6)-1 = 11.

F_B = 4n+2 = [BH](6)+2 = $B_6H_6^{2-}$. The closo boranes are known to be negatively charged (McGlinchey, et al, 2011).The ideal skeletal shape of $B_6H_6^{2-}$ is expected to be octahedral for $(Cp^*Ru)_2B_4H_8$. The expected skeletal shape is shown in F-5A. However, the observed shape is a capped square pyramid (Fehlner, 2002) shown below F-5B. This is due to k isomerism, that is, net k = k(square pyramid)+ k (capping) = 8+3 =11. This is the same as k=11 for octahedral shape. This is in agreement with the series, thus, M-5 square pyramid has corresponding series S = 4n+4(n=5), k =2n-2 =2(5)-2 =8; and the capping fragment $[BH]^{2+}$ has series S =4n-2(n=1), and k =2n+1 = 2(1)+1 = 3. The net cluster, F_B = B_5H_9+$[BH]^{2+}$ =B_6H_8 = $B_6H_6^{2-}$, has a net k =8+3 = 11.

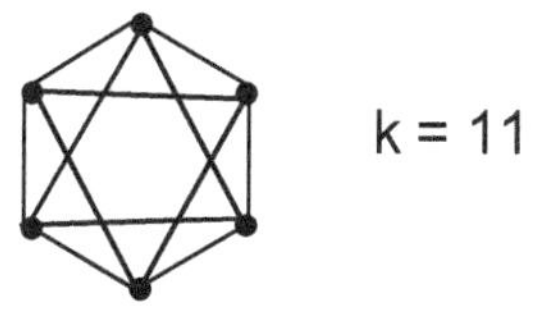

F-5 A. The expected ideal skeletal shape of $(Cp^*Ru)_2B_4H_8$

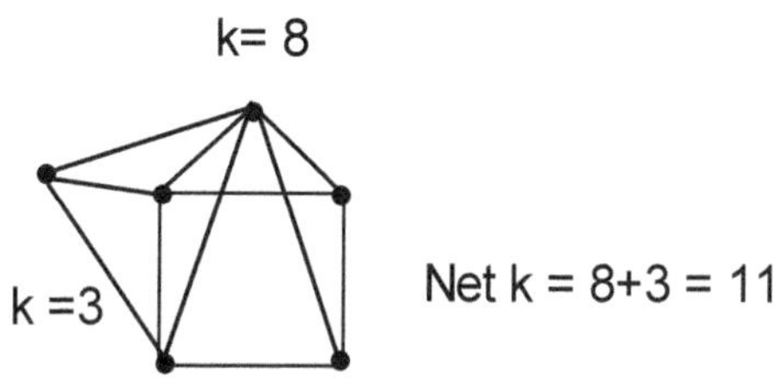

F-5B. Sketch of Ideal observed shape of $(Cp^*Ru)_2B_4H_8$

2.10 $Cp^*Co)B_4H_8$

1[Cp*Co]→1[5+9] =1[14+0]→1[14+0-10] = 1[4+0]→4n+0(n=1)

4[BH]→4n+0(n=4)

4H→0+4(n=0)

$$S = 4n+4(n=5)\text{; this is a nido cluster.}$$

$$F_B = 4n+4 = [BH](5)+4(H) = B_5H_5+4H = B_5H_9$$

$$F_{CH} = 4n+4(n=5) = [C](5)+4(H) = C_5H_4$$

This cluster is expected to have a shape similar to that of B_5H_9.

2.11 (Cp)$WH_3B_4H_8$*

$$1[Cp^*WH_3] = 1[5+6+3] = 1[14+0] \rightarrow 1[14+0-10] = 1[4+0] \rightarrow 4n+0(n=1)$$

$$4[BH] \rightarrow 4n+0(n=4)$$

$$4H \rightarrow 0+4(n=0)$$

$S = 4n+4(n=5)$; this is a nido cluster. $F_B = B_5H_9$, $F_{CH} = C_5H_4$, $k = 8$. Shape is a square pyramid

2.12 $(Cp^*Re)_2B_4H_8$

$$2[Cp^*ReH_2] \rightarrow 2[5+7+2] = 2[14+0] \rightarrow 4n+0(n=2)$$

$$4[BH] \rightarrow 4n+0(n=4)$$

$$[8-4-4](H) \rightarrow 0+0$$

$$S = 4n+0(n=6)$$

The cluster belongs to mono-capped series.

$F_B = 4n+0 = [BH](6)+0(H) = B_6H_6$; $F_{CH} = [C](6)+0(H) = C_6$.

This is a mono-capped cluster with symbol $Cp = C^{1}C[M-5]$. It is a mono-capped trigonal bipyramid (F-8A). This is in agreement with the observed cluster 2.10 in F-8B.

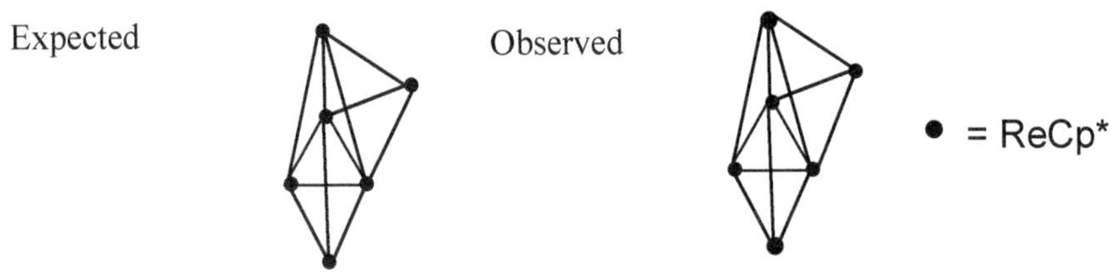

F-8A F-8B

F-8A. Expected ideal skeletal shape of $(Cp^*Re)_2B_4H_8$

F-8B. Observed ideal skeletal shape of $(Cp^*Re)_2B_4H_8$

2.13 $(Cp^*Cr)_2B_4H_8$

$$2[Cp^*CrH_3] \rightarrow 2[5+6+3] = 2[14+0] \rightarrow 2[14+0-10] = 2[4+0] \rightarrow 4n+0(n=2)$$

$$4[BH] \rightarrow 4[4+0] \rightarrow 4n+0(n=4)$$

$$[8-6-4](H) = -2(H) \rightarrow 0-2(n=0)$$

$$S = 4n-2(n=6)$$

This is a bi-capped tetrahedral, $Cp = C^2C[M-4]$. The skeletal sketch of $(Cp^*Cr)_2B_4H_8$ is shown in F-9A. The observed similar shape (Fehlner, 2002) is shown in F-9B.

Cp = $C^2C[M\text{-}4]$ Cp = $C^2C[M\text{-}4]$

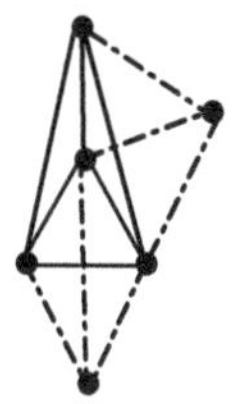

F-9A

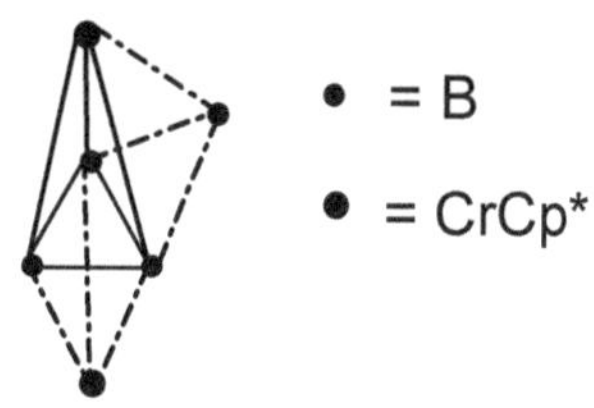

F-9B

Dotted lines denote capping

F-9A. Expected ideal skeletal shape of $(Cp^*Cr)_2B_4H_8$

F-9B. Observed ideal skeletal shape of $(Cp^*Cr)_2B_4H_8$

2.14 $(Cp^*W)_3(H)B_8H_8$

$$3[Cp^*WH_3]\rightarrow 3[5+6+3]=3[14+0]\rightarrow 3[14+0-10]=3[4+0]\rightarrow 4n+0(n=3)$$
$$8[BH]\rightarrow 8[3+1]=8[4+0]\rightarrow 4n+0(n=8)$$
$$[9-9-8](H)=-8(H)\rightarrow 0-8(n=0)$$
$$S=4n-8(n=11)$$
$$Cp=C^1+C^4=C^5C[M-6]$$

This means the cluster is a penta-capped octahedral. The predicted skeletal sketch is shown in F-10. This is almost similar to the observed shape (Fehlner, 2002).

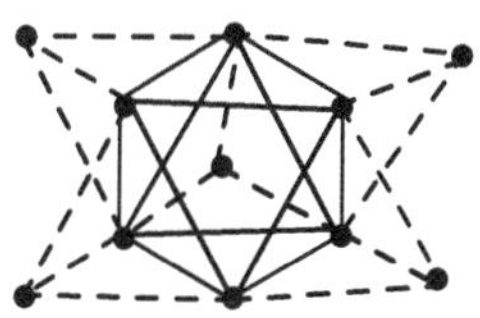

Dotted lines denote capping

F-10. Sketch of ideal skeletal shape of penta-capped $(Cp^*W)_3(H)B_8H_8$

2.15 $Cp^*_2Re_2H_2B_4H_8$

$$2(Cp^*ReH_2)=2[5+7+2]=2[14+0]\rightarrow 2[14+0-10]=2[4+0]\rightarrow 4n+0(n=2)$$
$$4[BH]=4[3+1]=4[4+0]\rightarrow 4n+0(n=4)$$
$$[10-4-4](H)=2(H)\rightarrow 0+2(n=0)$$
$$S=4n+2(n=6),\text{ Closo system.}$$

This corresponds to $B_6H_6^{2-}$ cluster which has an octahedral shape. The expected shape is sketched in F-11A and the observed one is given in F-11B.

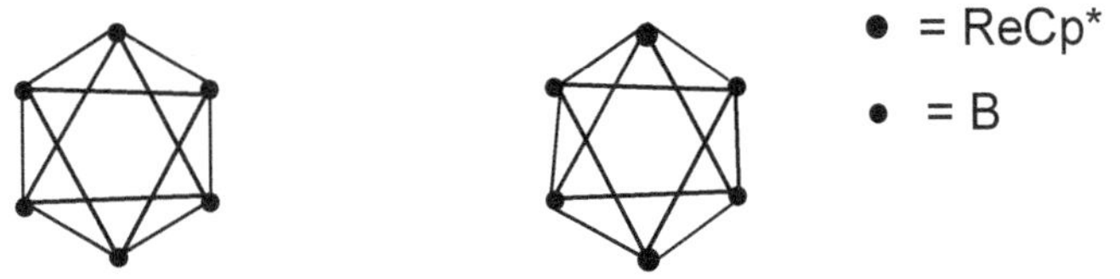

F-11A F-11B

F-11A. Sketch of expected ideal skeletal shape of $(Cp^*Re)_2(H)_2B_4H_8$

F-11B. Sketch of the observed skeletal shape of $(Cp^*Re)_2(H)_2B_4H_8$

2.16 $Cp^*_2Re_2B_6H_6$

$$2[Cp^*ReH_2]=2[5+7+2]=2[14+0]=2[14+0-10]=2[4+0] \rightarrow 4n+0(n=2)$$
$$6[BH]= 6[3+1]=6[4+0] \rightarrow 4n+0(n=6)$$
$$[6-4-6](H) = -4H \rightarrow 0-4(n=0)$$
$$S = 4n-4(n=8)$$
$$Cp = C^3C[M-5]$$

This means that the cluster is expected to have an inner shape of trigonal bipyramid with three atoms in capping positions.

The expected ideal shape tri-capped trigonal bipyramid skeletal shape is shown in F-12A. This in agreement with the observed shape (Fehlner, 2002) of $Cp^*_2Re_2B_6H_6$ sketched in F-12B.

$Cp = C^3C[M\text{-}5]$ $Cp = C^3C[M\text{-}5]$

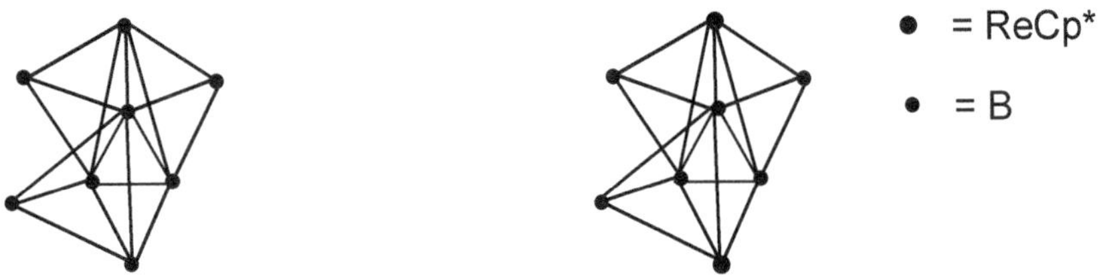

F-12A F-12B

F-12A. Sketch of expected ideal skeletal shape of $(Cp^*Re)_2B_6H_6$

F-12B. Sketch of observed skeletal shape of $(Cp^*Re)_2B_6H_6$

2.17 $Cp^*_2Re_2B_7H_7$

$$2[Cp^*ReH_2]=2[5+7+2]=2[14+0]=2[14+0-10]=2[4+0] \rightarrow 4n+0(n=2)$$
$$7[BH]= 7[3+1]=7[4+0] \rightarrow 4n+0(n=7)$$
$$[7-4-7](H) = -4H \rightarrow 0-4(n=0)$$
$$S = 4n-4(n=9)$$
$$Cp = C^3C[M-6]$$

Thus, $Cp^*_2Re_2B_7H_7$ expected to be a tri-capped octahedral complex. The proposed skeletal sketch is given in F-13A. The shape is similar to the one reported in literature (Fehlner, 2002) and its sketch is shown in F-13B.

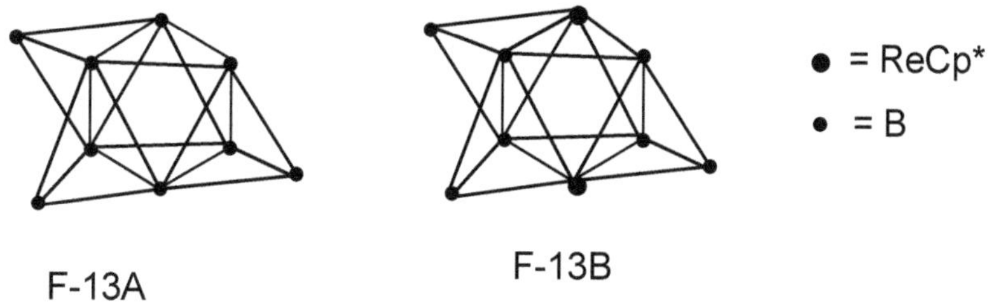

F-13A. Sketch of expected ideal skeletal shape of $(Cp^*Re)_2B_7H_7$

F-13B. Sketch of the observed skeletal shape of $(Cp^*Re)_2B_7H_7$

2.18 Categorization of Metalloboranes

More than than one hundred metalloborane clusters were categorized using the 4n series method. The results are shown in Tables 1 and 2. Table 2 lists mainly the capped clusters for ease of reference. The clusters form a wide range of series starting with S =4n+6 (arachno) and ending with 4n-8 (penta-capped). The range covers eight families of clusters. This is in contrast to the known stable boranes which comprise only of three main family groups namely, 4n+6 (arachno), 4n+4 (nido) and 4n+2 (closo) with almost no known stable capped boranes (Kiremire, 2015b). According to the 4n series method, the capping series start at 4n+0 and progress to higher capped series. Since capping is prevalent in metalloborane and rare in known borane clusters, it implies that metalloborane clusters indirectly stabilize and promote the existence of fragile capped boranes which are unable to exist on their own.

Table 1. Categorization of Metalloboranes

Complex	Series	k value	n Value	Classification	Equivalent Borane	Equivalent Hydrocarbon
	S = 4n+q	k = 2n-(q/2)			cluster	cluster
$Mn(CO)_4(B_3H_8)$	4n+6	5	4	Arachno	B_4H_{10}	C_4H_6
$(Cp^*Ir)H_2B_3H_7$	4n+6	5	4	Arachno	B_4H_{10}	C_4H_6
$Cp^*ReH_3B_3H_8$	4n+6	5	4	Arachno	B_4H_{10}	C_4H_6
$Mn(CO)_3(B_3H_8)$	4n+4	6	4	Nido	B_4H_8	C_4H_4
$B_5H_9L_2$	4n+8	6	5	Hypho	B_5H_{13}	C_5H_8
$Cp^*FeB_4H_{11}$	4n+6	7	5	Arachno	B_5H_{11}	C_5H_6
$Ir(CO)L_2(B_4H_9)$	4n+6	7	5	Arachno	B_5H_{11}	C_5H_6
$R_2C_2B_3H_5^{2-}$	4n+6	7	5	Arachno	B_5H_{11}	C_5H_6
$C_2B_3H_7$	4n+4	8	5	Nido	B_5H_9	C_5H_4
$(Cp^*RuH)_2B_3H_7$	4n+4	8	5	Nido	B_5H_9	C_5H_4
$Fe(CO)_3B_4H_8$	4n+4	8	5	Nido	B_5H_9	C_5H_4
$(CpCo)\ B_4H_8$	4n+4	8	5	Nido	B_5H_9	C_5H_4
$Fe(CO)_3C_4H_4$	4n+4	8	5	Nido	B_5H_9	C_5H_4
$(Cp^*RuH)_2B_3H_7$	4n+4	8	5	Nido	B_5H_9	C_5H_4
$(Cp^*Co)B_4H_8$	4n+4	8	5	Nido	B_5H_9	C_5H_4
$(Cp^*Co)_2B_3H_7$	4n+4	8	5	Nido	B_5H_9	C_5H_4
$[Fe(CO)_3]_2B_3H_7$	4n+4	8	5	Nido	B_5H_9	C_5H_4
$(Cp^*Rh)_2B_3H_7$	4n+4	8	5	Nido	B_5H_9	C_5H_4
$C_5H_5^+$	4n+4	8	5	Nido	B_5H_9	C_5H_4
$B_4H_8Fe(CO)_3$	4n+4	8	5	Nido	B_5H_9	C_5H_4
$Bi_3Mo_2(CO)_6^{3-}$	4n-4	12	5	$C^3C[M-2]$	B_5H_1	C_5-4H

Table 1. Continued

Complex	Series	k value	n Value	Classification	Equivalent Borane	Equivalent Hydrocarbon
$(CpFe)B_5H_{10}$	4n+4	10	6	Nido	B_6H_{10}	C_6H_4
(Cp^*Ir) $B_3H_7Co_2(CO)_5$	4n+4	10	6	Nido	B_6H_{10}	C_6H_4
$(Cp^*RuH)_2Me_2C_2B_2$ H_4	4n+4	10	6	Nido	B_6H_{10}	C_6H_4
$(Cp^*Ru)_2\ HB_4H_9$	4n+4	10	6	Nido	B_6H_{10}	C_6H_4
$(Cp^*Ir)_2B_4H_8$	4n+4	10	6	Nido	B_6H_{10}	C_6H_4
$(Cp^*RuH)_2$ $C_2B_2H_4Me_2$	4n+4	10	6	Nido	B_6H_{10}	C_6H_4
$(Cp^*Ir)_2B_4H_8$	4n+4	10	6	Nido	B_6H_{10}	C_6H_4
$R_2C_2B_4H_4^{2-}$	4n+4	10	6	Nido	B_6H_{10}	C_6H_4
$IrL_2(CO)(B_5H_8)$	4n+4	10	6	Nido	B_6H_{10}	C_6H_4
$CuL_2B_5H_8$	4n+4	10	6	Nido	B_6H_{10}	C_6H_4
$(Cp^*Ru)_2B_3H_6Co(C$ $O)_4$	4n+4	10	6	Nido	B_6H_{10}	C_6H_4
$(Cp^*Co)_2B_2H_2S_2$	4n+4	10	6	Nido	B_6H_{10}	C_6H_4
$C_2B_4H_8$	4n+4	10	6	Nido	B_6H_{10}	C_6H_4
$(Cp^*Ru)_2B_4H_{10}$	4n+4	10	6	Nido	B_6H_{10}	C_6H_4
CB_5H_9	4n+4	10	6	Nido	B_6H_{10}	C_6H_4

Table 1. Continued

Complex	Series	k value	n Value	Classification	Equivalent Borane	Equivalent Hydrocarbon
$(Cp*Re)_2H_2B_4H_8$	4n+2	11	6	Closo	$B_6H_6^{2-}$	C_6H_2
$H_2Os_6(CO)_{18}$	4n+2	11	6	Closo	$B_6H_6^{2-}$	C_6H_2
$(CpCo)_3B_3H_5$	4n+2	11	6	Closo	$B_6H_6^{2-}$	C_6H_2
$(Cp*Co)_2B_4H_2Br_4$	4n+2	11	6	Closo	$B_6H_6^{2-}$	C_6H_2
$(Cp*Ru)_2B_4H_8$	4n+2	11	6	Closo	$B_6H_6^{2-}$	C_6H_2
$(Cp*Ru)_3B_3H_8$	4n+2	11	6	Closo	$B_6H_6^{2-}$	C_6H_2
$(Cp*Ir)\ B_3H_5Co_2(CO)_5$	4n+2	11	6	Closo	$B_6H_6^{2-}$	C_6H_2
$(Cp*Ru)_2B_4H_8$	4n+2	11	6	Closo	$B_6H_6^{2-}$	C_6H_2
$(CpCo)_3B_3H_3(CO)$	4n+2	11	6	Closo	$B_6H_6^{2-}$	C_6H_2
$(Cp*Co)_2B_4H_6$	4n+2	11	6	Closo	$B_6H_6^{2-}$	C_6H_2
$(Cp*ReH_2)\,(CO)B_4H_6$	4n+0	12	6	$C^1C[M-5]$	B_6H_6	C_6
$(CpCo)_2B_2H_5Mo_2(CO)_6(I)$	4n+0	12	6	$C^1C[M-5]$	B_6H_6	C_6
$(Cp*ReH_2)_2B_4H_4$	4n+0	12	6	$C^1C[M-5]$	B_6H_6	C_6
$(Cp*Ru)_2B_3H_4Co(CO)_3$	4n+0	12	6	$C^1C[M-5]$	B_6H_6	C_6
$(Cp*Re)_2B_4H_8$	4n+0	12	6	$C^1C[M-5]$	B_6H_6	C_6
$(Cp*Cr)_2B_4H_8$	4n-2	13	6	$C^2C[M-4]$	B_6H_4	C_6-2H
$(Cp*Cr)_2B_4H_8$	4n-2	13	6	$C^2C[M-4]$	B_6H_4	C_6-2H
$(Cp*Cr)_2(CO)_2B_4H_2$	4n-4	14	6	$C^3C[M-3]$	B_6H_2	C_6-4H
$(Cp*Re)_2B_4H_4$	4n-4	14	6	$C^3C[M-3]$	B_6H_2	C_6-4H

Complex	Series	k value	n Value	Classification	Equivalent Borane	Equivalent Hydrocarbon
$Co_5(CO)_{14}B_2H$	4n+2	13	7	Closo	$B_7H_7^{2-}$	C_7H_2
$Bi_3Ni_4(CO)_6^{2-}$	4n+2	13	7	Closo	$B_7H_7^{2-}$	C_7H_2
$(Cp*Co)_3B_4H_4$	4n+0	14	7	$C^1C[M-6]$	B_7H_7	C_7
$(Cp*Mo)_2B_5H_9$	4n-2	15	7	$C^2C[M-5]$	B_7H_5	C_7-2H
$(Cp*Cr)_2B_4H_7Co(CO)_3$	4n-2	15	7	$C^2C[M-5]$	B_7H_5	C_7-2H
$(Cp*Re)_2B_6H_2Cl_5$	4n-2	15	7	$C^2C[M-5]$	B_7H_5	C_7-2H
$(Cp*Ru)_3Co(CO)_3(BH)_3$	4n-2	15	7	$C^2C[M-5]$	B_7H_5	C_7-2H
$Me_4C_4B_4H_4$	4n+4	14	8	Nido	B_8H_{12}	C_8H_4
$(Cp*Ni)_4B_4H_4$	4n+4	14	8	Nido	B_8H_{12}	C_8H_4
$Bi_4Ni_4(CO)_6^{2-}$	4n+2	15	8	Closo	$B_8H_8^{2-}$	C_8H_2
$(CpCo)_4B_4H_4$	4n+0	16	8	$C^1C[M-7]$	B_8H_8	C_8
$(CpCo)_4B_4H_4$	4n+0	16	8	$C^1C[M-7]$	B_8H_8	C_8
$(Cp*Re)_2B_4H_4\ Co_2(CO)_5$	4n-4	18	8	$C^3C[M-5]$	B_8H_4	C_8-4H
$(Cp*Re)_2B_6H_6$	4n-4	18	8	$C^3C[M-5]$	B_8H_4	C_8-4H
$(Cp*Re)_2B_6H_4Cl_2$	4n-4	18	8	$C^3C[M-5]$	B_4H_4	C_8-4H
$(Cp*Rh)_2\,(Co)_3(CO)_8B_3H(Cl)$	4n+0	16	8	$C^1C[M-7]$	B_8H_8	C_8
$Cp*Ni)_4B_5H_5$	4n+4	16	9	Nido	B_9H_{13}	C_9H_4
$(CpNi)_3CB_5H_6$	4n+4	16	9	Nido	B_9H_{13}	C_9H_4
$Bi_3Ni_6(CO)_9^{2-}$	4n+0	18	9	$C^1C[M-8]$	B_9H_9	C_9
$(Cp*Re)_2B_7H_{11}$	4n+0	18	9	$C^1C[M-8]$	B_9H_9	C_9
$(Cp*Re)_2B_7H_7$	4n-4	20	9	$C^3C[M-6]$	B_9H_5	C_9-4H

Complex	Series	k value	n Value	Classification	Equivalent Borane	Equivalent Hydrocarbon
$(C_6H_6)RuB_9H_9^{2-}$	4n+2	19	10	Closo	$B_{10}H_{10}^{2-}$	$C_{10}H_2$
$(C_6H_6)RuB_9H_9$	4n+0	20	10	$C^1C[M-9]$	$B_{10}H_{10}$	C_{10}
$(CpCo)_3B_6H_7Co(CO)_2$	4n+0	20	10	$C^1C[M-9]$	$B_{10}H_{10}$	C_{10}
$(Cp*Co)_4B_6H_6$	4n+0	20	10	$C^1C[M-9]$	$B_{10}H_{10}$	C_{10}
$(Cp*Ru)_2(C_6H_6)RuB_7H_7$	4n-2	21	10	$C^2C[M-8]$	$B_{10}H_8$	C_{10}-2H
$(Cp*Ru)_2(C_6H_6)RuB_7H_7$	4n-2	21	10	$C^2C[M-8]$	$B_{10}H_8$	C_{10}-2H
$(Cp*Re)_2B_8H_8$	4n-4	22	10	$C^3C[M-7]$	$B_{10}H_6$	C_{10}-4H
$B_{11}H_{14}^{-}$	4n+4	20	11	Nido	$B_{11}H_{15}$	$C_{11}H_4$
$Ir(CO)(L_2)(S)B_9H_{10}$	4n+4	20	11	Nido	$B_{11}H_{15}$	$C_{11}H_4$
$B_{10}H_{12}TlMe^{2-}$	4n+4	20	11	Nido	$B_{11}H_{15}$	$C_{11}H_4$
$(Cp*Re)_2B_9H_9$	4n-4	24	11	$C^3C[M-8]$	$B_{11}H_7$	C_{11}-4H
$(Cp*W)_3HB_8H_8$	4n-8	26	11	$C^5C[M-6]$	$B_{11}H_5$	C_{11}-8H
$(CpNi)_2B_{10}H_{10}$	4n+2	23	12	Closo	$B_{12}H_{12}^{2-}$	$C_{12}H_2$
$(Cp*Re)_2B_{10}H_{10}$	4n-4	26	12	$C^3C[M-9]$	$B_{12}H_8$	C_{12}-4H
$(Cp*Re)_2B_{10}H_{10}$	4n-4	26	12	$C^3C[M-9]$	$B_{12}H_8$	C_{12}-4H

Table 1. Continued

Complex	Series	k value	n Value	Classification	Equivalent Borane	Equivalent Hydrocarbon
$(Cp^*Co)_2B_4H_6$	4n+2	11	6	Closo	$B_6H_6^{2-}$	C_6H_2
$(Cp^*ReH_2)(CO)B_4H_6$	4n+0	12	6	$C^1C[M-5]$	B_6H_6	C_6
$(CpCo)_2B_2H_5Mo_2(CO)_6(I)$	4n+0	12	6	$C^1C[M-5]$	B_6H_6	C_6
$(Cp^*ReH_2)_2B_4H_4$	4n+0	12	6	$C^1C[M-5]$	B_6H_6	C_6
$(Cp^*Ru)_2B_3H_4Co(CO)_3$	4n+0	12	6	$C^1C[M-5]$	B_6H_6	C_6
$(Cp^*Re)_2B_4H_8$	4n+0	12	6	$C^1C[M-5]$	B_6H_6	C_6
$(Cp^*Cr)_2B_4H_8$	4n-2	13	6	$C^2C[M-4]$	B_6H_4	C_6-2H
$(Cp^*Cr)_2B_4H_8$	4n-2	13	6	$C^2C[M-4]$	B_6H_4	C_6-2H
$(Cp^*Cr)_2(CO)_2B_4H_2$	4n-4	14	6	$C^3C[M-3]$	B_6H_2	C_6-4H
$(Cp^*Re)_2B_4H_4$	4n-4	14	6	$C^3C[M-3]$	B_6H_2	C_6-4H
$(Cp^*Co)_3B_4H_4$	4n+0	14	7	$C^1C[M-6]$	B_7H_7	C_7
$(Cp^*Mo)_2B_5H_9$	4n-2	15	7	$C^2C[M-5]$	B_7H_5	C_7-2H
$(Cp^*Cr)_2B_4H_7Co(CO)_3$	4n-2	15	7	$C^2C[M-5]$	B_7H_5	C_7-2H
$(Cp^*Re)_2B_6H_2Cl_5$	4n-2	15	7	$C^2C[M-5]$	B_7H_5	C_7-2H
$(Cp^*Ru)_3Co(CO)_3(BH)_3$	4n-2	15	7	$C^2C[M-5]$	B_7H_5	C_7-2H
$(CpCo)_4B_4H_4$	4n+0	16	8	$C^1C[M-7]$	B_8H_8	C_8
$(CpCo)_4B_4H_4$	4n+0	16	8	$C^1C[M-7]$	B_8H_8	C_8
$(Cp^*Re)_2B_4H_4\ Co_2(CO)_5$	4n-4	18	8	$C^3C[M-5]$	B_8H_4	C_8-4H
$(Cp^*Re)_2B_6H_6$	4n-4	18	8	$C^3C[M-5]$	B_8H_4	C_8-4H
$(Cp^*Re)_2B_6H_4Cl_2$	4n-4	18	8	$C^3C[M-5]$	B_4H_4	C_8-4H
$(Cp^*Rh)_2(Co)_3(CO)_8B_3H(Cl)$	4n+0	16	8	$C^1C[M-7]$	B_8H_8	C_8

Table 2. Display of mainly capped series

Complex	Series	k value	n Value	Classification	Equivalent Borane	Equivalent Hydrocarbon
$(Cp^*Re)_2B_7H_{11}$	4n+0	18	9	$C^1C[M-8]$	B_9H_9	C_9
$(Cp^*Re)_2B_7H_7$	4n-4	20	9	$C^3C[M-6]$	B_9H_5	C_9-4H
$(C_6H_6)RuB_9H_9$	4n+0	20	10	$C^1C[M-9]$	$B_{10}H_{10}$	C_{10}
$(CpCo)_3B_6H_7Co(CO)_2$	4n+0	20	10	$C^1C[M-9]$	$B_{10}H_{10}$	C_{10}
$(Cp^*Co)_4B_6H_6$	4n+0	20	10	$C^1C[M-9]$	$B_{10}H_{10}$	C_{10}
$(Cp^*Ru)_2(C_6H_6)RuB_7H_7$	4n-2	21	10	$C^2C[M-8]$	$B_{10}H_8$	C_{10}-2H
$(Cp^*Ru)_2(C_6H_6)RuB_7H_7$	4n-2	21	10	$C^2C[M-8]$	$B_{10}H_8$	C_{10}-2H
$(Cp^*Re)_2B_8H_8$	4n-4	22	10	$C^3C[M-7]$	$B_{10}H_6$	C_{10}-4H
$(Cp^*Re)_2B_9H_9$	4n-4	24	11	$C^3C[M-8]$	$B_{11}H_7$	C_{11}-4H
$(Cp^*W)_3HB_8H_8$	4n-8	26	11	$C^5C[M-6]$	$B_{11}H_5$	C_{11}-8H
$(CpNi)_2B_{10}H_{10}$	4n+2	23	12	Closo	$B_{12}H_{12}^{2-}$	$C_{12}H_2$
$(Cp^*Re)_2B_{10}H_{10}$	4n-4	26	12	$C^3C[M-9]$	$B_{12}H_8$	C_{12}-4H
$(Cp^*Re)_2B_{10}H_{10}$	4n-4	26	12	$C^3C[M-9]$	$B_{12}H_8$	C_{12}-4H

3. Conclusion

Capped clusters are relatively common among transition metal carbonyl complexes. On the other hand, such capped clusters are rare in stable borane clusters. In this work, by using the 4n series method, it has been found that capped clusters are prevalent in metalloborane complexes. The capped clusters ranged from 4n+0 to 4n-8 series. This clearly illustrates the fact that metal fragments have a tendency of stabilizing the fragile capped boranes. The 4n series method is extremely useful in categorizing skeletal atoms, molecules, fragments and clusters.

Acknowledgement

The financial support by the University of Nambia and NAMSOV, Namibia is greatly appreciated. Gratitude is also extended to Mrs Merab Kambamu Kiremire for assisting in proof reading the paper.

References

Callahan, K. P., & Hawthorne, M. F. (1974). New Chemistry of Metallocarboranes and Metalloboranes. *Pure and Appl. Chem., 39*(4), 475-495. http://dx.doi.org/10.1351/pac197439040475

Cox, D. N., & Mingos, D. M. P. (1981). Extended Hückel Molecular-Orbital Calculations on Dodecahedral Metalloboranes which do not conform to the Polyhedral Skeletal Electron-Pair Theory. *J. Chem. Soc., Dalton*, 1788-1797.

Cox, D. N., & Mingos, D. M. P. (1981). Extended Hückel Molecular-Orbital Calculations on Dodecahedral Metalloboranes which do not conform to the Polyhedral Skeletal Electron-Pair Theory. *J. Chem. Soc., Dalton,* 1788-1797. http://dx.doi.org/10.1039/DT9810001788

Fehlner, T. P. (2002). A Systematic Approach to Varied Roles of Transition Metals in Metallaborane Chemistry. *PINSA,*

68, A(6), 579-596.

Grimes, R. N. (1974). Recent Studies of Metalloborane Cage Compounds Derived from Small Carboranes. *Pure and Appl. Chem., 39*(4), 455-474. http://dx.doi.org/10.1351/pac197439040455

Hoffmann, R. (1982). Building Bridges between Inorganic and Organic Chemistry. *Angew. Chem. Int. Ed. Engl., 21,* 711-724. http://dx.doi.org/10.1002/anie.198207113

Kiremire, E. M. (2015a). Isolobal Series of Chemical Fragments. *Orient. J. Chem., 31*(spl. Edn), 59-70.

Kiremire, E. M. R. (2015b). Capping and Decapping Series of Boranes. *Int. J. Chem.* 7(2), 186-197. http://dx.doi.org/10.5539/ijc.v7n2p186

McGlinnchey, M. J., & Hopf, H. (2011). Reciprocal polyhedral and the Euler relationship: Cage hydrocarbons, C_nH_n and Closo-boranes $B_xH_x^{2-}$. *Beilstein J. Org. Chem., 7,* 222-233. http://dx.doi.org/10.3762/bjoc.7.30

Mingos, D. M. P. (1972). A General Theory for Cluster and Ring Compounds of the Main Group and Transition Elements. *Nature(London), Phys. Sci., 236,* 99-102. http://dx.doi.org/10.1038/physci236099a0

Mingos, D. M. P. (1984). *Polyhedral Skeletal Electron Pair Approach. Acc. Chem. Res., 17*(9), 311-319. http://dx.doi.org/10.1021/ar00105a003

Mingos, D. M. P. (1991). Theoretical aspects of metal cluster chemistry. *Pure and Appl. Chem., 83*(6), 807-812. http://dx.doi.org/10.1351/pac199163060807

Rudolph, R. W. (1976). Boranes and heteroboranes: a paradigm for the electron requirements of clusters? *Acc. Chem. Res., 9*(12), 446-452. http://dx.doi.org/10.1021/ar50108a004

Wade, K. (1971). The structural significance of the number of skeletal bonding electron-pairs in carboranes, the higher boranes and borane ions and various transition metal carbonyl cluster compounds. *Chem. Commun.,* 792-793. http://dx.doi.org/10.1039/c29710000792

3

Synthesis and Properties of Poly(ferrocenylsilane)s Containing Si-Vinylene Units in the Main Chain

Naofumi Naga[1], Yoshie Saito[1]

[1]Department of Applied Chemistry, Materials Science Course, College of Engineering, Shibaura Institute of Technology, 3-7-5 Toyosu, Koto-ku, Tokyo 135-8548, Japan

Correspondence: Naofumi Naga, Department of Applied Chemistry, Materials Science Course, College of Engineering, Shibaura Institute of Technology, 3-7-5 Toyosu, Koto-ku, Tokyo 135-8548, Japan. E-mail: nnaga@sic.shibaura-it.ac.jp

Abstract

Poly(ferrocenylsilane)s containing Si-vinylene unit in the main chain have been synthesized by Mizoroki-Heck reaction of 1,1'-bis(dimethylvinylsilyl)ferrocene (MVFS) and dibromo aryl compounds using palladium diacetate as a catalyst. The alternating copolymers with relatively low molecular weight were obtained in good yields. Cyclic voltammetry of the copolymers in CH_2Cl_2 solutions showed redox potential derived from the ferrocene units in the copolymers. The voltammograms the copolymers indicated that almost no difference in numbers of electron transfer compared with MVFS. By contrast, diffusion coefficients of the ferrocene units in the copolymers were smaller than that of MVSF. The copolymers showed reversible electrochromism from yellow to blue-green due to the redox of ferrocene units in the main chain by 3V of application.

Keywords: poly(ferrocenylsilane), vinylene unit, Mizoroki-Heck reaction, copolymerization

1. Introduction

Metallocene containing polymers have been developed because of their high potentials such as redox, magnetic, photo-physical, and catalytic properties (for example Abd-El-Aziz, Agatemor, & Etkin, 2014). Poly(ferrocene) is one of the most synthesized and studied metallocene containing polymers due to their characteristic features derived from the ferrocene units in the polymer. The main chain type poly(ferrocene)s have been synthesized by ring opening polymerization, condensation polymerization, coupling reaction, and so on (for examples Bellas & Rehahn, 2007; Miles, Ward, & Foucher, 2009; Miles, Ward, & Foucher, 2010; Wang, Huo, Ding, & Tan, 2008; Huo, Wang, Yu, Deng, Zhou, & Yang, 2007; Cazacu, Munteanu, Racles, Vlad, & Marcu, 2006; Caracu, Vlad, Marcu, Racles, Airinei, & Munteanu, 2006; Southard & Curti, 2001). Poly(ferrocenylsilane)s have been widely synthesized by ring opening polymerization of Si bridged ferrocene compounds, as shown in Scheme 1 (for example Bellas & Rehahn, 2007). Modification of substituted groups, R_1 and R_2 in Scheme 1, of the Si bridge of ferrocene compounds or resulting polymers, and copolymerization are usable to develop the highly functionalized poly(ferrocenylsilane)s (for examples Zechel, Hultzsch, Rulkens, Balaishis, Pudelski, Lough, & Manners, 1996; Nanjo, Cyr, Liu, Sargent, & Manners, 2008; Qian, Lu, Cambridge, Guerin, Manners, & Winnik, 2012; Sui, Hempenius, & Vancso, 2012; Wang, Wang, Wang, Chen, Yu, Wang, & Wang, 2006; Gilroy, Rupar, Whittell, Chabanne, Terrill, Winnik, Manners, & Richardson, 2011; Roerdink, van Zanten, Hempenius, Zhoug, Feijen, & Vancso, 2007; Ren, Shaushan, Liu, & Tong, 2007; Bunz, Maker, & Porz, 2011).

Incorporation of the ferrocene units in the main chain of π-conjugated polymers has been investigated by metathesis polymerization, Pd catalyzed coupling polymerization, and click reaction (Bueretea & Tilley, 1997; Heo, Park, & Lee, 2005; Weyehardt & Plenio, 2008; Masson, Lugh, & Manners, 2008; Lang, Voll, Inglis, Dingenouts, Goldmann, Barner, & Barner-Kowollik, 2011; Choi, Lee, Joo, Lee, Lee, & Chae, 2007; Yamamoto, Morikita, Maruyama, Kubota, & Katada, 1997; He, Gadt, Jones, Scholes, Manners, & Winnik, 2009). Sheridan et al. synthesized poly(ferrocenylsilane)s having conjugation units in the main chain using hydroslilylation reaction of 1,1'-bis(dimethylsilyl)ferrocene and diethynylbenzenes (Jain, Lalancette, & Sheridan, 2004). The resulting copolymers have ferrocene, silyl, and phenylene-vinylene units in the polymer back bone.

We developed fluorene- and carbazole-based alternating copolymers containing Si-vinylene units in the main chain (Naga, Tagaya, Noda, Imai, & Tomoda, 2008; Naga, Ohkura, Tagaya, & Tomoda, 2011). These copolymers were synthesized by alternating copolymerization of dibromofluorene or dibromocarbazole with Si containing divinyl or diallyl compounds using Mizoroki-Heck reaction with a Pd catalyst. Incorporation of Si and unsaturated vinylene units in the main chain of the conjugated polymers was effective to improve solubility of the polymers in organic solvents. Furthermore, the copolymerization is usable to control the photophysical properties of the conjugated polymers. We came to an idea to synthesize poly(ferrocenylsilane)s containing Si-vinylene units in the main chain using the alternating copolymerization. This paper reports the alternating copolymerization of 1,1-bis(dimethylvinylsilyl)ferrocene (MVSF) with aromatic dibromo compounds ($ArBr_2$) using Mizoroki-Heck reaction with a Pd catalyst, as shown in Scheme 2, and the photophysical and electro chemical properties of the resulting copolymers.

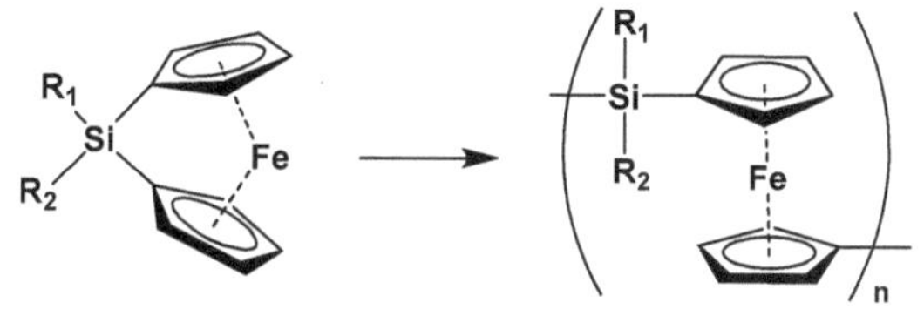

Scheme 1. Synthesis of poly(ferrocenylsilane) by ring-opening polymerization of bridged ferrocene.

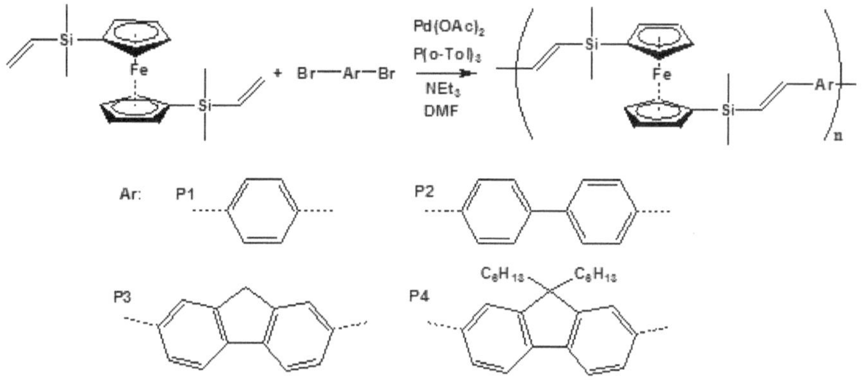

Scheme 2. Synthesis of poly(ferrocenylsilane)s containing Si-vinylene unit in the main chain by Mizoroki-Heck reaction of bis(dimethylvinylsilyl)ferrocene (MVFS) and dibromo aryl compounds.

2. Method

2.1 Materials

Ferrocene (Tokyo Chemical Industry Co., Ltd.), n-butyllithium (1.63 M n-hexane solution, Kanto Chemical Co. Ltd.), N,N,N',N'-tetramethylethylenediamine (TMEDA, Kanto Chemical Co. Ltd.), and chlorodimethylvinylsilane (AZmax.co) were commercially obtained, and used without further purification. n-Hexane was dried over calcium hydride under refluxing for 6 h and distilled before use under nitrogen atmosphere. Palladium(II) acetate ($Pd(OAc)_2$) (Sigma-Aldrich Co. LLC.), tri-*o*-tolylphosphine ($P(o\text{-}Tol)_3$) (Kanto Chemical Co. Ltd.), triethylamine (NEt_3) (Kanto Chemical Co. Ltd.), and N,N-dimethylformamide (DMF) (dehydrated grade, Kanto Chemical Co. Ltd.), were commercially obtained and used as received. *p*-Dibromobenzene (Bz, Kanto Chemical Co. Ltd.), 4,4'-dibromobiphenyl (BPh, Sigma-Aldrich Co. LLC.), 2,7-dibromofluorene (Flu, Wako Pure Chemical Industries, Ltd.), and 9,9'-dihexyl-2,7-dibromofluorene (HFlu, Sigma-Aldrich Co. LLC.) were commercially obtained and used without further purification.

Indium tin oxide (ITO) coated glass slide (surface resistivity: 8-12 Ω/sq) was purchased from Sigma-Aldrich Co. LLC. Chloroform and CH_2Cl_2 (spectroscopy grade) were commercially obtained from Dojindo Molecular Technologies, Inc. Tetrabutylammonium hexafluorophosphate was commercially obtained from Tokyo Chemical Industry Co., Ltd., and used as received.

2.2 Synthesis of 1,1-bis(dimethylvinylsilyl)Ferrocene (MVSF)

MVSF was synthesized according to the previous report (Majchrzak, Marciniec, Kubicki, & Pawelczyk, 2005). A n-hexane solution of n-butyllithium (40.0 mmol, 1.63 M, 24.5 mL) was added dropwise to a hexane (80 mL) solution of TMEDA (40.8 mmol) and ferrocene (19.9 mmoL) at room temperature under nitrogen atmosphere, and the reaction system was stirred for 20 h. The reaction system was cooled to -78°C, and chlorodimethylvinylsilane (41.8 mmol) was added dropwise to the reaction solution, and stirred for 25 min. The reaction system was gradually warmed to room temperature, and stirred for 8 h. The solvent was evaporated, and the products was purified by silica gel column using n-hexane as an eluent.

2.3 Synthesis of Poly(ferrocenylsilane)s Containing Si-vinylene Unit in the Main Chain

The copolymerization was carried out in a 100 mL glass reactor equipped with a magnetic stirrer. Equivalent molar of MVSF (1.25 mmol) and a $ArBr_2$ (1.25 mmol) were added to the reactor under nitrogen atmosphere. P(*o*-Tol)$_3$ (0.25 mmol), NEt_3 (3.13 mmol), and a DMF solution (4.0 mL) of $Pd(OAc)_2$ (0.05 mmol) were introduced to the reactor. The copolymerization was conducted at 100ºC for 24 h. The reaction was terminated by adding a small amount of methanol. The polymer was precipitated in a large excess of methanol and recovered by decantation. The copolymer obtained was dissolved in chloroform and re-precipitated in methanol. The precipitate was collected by decantation and dried in vacuo at 60°C for 6 h.

2.4 Analytical Procedures

^{1}H NMR spectra of the copolymers were recorded at room temperature on a JEOL-JNM-LA300 spectrometer in pulse Fourier transform mode. The sample solution was made in $CDCl_3$ as a solvent and the resonance of $CDCl_3$ (7.24 ppm) was used as an internal reference. Molecular weight and molecular weight distribution of the copolymers were measured at 40ºC by means of gel-permeation chromatography, SHIMADZU Prominence GPC System, using chloroform as a solvent, and calibrated with standard polystyrene samples. UV-vis absorption spectroscopy of the copolymers was conducted with SHIMADZU UV-1200 in a chloroform or CH_2Cl_2 solution. Photoluminescence (PL) spectroscopy was investigated with a SHIMADZU RF-1500 in a chloroform solution of the copolymer. Cyclic voltammetry was conducted with Electrochemical Measurement System HZ-5000 (Hokuto Denko) using Ag|AgCl reference electrode, glassy carbon working electrode, and Pt counter electrode at a scan rate of 0.1 V/s. Voltage for electrochlomism observation was applied by GP 025-5 (Takasago Ltd.) ranged from 0 to 3.0 V at a scan rate of 10 mV/s.

3. Results and Discussion

3.1 Synthesis and Structure of Poly(ferrocenylsilane)s

Copolymerizations of MVSF with $ArBr_2$ compounds have been investigated with $Pd(OAc)_2$ in DMF at 100ºC. The results are summarized in Table 1. Although the molecular weights of the copolymers were relatively low, the corresponding copolymers were obtained in good yield.

Table 1. Synthesis of alternating copolymers of MVFS with $ArBr_2$, and their structure and optical properties

Sample	$ArBr_2$	Yield [wt%]	M_n [g/mol]	M_w/M_n	Absorption[a] [nm]	Emission[a,b] [nm]
P1	Bz	46	2000	1.9	275	381, 416
P2	BPh	37	2100	2.0	275	386, 418
P3	Flu	53	2900	3.1	275	385, 420
P4	HFlu	75	3000	6.0	275	380, 420
MVSF					275	379, 418
Ferrocene					275	381, 416

[a] Evaluated in $CHCl_3$ (10^{-8} mol/L of ferrocene unit). [b] Emission was exited at λ_{max} of absorption.

Structure of the copolymers was studied by ^{1}H NMR spectroscopy. Figure 1 shows ^{1}H NMR spectrum of P1 prepared by the copolymerization of MVFS and Bz. The signals derived from methyl, cyclopentadienyl, vinylene, and phenyl groups are observed at 0.3-0.4, 4.1-4.4, 6.6-6.9, and 7.4 ppm, respectively. The peaks assignments of the ^{1}H NMR spectra of the obtained copolymers are listed below. P1 (ppm) : 0.3-0.4 (d, 12H), 4.1-4.4 (m, 8H), 6.6-6.9 (m, 4H), 7.4 (s, 6H); P2 (ppm) : 0.3-0.4 (d, 12H), 3.9 (s, 2H), 4.2-4.4 (m, 8H), 6.6-6.9 (m, 4H), 7.6 (m, 8H); P3 (ppm) : 0.3-0.4 (d, 12H), 4.1-4.4 (m, 8H), 6.6-6.9 (m, 4H), 7.4-7.7 (m, 6H); P4 (ppm) : 0.3-0.4 (m, 18H), 0.6-1.1 (m, 20H), 4.1-4.4 (m, 8H), 6.6-6.9 (m, 4H), 7.4-7.7 (m, 6H). These chemical shift assignments and quantitative analyses of the peaks intensity cleared that the copolymerizations yielded the corresponding copolymers.

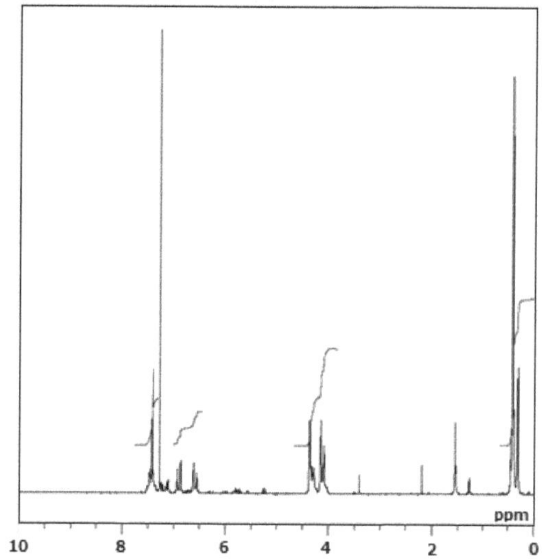

Figure 1. ^{1}H NMR spectrum of P1.

3.2 Optical Properties

UV-vis and PL spectra of the copolymers were acquired in chloroform. Figure 2 shows absorption and emission spectra of P1. The wavelengths UV-vis and PL spectra of the copolymers are summarized in Table 1. All the copolymers showed an absorption peak at 275 nm, and emission peaks at around 380 and 420 nm. These wavelengths were almost same those of MVSF or ferrocene. The absorption and emission of the copolymers are derived from ferrocene units, and their wavelengths are independent of structure of the aryl units.

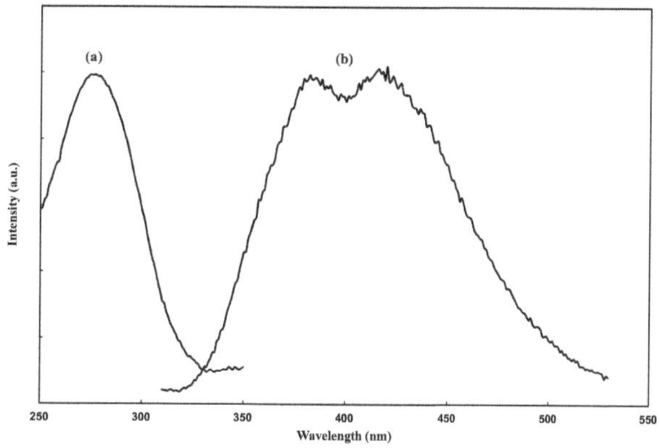

Figure 2. UV-vis (a) and PL (b) spectra of P1 in $CHCl_3$ solution, 10^{-8} mol/L of ferrocene unit, emission was exited at λ_{max} of absorption of 275 nm.

3.3 Electrochemical Properties

Wang et al. reported effect of molecular structure of poly(ferrocenlysilane)s on electrochemical behavior (Wang, Wang, Wang, Chen, & Chen, 2006; Wang, Wang, Wang, & Chen, 2007). Electrochemical properties of the present copolymers in CH_2Cl_2 solution containing 0.1 M of tetrabutylammonium hexafluorophosphate were investigated with cyclic voltammetry. The results are summarized in Figure 3, and Table 2. The copolymers showed redox potential, derived from ferrocene units in the main chain. All the copolymers and MVSF, which have dimethylsilylene-vinylene units, showed higher $E_{1/2}$ values than that of ferrocene, indicating shift of the voltammograms to anode due to the electron withdrawing of the ferrocene units by the vinylene units. ΔE values of the copolymers (0.08 -0.11 V) were almost same with those of the MVFS. The results indicate almost no difference in numbers of electron transfer between the copolymers and MVFS. By contrast, I_{pa} and I_{pc} of the copolymers were smaller than those of MVFS. Small diffusion coefficient of the ferrocene units in the copolymers, which was caused by large molecular weight, should decrease the I_{pa} and I_{pc} values of the copolymers.

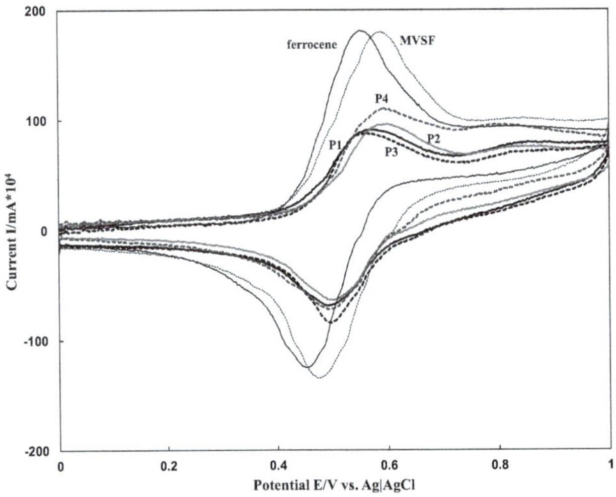

Figure 3. Cyclic voltammograms of alternating copolymers of MVFS with $ArBr_2$, P1: black bold line, P2: black dot line, P3: gray bold line, P4: gray dot line, MVFS: thin dot line, and ferrocene: thin line, in CH_2Cl_2 containing 0.5 mM of ferrocene unit and 0.1 M of tetrabutylammonium hexafluorophosphate, at a scan rate of 0.1 V/s.

Table 2. Electrochemical properties of MVFS/$ArBr_2$ alternating copolymers [a]

Sample	E_{pa} [V]	I_{pa} [mA*10^4]	E_{pc} [V]	I_{pc} [mA*10^4]	$E_{1/2}$ [V]	ΔE [V]
P1	0.57	91.1	0.49	-68.7	0.53	0.08
P2	0.60	96.2	0.50	-63.8	0.55	0.10
P3	0.57	87.7	0.49	-84.2	0.53	0.08
P4	0.60	110.1	0.49	-72.2	0.55	0.11
Ferrocene	0.55	181.2	0.46	-125.3	0.50	0.10
MVSF	0.59	180.1	0.48	-134.7	0.53	0.11

[a] Determined by cyclic voltammetry of CH_2Cl_2 solution of the copolymers containing 0.5 mM of ferrocene unit and 0.1 M electrolyte (tetrabutylammonium hexafluorophosphate) at scan rate of 0.1 V/s. E_{pa} = peak anodic potential; I_{pa} = peak anodic current, E_{pc} = peak cathodic potential; I_{pc} = peak cathodic current; $E_{1/2} = (E_{pa}+E_{pc})/2$, ΔE = difference between E_{pa} and E_{pc}.

3.4 Electrochromism

Electrochromism of the polymer in CH_2Cl_2 solution was investigated in an ITO cell shown in Figure 4. UV-vis spectra of the P2 solution containing tetrabutylammonium hexafluorophosphate in the ITO cell without or with applied voltage are shown in Figure 5. The application of 3.0 V increased intensity of the absorption ranged from 200-450 nm. Figure 6 shows photos of the ITO cell containing the P2 solution. The change of UV-vis spectrum induced electrochromism of the polymer solution from yellow to blue-green derived from oxidation of ferrocene units in the polymer chain. The electrochromism was reversible for several times of electronic cycles. The copolymers of P1, P2, and P4 showed similar electrochromism behavior.

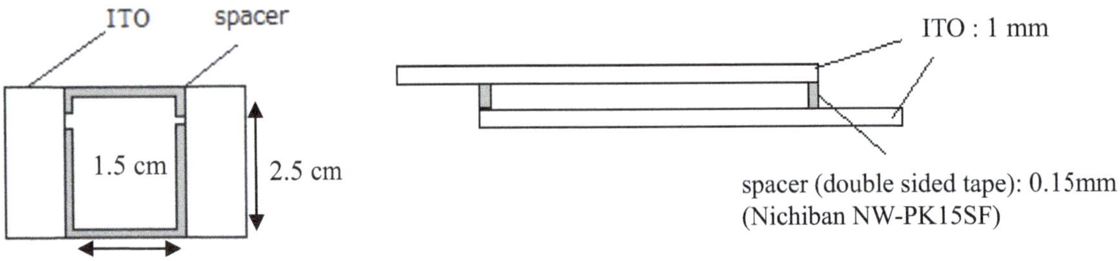

Figure 4. Sandwich-type ITO electrode cell.

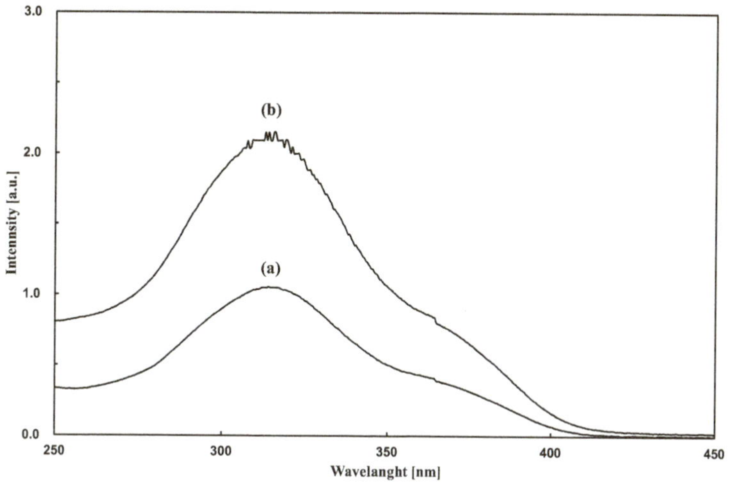

Figure 5. UV-vis spectra of CH_2Cl_2 solution of P2, (a) without voltage, (b) with 3.0 V containing 1.0 M of ferrocene unit and 0.2 M of tetrabutylammonium hexafluorophosphate.

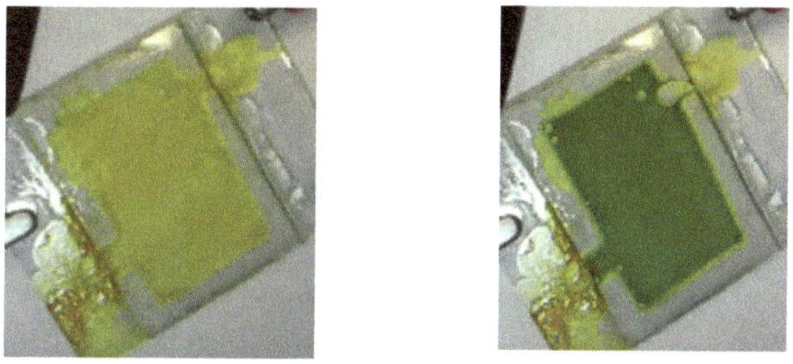

(a) (b)

Figure 6. Photos of CH_2Cl_2 solution of P2 in ITO cell, (a) without voltage, (b) with 3.0 V, containing 1.0 M of ferrocene unit and 0.2 M of tetrabutylammonium hexafluorophosphate.

4. Conclusions

Poly(ferrocenylsilane)s containing Si-vinylene unit in the main chain were obtained with Mizoroki-Heck reaction of bis(dimethylvinylsilyl)ferrocene and dibromo aryl compounds using $Pd(OAc)_2$ catalyst. The copolymers showed good solubility in $CHCl_3$ and CH_2Cl_2. The $CHCl_3$ solutions of the copolymers showed absorption and photo luminescence derived from the ferrocene units in the copolymers. The absorption and luminescence wavelengths were almost independent of the molecular structure of the aryl compounds. The cyclic voltammograms of the alternating copolymers indicated the electron withdrawing of the ferroecene units by the vinylene units, and small diffusion coefficient of the ferrocene units in the copolymers caused by low mobility derived from large molecular weight. The copolymers also showed reversible electrochromism due to the redox of the ferrocene units in the main chain.

Although, the present copolymerization should be one of the useful methods to design and synthesize the poly(ferrocenylsilane)s containing conjugation, the molecular weights of the copolymers were too low to form the films. Furthermore, the optical properties of the copolymers were independent with the structure of Ar units. The ferrocene units are dominant of the optical properties of the copolymers. As the next step, we are trying to synthesize the high molecular weight poly(ferrocenylsilane)s containing long conjugation, which can control the optical properties of the resulting copolymers, and the results will be reported elsewhere.

References

Abd-El-Aziz, A., Agatemor, C., & Etkin, N. (2014). Sandwich Complex-Containing Macromolecules: Property Tunability Through Versatile Synthesis. *Macromol. Rapid Commun. 35*, 513-559. http://onlinelibrary.wiley.com/doi/10.1002/marc.201300826/full

Bellas, V., & Rehahn, M. (2007). Polyferrocenylsilane-Based Polymers Systems. *Angew. Chem. Int. Ed., 46*, 5082-5104. http://onlinelibrary.wiley.com/wol1/doi/10.1002/anie.200604420/full

Bueretea, M. A., & Tilley, T. D. (1997). Poly(ferrocenylenevinylene) from Ring-Opening Metathesis Polymerization of *ansa*-(Vinylene)ferrocene. *Organometallics, 16*, 1507-1510. http://pubs.acs.org/doi/abs/10.1021/om960940l

Bunz, U. H. F., Maker, D., & Porz, M. (2011). Alkene Metathesis–A Tool for the Synthesis of Conjugated Polymers. *Macromol. Rapid Commun., 33*, 886-910. http://onlinelibrary.wiley.com/doi/10.1002/marc.201200001/full

Caracu, M., Vlad, A., Marcu, M., Racles, C., Airinei, A., & Munteanu, G. (2006). New Organometallic Polymers by Polycondensation of Ferrocene and Siloxane Derivatives. *Macromolecules, 39*, 3786-3793. http://pubs.acs.org/doi/abs/10.1021/ma052030y

Cazacu, M., Munteanu, G., Racles, C., Vlad, A., & Marcu, M. (2006). New ferrocene-containing structures: Poly(silyl ester)s. *J. Organomet. Chem., 691*, 3700-3707. http://www.sciencedirect.com/science/article/pii/S0022328X06003809

Choi, T. L., Lee, K. H., Joo, W. J., Lee, S., Lee, T. W., & Chae, M. Y. (2007). Synthesis and nonvolatile memory behavior of redox-active conjugated polymer containing ferrocene. *J. Am. Chem. Soc., 129*, 9842-9843. http://pubs.acs.org/doi/abs/10.1021/ja0717459

Gilroy, J. B., Rupar, P. A., Whittell, G. R., Chabanne, L., Terrill, N. J., Winnik, M. A., Manners, I., & Richardson, R. M. (2011). Probing the Structure of the Crystalline Core of Field-Aligned, Monodisperse, Cylindrical Polyisoprene-*block*-Polyferrocenylsilane Micelles in Solution Using Synchrotron Small- and Wide-Angle X-ray Scattering. *J. Am. Chem. Soc., 133*, 17056-17062. http://pubs.acs.org/doi/abs/10.1021/ja207417z

He, F., Gadt, T., Jones, M., Scholes, G. D., Manners, I. & Winnik, M. A. (2009). Synthesis and Self-Assembly of Fluorescent Micelles from Poly(ferrocenyldimethylsilane-b-2-vinylpyridine-b-2,5-di(2'-ethylhexyloxy)-1,4-phenyl-vinylene)Triblock Copolymer. *Macromolecules, 42*, 7953-7960. http://pubs.acs.org/doi/abs/10.1021/ma900990y

Heo, R. W., Park, J. S., & Lee, T. R. (2005). Synthesis and Ring-Opening Metathesis Polymerization of Aryl-Substituted 1,1'-(1,3-Butadiennylene)ferrocenes. *Macromolecules, 38*, 2564-2573. http://pubs.acs.org/doi/abs/10.1021/ma047826w

Huo, J., Wang, L., Yu, H. J., Deng, L. B., Zhou, J. F., & Yang, Q. (2007). Synthesis and Electrochemical Behavior of Linear Oligo(ferrocenylsilane) and Hyperbranched Poly(ferrocenylsilane). *J. Polym. Sci.: Part B: Polym. Phys., 45*, 2880-2889. http://onlinelibrary.wiley.com/wol1/doi/10.1002/polb.21179/abstract

Jain, R., Lalancette, R. A., & Sheridan, J. B. (2004). Poly{ferrocecne(phenylene)bis(silylenevinylene)}s via Platinum- and Rhodium-Catalyzed Hydrosilylation of Diethynylbenzenes with 1,1'-Bis(dimethylsilyl)ferrocene, *Organometallics, 24, 1458-1467*. http://pubs.acs.org/doi/abs/10.1021/om049166p

Lang, C., Voll, D., Inglis, A. J., Dingenouts, N., Goldmann, A. S., Barner, L., & Barner-Kowollik, C. (2011). An Access Route to Polyferrocenes via Modular Conjugation. *Macromol. Chem. Phys., 212*, 831-839. http://onlinelibrary.wiley.com/doi/10.1002/macp.201100010/full

Majchrzak, M., Marciniec, B., Kubicki, M., & Pawelczyk, A. (2005). Highly Selective Synthesis and Structural Characterization of Bis(silyl)-[3]-ferrocenopane Derivatives. *Organometallics, 24,* 3731-3736. http://pubs.acs.org/doi/abs/10.1021/om050194x#om050194xAF1

Masson, G., Lugh, A. J., & Manners, I. (2008). Soluble Poly(ferrocenylene vinylene) with *t*-Butyl Substituents on the Cyclopentadienyl Ligands via Ring-Opening Metathesis Polymerization. *Macromolecules, 41*, 539-547. http://pubs.acs.org/doi/abs/10.1021/ma071034v

Miles, D., Ward, J., & Foucher, D. A. (2009). Polyferrocenyldisiloxane from the Platinum-Catalyzed Reactions of Tertiary Bis(dimethylsilyl)ferrocene in a Polar Aprotic Solvent. *Macromolecules, 42*, 9199-9203. http://pubs.acs.org/doi/pdf/10.1021/ma9018608

Miles, D., Ward, J., & Foucher, D. A. (2010). Unexpected Synthesis of Oligomeric (Dimethylsilyl)-Bridged Ferrocenes from the Desilylative Coupling of Tertiary 1,1'-Bis(dimethylsilyl)ferrocene. *Organometallics, 29*, 1057-1060. http://pubs.acs.org/doi/abs/10.1021/om100041k

Naga, N., Ohkura, Y., Tagaya, N., & Tomoda, H. (2011). Synthesis and Chemosensing Behavior of Fluorene-Based Alternating Copolymers Containing Ether Side Chains and Si-Vinylene Units in the Main Chain. *Journal of Polymer Science: Part A: Polymer Chemistry, 49*, 4935-4940. http://onlinelibrary.wiley.com/wol1/doi/10.1002/pola.24943/abstract

Naga, N., Tagaya, N., Noda, H., Imai, T., & Tomoda, H. (2008). Synthesis and Properties of Fluorene or Carbazole-Based Alternating Copolymers Containing Si and Vinylene Units in the Main Chain. *Journal of Polymer Science: Part A: Polymer Chemistry, 46*, 4513-4521. http://onlinelibrary.wiley.com/wol1/doi/10.1002/pola.22788/abstract

Nanjo, M., Cyr, P. W., Liu, K., Sargent, E. H., & Manners, I. (2008). Donor-Acceptor C_{60}-Containing Polyferrocenylsilanes: Synthesis, Characterization, and Applications in Photodiode Devices. *Adv. Func. Mater., 18*, 470-477. http://onlinelibrary.wiley.com/wol1/doi/10.1002/adfm.200700315/abstract

Qian, J., Lu, Y., Cambridge, G., Guerin, G., Manners, I., & Winnik, M. A. (2012). Polyferrocenylsilane Crystals in Nanoconfinement: Fragmentation, Dissolution, and Regrowth of Cylindrical Block Copolymer Micelles with a Crystalline Core. *Macromolecules, 45*, 8363-8372. http://pubs.acs.org/doi/abs/10.1021/ma3014788

Ren, B., Shaushan, L., Liu, X., & Tong, Z. (2007). Synthesis and Characterization of Poly(ferrocenylsilanes) with Coumarin Side Groups and Their Photochemical Reactivity and Electrochemical Behavior. *Macromolecules, 40,* 4501-4508. http://pubs.acs.org/doi/abs/10.1021/ma062502o

Roerdink, M., van Zanten, T. S., Hempenius, M. A., Zhoug, Z., Feijen, J., & Vancso, J. (2007). Poly(ferrocenylsilane)-*block*-Polylactide Block Copolymers. *Macromol. Rapid Commun., 28*, 2125-2130. http://onlinelibrary.wiley.com/doi/10.1002/marc.200700364/full

Southard, G. E., & Curtis, M. D. (2001). Synthesis and Characterization of Soluble Poly(ferrocenylenearylene)s from Condensation of Dilithio Bis(alkylcyclopentadienide)arenes with Iron(II) Halides: A general Route to Conjugated Poly(metallocene)s. *Organometallics, 20*, 508-522. http://pubs.acs.org/doi/abs/10.1021/om000722z

Sui, X., Hempenius, M. A., & Vancso, G. J. (2012). Redox-Active Cross-Linkable Poly(ionic liqud)s. *J. Am. Chem. Soc., 134*, 4023-4025. http://pubs.acs.org/doi/abs/10.1021/ja211662k

Wang, H. Y., Huo, J., Ding, J., & Tan, Q. (2008). Synthesis and curing behavior of a novel ferrocene-based epoxy compound. *J. Appl. Polym. Sci., 110*, 1594-1599. http://onlinelibrary.wiley.com/wol1/doi/10.1002/app.28584/abstract

Wang, J. J., Wang, L., Wang, X. J., Chen, T., & Chen, Y. (2006). Synthesis of poly(ferrocenylsilanes) with different molecular weight and their electrochemical behavior. *J. Appl. Polym. Sci., 100,* 473-477. http://onlinelibrary.wiley.com/wol1/doi/10.1002/app.23168/abstract

Wang, J. J., Wang, L., Wang, X. J., Chen, T., Yu, H. J., Wang, W., & Wang, C. L. (2006). Preparation of polyferrocenylsilane via ROP, synthesis of polyferrocenylsilane with methacrylate side chain and its photochemical cross-linking properties. *Materials Letters, 60*, 1416-1419. http://www.sciencedirect.com/science/article/pii/S0167577X05011638

Wang, X. J., Wang, L., Wang, J. J., & Chen, T. (2007). Electrochemical behaviors of poly(ferrocenylsilane) solutions. *J. Appl. Polym. Sci., 103, 789-794.* http://onlinelibrary.wiley.com/wol1/doi/10.1002/app.25072/abstract

Weyehardt, H., & Plenio, H. (2008). Acyclic Diene Metathesis Polymerization of Divinylarenes and Divinylferrocenes with Grubbs-Type Olefin Metathesis Catalysts. *Organometallics, 27*, 1479-1485. http://pubs.acs.org/doi/abs/10.1021/om701277p

Yamamoto, T., Morikita, T., Maruyama, T., Kubota, K., & Katada, M. (1997). Poly(aryleneethynylene) Type Polymers Containing a Ferrocene Unit in the-Conjugated Main Chain. Preparation, Optical Properties, Redox Behavior, and Mossbauer Spectroscopic Analysis. *Macromolecules, 30*, 5390-5396. http://pubs.acs.org/doi/abs/10.1021/ma9704368

Zechel, D. L., Hultzsch, K. C., Rulkens, R., Balaishis, D., Ni, Y., Pudelski, J. K., Lough, A. J., & Manners, I. (1996). Thermal and Transition-Metal-Catalyzed Ring-Opening Polymerization (ROP) of [1]Silaferrocenophanes with Chlorine Substituents at Silicon: A Route to Tunable Poly(ferrocenylsilanes). *Organometallics, 15*, 1972-1978. http://pubs.acs.org/doi/abs/10.1021/om950911v

4

On the Extraction of Vanadium (V) by Alkanols: A comparative Study of 1-pentanol and 3-methyl-1-butanol Isomers as Extractants for Vanadium (V) Extraction from Hydrochloric Acid Solutions

James O. Ojo[1], David A. Oyegoke[1]

[1]Department of Chemistry, Federal University of Technology, Akure, Ondo State, Nigeria

Correspondence: James O. Ojo, Department of Chemistry, Federal University of Technology, Akure, P.M.B. 704, Akure, Ondo State, Nigeria. E-mail: oluwasanjo85@yahoo.com.

Abstract

Further to the previous study on the extractive recovery of V(V) from HCl solution by some alkanols, the extraction of V(V) from HCl solutions by 1-pentanol and 3-methyl -1-butanol (isomers) dissolved in n-heptane were investigated in the present study, still propelled by the need to proffering viable alternatives to the scarce and relatively costly organophosphorous extractants, and also compare the extraction efficiency of straight- and branched-chain extractants. At the studied pH range (0.5-4.0) with 1-penttanol as extractant, the percentage V(V) extraction (extraction efficiency, E%) reached optimum E%= 40.0 at pH of 3.0 and for 3-methyl-1-butanol, optimum E% was 36.7 at pH 2.0. Increasing concentration of HCl at the studied range (1.0 – 6.0 mol L^{-1}) was found to have an appreciable effect on percentage V(V) extraction. With both extractants, E% peaked at 5.0 and 6.0 mol L^{-1}HCl with values 70.0 and 66.7% for 1-pentanol and 3-methyl-1-butanol respectively.

With increase in the concentration of extractants, both extractants produced a trend of increased/decreased in E% for V(V), suggesting the extraction of two different species. The optimal percentage extraction values were 70.0 and 66.7% observed at log [1-pentanol] = 0.86 mol L^{-1} and log[3-methyl-1-butanol]=0.86 mol L^{-1}, in that order. The determined enthalpy change. ΔH, for 1-pentanol and 3-methyl-1-butanol systems were -80.4/120 kJ mol^{-1} and -203.0/254.7 kJ mol^{-1} respectively, indicating exothermic/endothermic extraction processes. The isomer 3- methyl -1-butanol gave a higher extraction efficiency after a three-stage extraction (E% = 92.5) than 1-pentanol (E% = 83.4). Recycling experiment revealed both to be recyclable. From slope analysis, analytical and spectra data, the extracted complexes have been formulated as $VO_2Cl.nROH$ and $(VO_2Cl)_m.nROH$ (where m>> n) at low and high concentrations of extractant, respectively. On the whole, 3-methyl-1-butanol was assessed as fairly better as an extractant for vanadium (V) at lower temperature experimental conditions. The effect of methyl branched-chain played an insignificant role on the extraction.

Keywords: vanadium (V), extraction, 1-pentanol, 3-methyl-1-butanol, isomers, slope analysis, enthalpy

1. Introduction

Vanadium applications in various fields of human endeavours particularly modern technology, as well as its extraction in form of cationic vanadium (V) have been reported in recent work (Ojo and Oyegoke, 2013). The use of alkanols to extract valuable metals such as niobium (V) and tantalum (V), aluminium (III), gallium (III) and indium (III) was reviewed as well (Yamada, Hayashi & Yasui, 2006).

In the previous work reported by the current authors, vanadium (V) was extracted from HCl solutions by 1–butanol and 1–octanol with remarkable efficiencies, (Ojo & Oyegoke, 2013). It is a common knowledge that several alkanols such as 1–pentanol (Figure 1) and 3-methyl-1-butanol (Figure 2) share similar electronic and solvating properties with those previously investigated, and at least superficially, are expected to have the potentiality for V (V) extraction.

In addition to this, isomerization in compounds as in the case of 3–methyl–1–butanol, (branched–chain) and the straight–chain 1–pentanol expectedly should affect covalent and lipophillic properties of compounds, with the tendency to cause a disparity in solvating power and invariably the extraction efficiency. The presence of methyl branched–chain is also expected to bring a new dimension of disparity in steric hindrance of the extractants, and probably influence the extraction efficiency. It is in the light of this that 1–pentanol (straight–chain) and 3-methyl-1-butanol (branched–chain) were investigated as extractants for V (V) from HCl solutions.

$CH_3 - CH_2 - CH_2 - CH_2 - CH_2 - OH$

Figure 1. The structure of 1 – pentanol

CH_3
|
$CH_3 - CH_2 - CH_2 - CH_2 - OH$

Figure 2. The structure of 3– methyl – 1 – butanol

2. Materials and Methods

2.1 Materials Used and Their Sources

The chemicals NH_4VO_3, KCl, $Na_2WO_4.2H_2O$, HCl, HNO_3, H_3PO_4, and the reagents 1–pentanol, 3-methyl-1-butanol and n – heptane were purchased from the British Drug House Limited (BDH). All the materials were of analytical grade except the reagents (solvents) and were used as purchased.

Ammonium metavanadate (3.51g, 0.03mol) with KCl (7.48g, 0.1mol) were dissolved in hydrochloric acid solutions of varied concentrations, and made up to the 1–litre mark as V (V) stock solution.

Solutions of 1–pentanol and 3-methyl-1-butanol of varied concentrations in n–heptane and of the ranges 5.43–9.05 mol L^{-1} and 5.47–9.11 mol L^{-1} respectively, were prepared by dissolving the pure alkanols in n–heptane and made up to the 1-litre mark with the diluent.

2.2 Extraction and Analytical Procedure

Extraction, separation and analysis (vanadium –phosphotungstate method) were carried out as previously reported except that pentanol and 3-methyl-1-butanol were used as extractants (Ojo & Oyegoke, 2013).

The effects of pH (in the range 0.5–4.0), concentration of HCl (1.0 – 7.0 mol L^{-1}), variation in temperature between 301 and 321 K, recycling of pentanol and 3-methyl-1-butanol, batch extraction stages were investigated. The amount of vanadium extracted was determined by UV–visible spectrophotometry of the yellow phosphotungstate–vanadium complex at 400 nm (Vogel, 1998). The electronic spectra of the pure 1-pentanol and 3-methyl-1-butanol and their organic phases of V (V) were recorded on the Alpha spectrophotometer in the range 200–900 nm and 4000-650 cm^{-1} at the Obafemi Awolowo University, Ile–Ife, and Bowen University, Iwo, Nigeria, respectively.

3. Results and Discussion

3.1 The Effect of pH

The effect of pH on percentage V (V) extraction (E%) from HCl solutions with 1–pentanol and 3–methyl – 1 – butanol is presented in Figure 3. The E% for V (V) extraction using 1–pentanol and 3-methyl-1-butanol follow an irregular trend characterized by increased/decreased, with optimal values attained at pH of 3.0 (E% = 40.0) and 2.0 (E% =36.7) for pentanol and 3-methyl-1-butanol respectively. The observed trend was due to the prevalence of diverse kinds of extractable/unextractable V (V) species such as VO_2^+, H_2VO_4 and $H_2V_{10}O_{28}^{4-}$ which is consistent with what some authors reported in the organophosphorous extraction of V (V) from aqueous acids (Kurbatova & Kurbatov, 2006; Greenwood & Earnshaw, 1997; Ojo 2010; J. Ojo, Adebayo & I. Ojo, 2012), and also the alkanol extraction of V (V) by current authors (Ojo and Oyegoke, 2013). The unappreciable difference in the extraction efficiency (E%) of both extractants at this pH region might be due to the observed closeness in their hydrolytic stability(from preliminary work), implying that linear chain–length plays no role on E% under this experimental condition.

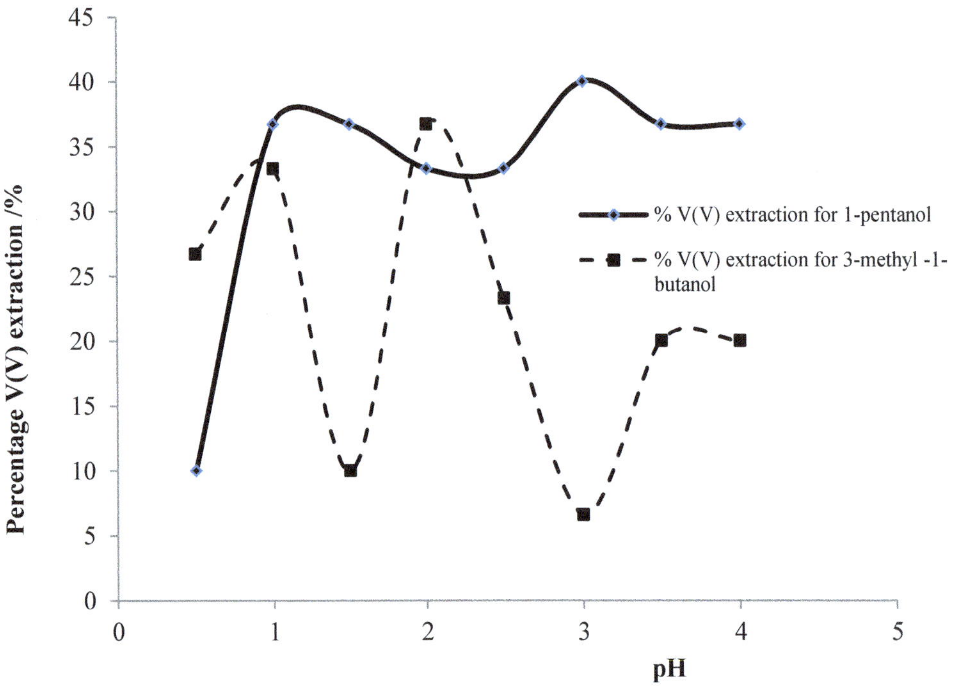

Figure 3. Effect of pH on the extraction of V(V) from HCl solutions with 1-pentanol and 3-methyl-1-butanol [V(V)] = 0.03 mol L^{-1}; [1 – pentanol] = 5.89 mol L^{-1}; [3-methyl-1-butanol] = 6.38 mol L^{-1}; Temperature = 301 K

3.2 The Effect of Concentration of HCl

The effect of concentration of HCl investigated at the range 1.0–7.0 mol L^{-1} on V (V) extraction with 1–pentanol and 3-methyl-1-butanol is presented in Figure 4. The E% increased gradually from 30.0% at [HCl] = 1.0 mol L^{-1} to 70.0% at [HCl] = 5.0 mol L^{-1} followed by a decrease (E% = 46.7) with the use of 1–pentanol as extractant.While for 3-methyl-1-butanol, the trend was increased, decreased and increased. The least E% was 20.0% observed at [HCl] = 1.0 mol L^{-1} and 66.7% at [HCl] = 6.0 mol L^{-1}.

The observed general increase in E% at increased concentration of HCl for both extractants could be explained on the basis of increased desolvation of coordinated water from the V (V) hydration sphere, making the pervanadyl specie (VO_2^{+}) available for binding to the alkanolic oxygen, as similarly reported by the current authors on V (V) extraction by 1- butanol and 1-octanol and other workers (Ojo & Adeeyinwo, 2011; Thomas & Reddy, 2003; Ojo & Oyegoke, 2013).

It is worth noting that the E% values of 1- pentanol are generally higher (although minimally) than 3-methyl-1-butanol at the studied concentration range which could not be explained on the basis of solubility of both extractants in aqueous solutions, as they have nearly the same solubilities, 2.3 g /100 g and 2.0 g/100 g H_2O for 1- pentanol and 3-methyl-1-butanol respectively (Morrison and Boyd, 2009). And for the fact that both extractants showed good phase separation at nearly the same rate (from preliminary work), differential interfacial and coalescent properties could very scarcely account for the disparities in E%. The plausible reason that could be advanced is the steric hindrance presented by 3-methyl-1-butanol with branched methyl group, which may relatively hinder the binding of the VO_2^+ specie to the alkanolic oxygen.

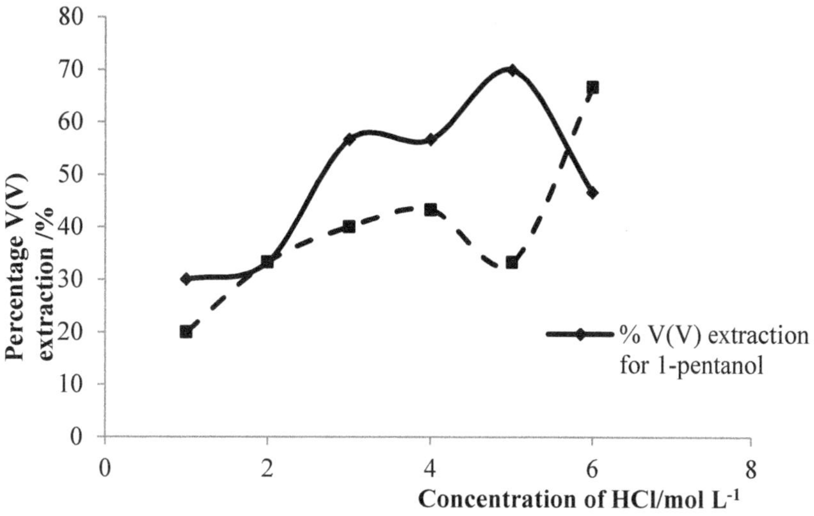

Figure 4. Effect of concentration of HCl solutions on percentage V(V) extraction [V(V)] = 0.03 mol L^{-1}; [1 – pentanol] = 5.89 mol L^{-1}; [3-methyl-1-butanol] = 6.38 mol L^{-1}; Temperature = 301 K

3.3 Effect of Concentration of Extractants

The effect of concentration of 1–pentanol and 3-methyl-1-butanol is depicted by Figure 5.From the slope analysis curves of the extractants, two linear portions were observed (at lower and higher concentrations). It was observed that increased extractant concentration produced an increased E% up to an optimum value, and subsequently decreased.

At log [1–pentanol] = 0.73, log D = 0 (E% = 50.0) which increased to an optimum of log D = 0.60 (E% = 86.5) at log [1-pentanol] = 0.86. As regards 3-methyl-1-butanol, at log [3-methyl-1-butanol] = 0.74, log D = 0.30 (E% = 66.7) which increased to an optimum of log D = 0.37 (E% = 70.0) at log [3-methyl-1-butanol] = 0.86 and subsequently decreased.

The difference in percentage extraction of V (V) by both extractants at lower concentration of extractants is minimal but becomes appreciable as the concentration increases.

This may be explained by the fact that angular VO_2^+ ion is not encountering much steric hinderance from the steric–prone methyl branch of 3-methyl – 1 – butanol at lower extractant concentration (lower amount of molecules) but becomes hindered at higher concentration (higher amount of molecules). This observation is in agreement with our reported work where the linear nature of the vanadyl ion (VO^{2+}) was viewed to have played a vital role on its better extraction than the steric–prone angular pervanady ion, VO_2^+ from HCl solution by di (2–ethylhexyl) phosphoric acid. (J. Ojo, Adebayo and I. Ojo, 2012).

If assumed that VO_2^+ is the predominant extractable specie, its extraction by both alkanols can be generally represented by Equation 1 (Ojo and Oyegoke, 2013).

$$VO_2^+{}_{(aq)} + Cl^-{}_{(aq)} + nROH_{(org)} \rightleftharpoons VO_2Cl \cdot nROH_{(org)} \quad (1)$$

Then,

$$K_{ex} = [VO_2Cl^- nROH]_{org}/[VO_2^+]_{aq} \cdot [Cl^-]_{aq} \cdot [ROH]_{org}^n \quad (2)$$

$$D = [V0_2Cl \cdot nROH]_{org}/[VO_2^+]_{aq} \quad (3)$$

Where "aq", "org", K_{ex} and D represent aqueous and organic phases, and extraction equilibrium constant and distribution coefficient, respectively.

Substituting, we have

$$K_{ex} = D/[ROH]^n \cdot [Cl^-] \quad (4)$$

Taking logarithm of both sides, while keeping $[Cl^-]$ constant, we have

$$\log D = \log K_{ex} + n\log[ROH] \quad (5)$$

A plot of log D against log [ROH] should expectedly yield a straight line of slope equal to 'n' the number of solvated molecules. However, the two linear portions observed on the plots of both 1–pentanol and 3–methyl-1-butanol at lower and higher extractant concentrations imply that two different species were extracted.

With 1–pentanol, at the ascending portion of the curve (lower concentration), the determined slope, 'n' was equal to six and for the descending portion of the curve, it was -4.6, which implies that six molecules of 1–pentanol were coordinated at lower concentration, while polymeric V (V) specie is coordinated to one molecule of 1–pentanol at high concentration which does not agree with Equation (1).

With respect to 3–methyl-1-butanol, the determined slope 'n' was equal to 1.0 and – 0.8 at lower and higher extractant concentrations, respectively, implying that one molecule of 3–methyl-1-butanol was coordinated to V (V) at lower concentrations while polymeric V (V) was coordinated to one molecule of 3–methyl-1-butanol at higher concentrations, similar to observations made in Mo(VI)/HCl/TBP and V (V)/HCl/ROH systems (Ojo, Ipinmoroti & Adeeyinwo, 2008.) and these could not be rationalized by Equation (1).

The extraction at higher concentration of alkanols could therefore be represented by the Equation (6).

$$mVO_2^+ + mCl^- + nROH \rightleftharpoons (VO_2Cl)_m.nROH \qquad (6)$$

which is of the limit m >> n.

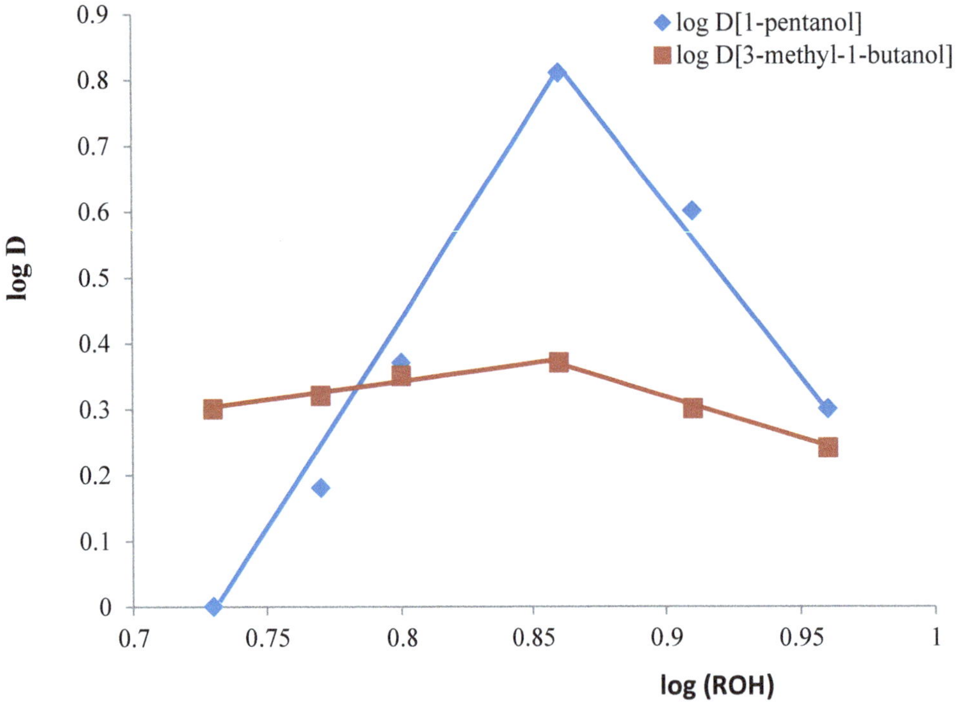

Figure 5. A plot of log D versus log [1-pentanol] or log [3-methyl-1-butanol] for the extraction of V(V) from HCl solutions.

[V(V)] = 0.03 mol L^{-1}; $[HCl]_{optimum}$: 5 mol L^{-1} (1-pentanol), 6 mol L^{-1} (3-methyl-1-butanol);

Temperature = 301 K

3.4 Effect of Temperature

The effect of temperature on the distribution coefficient of V(V) extraction from 5.0 and 6.0 mol L^{-1} HCl by 1-pentanol and 3-methyl-1-butanol respectively at the temperature range 301-321K is shown in Figures 6 and 7. Both extractants show a characteristic ascending and descending orders as the temperature increased.TheVan't Hoff equations is

$$\log D = -\Delta H/2.303RT + C \qquad (7)$$

where D, ΔH and C represent distribution coefficient, enthalpy change and constant for the extractive reaction respectively (Talla, Gaikward & Pawar, 2010). It could be explained (from fundamentals) that the two linear portions described by equation of a straight line are characteristic of mathematical step functions of positive and negative linear regressions of defined 1/T (that is 'x') limits ($3.22 \leq x < \infty$ to $-\infty < x < 3.22$ where $x = 1/T$). The plot of log D against 1/T x 1000 yielded two linear ascending /descending straight lines of the slopes equal to 4.2 and -6.3 and ΔH was determined to be -80.4 and 120 kJ mol^{-1} indicating exothermic and endothermic reaction at lower and higher temperatures with 1-pentanol. As regards 3-methyl-1-butanol the plot of log D versus 1/T x 1000 yielded slopes of 10.3

and -13.3 and ΔH values of -203.0 and 254.7 kJ mol^{-1}, the extraction could similarly be referred to as exothermic / endothermic.

These values of (ΔH) were much higher than were obtained for 1-butanol and 1-octanol extraction of V(V) reported in the previous work (Ojo and Oyegoke, 2013). The presence of dual energy profiles (exothermic/endothermic) as temperature changes in the extraction of V(V) by these alkanols are peculiar and exciting, which could be exploited to an advantage in making a choice of operating temperature if it were to be commercially exploited. This dual characteristic feature is similar to the presence of two activation energies (Arrhenius plots at low &high temperatures) observed in the kinetic study of Mo(VI) extraction from HNO_3 by tri-n—butyl phosphate reported by the current author (Ojo, 2007)

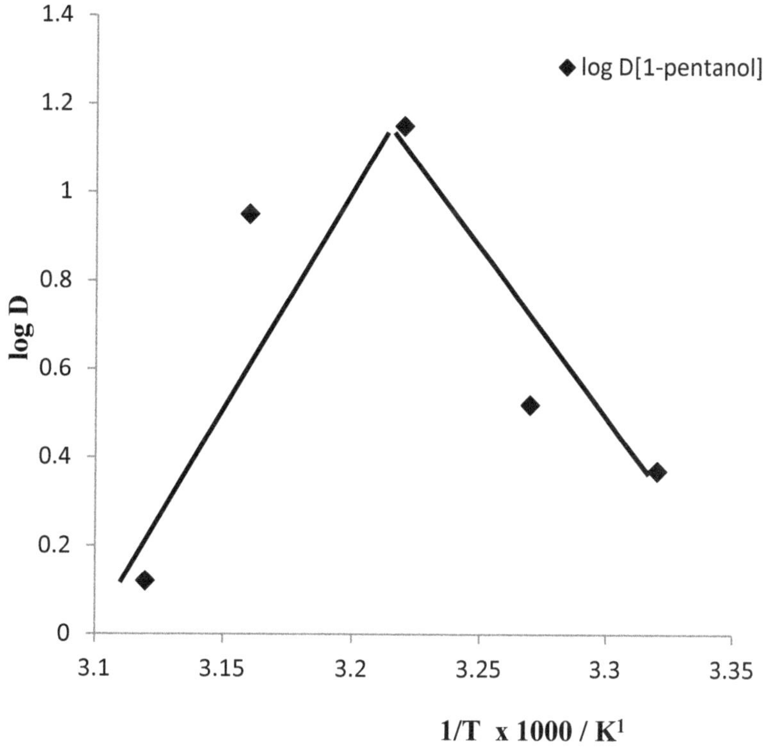

Figure 6. A plot of log D versus 1/T x 1000 for the extraction of V(V) from HCl solutions with 1 - pentanol[V(V)] = 0.03 mol L^{-1}; [1 – pentanol] = 5.89 mol L^{-1}; [HCl] = 5 mol L^{-1}.

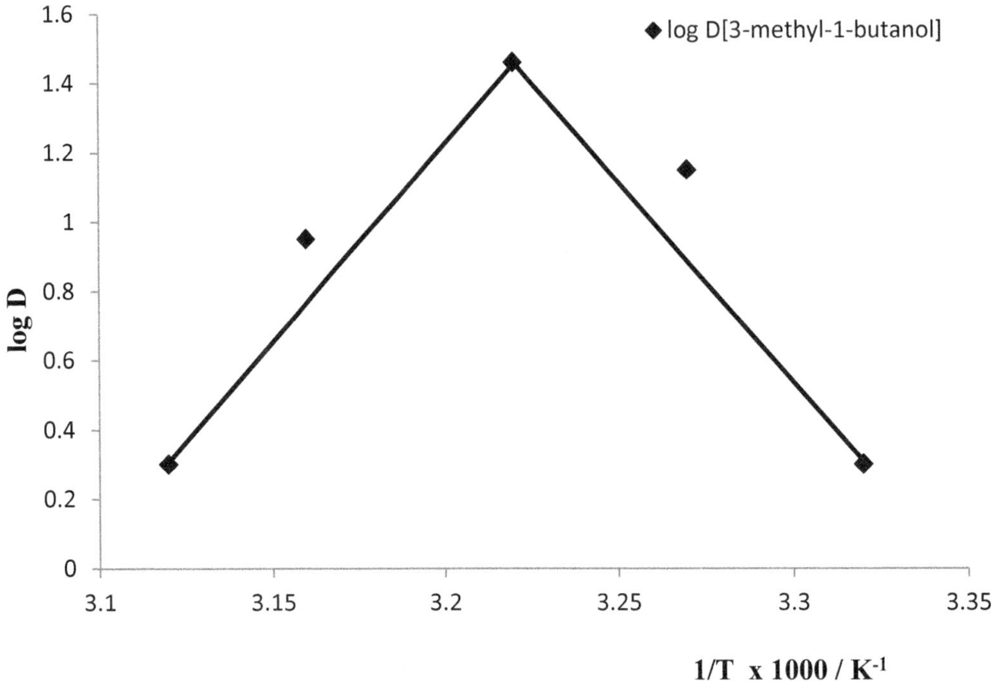

Figure 7. A plot of log D versus 1/T x 1000 for the extraction of V(V) from HCl solutions with 3 - methyl-1-butanol [V(V)] = 0.03 mol L^{-1}; [3-methyl-1-butanol] = 6.38 mol L^{-1}; [HCl} = 6.0 mol L^{-1}

3.5 Extraction Stages

The determination of number of extraction stages required for quantitative extraction of V(V) from 5.0 and 6.0 mol L^{-1} HCl by 1-pentanol and 3-methyl-1-butanol respectively is presented in Table 1. With the use of 1-pentanol the percentage of V(V) extracted after a three-stage extraction was 83.4% while for 3-methyl-1-butanol was 92.5% at likewise extraction stages implying the later was more efficient.

Table 1. The effect of number of extraction stages on percentage V(V) extraction with 1-pentanol and 3-methyl-1-butanol.

Stage	**1-pentanol**		**3-methyl-1-butanol**	
	V(V)conc/(g L^{-1})	**E(%)**	**V(V)conc/(g L^{-1})**	**E(%)**
1	0.458	70.0	0.509	67.3
2	0.356	76.7	0.229	85.0
3	0.254	83.4	0.115	92.5

Initial V(V) concentration = 1.529 g L^{-1}

3.6 Recycling

The effect of the number of recycling stages of 1-pentanol and 3-methyl-1-butanol on percentage extraction and recycling coefficient of V(V) in 5.0 and 6.0 mol L^{-1}HCl medium respectively is presented in Table 2.With 1-pentanol the distribution coefficient (D) and percentage extraction (E%) decrease minimally with three consecutive extractions from E%= 70,0 (at D=2.3) to 68.2(at D=2.1).In order to quantify the recycling effect, the recycling coefficient (R.C) was defined as "D" of nth extraction/D of initial extraction (or 1st stage). If R.C $\geq$ 1.0 the extractant is recyclable and if R.C< 1.0, it is weakly recyclable. The determined R.C values were greater than one (R.C. = 1.1) using 1-pentanol.

While for 3-methyl-1-butanol the E% =70.0 (at D = 2.3) was also greater than one (R.C. =1.2), hence, the extractants could be said to be recyclable (the latter being slightly higher) which was supported by the insignificant losses in volumes during equilibrations.

Table 2. Effect of number of recycling stages on distribution coefficient, percentage and recycling coefficient on V(V) extraction

Recycling stage	1-pentanol			3-methyl-1-butanol		
	D	E%	R.C[b]	D	E%	R.C[b]
1[c]	2.3	70.0	-	2.0	66.7	-
2	2.1	68.2	1.1	2.2	68.3	1.1
3	2.3	70.0	1.0	2.3	70.0	1.2

[b] = Recycling coefficient, 1[c]= initial extraction

3.7 Electronic Spectra

The electronic spectra bands of 1-pentanol dissolved in n-heptane, aqueous and organic phases of 1-pentanol-V(V) are presented in Table 3.

The electronic absorption spectra of 1-pentanol in n-heptane showed absorption bands at 29940 ($\mathcal{E} < 1.0\ M^{-1}cm^{-1}$) ascribable to $n \rightarrow \sigma^*$ transition. This band shifted positively (shorter wavelength) by about 3173 and 3617 cm^{-1} in the aqueous and organic phases of 1-pentanol-V(V) system respectively, suggesting a positive involvement of ligand-to-metal charge transfer (LMCT) between the bound Cl^- and V(V) lend credence to by the observed and typically high molar absorptivity ($\mathcal{E} = 246 - 2590\ M^{-1}\ cm^{-1}$). With 3-methyl-1-butanol dissolved in n-heptane, its absorption band was observed at 30674 ($\mathcal{E} < 1.0\ M^{-1}cm^{-1}$) ascribable to $n \rightarrow \sigma^*$ transition. The positive shift in absorption frequency (hypsochromic) was about 2659 cm^{-1} for both the aqueous and organic phases of 3-methyl-1-butanol- V(V) system and molar absorptivities were 3429 and 805 $M^{-1}\ cm^{-1}$ in that order , similarly, charge transfer was advanced (Ojo, Ipinmoroti & Adeeyinwo, 2007; J. Ojo, Adebayo, & I. Ojo, 2012; Ojo, 2013; Ojo & Oyegoke, 2013).

Table 3. Electronic spectra data for the extractants, aqueous and organic phases of V(V) in 1-pentanol and 3-methyl-1-butanol

Compound	Band (cm-1)	$\mathcal{E}$ / $M^{-1}cm^{-1}$	Assignment
1-pentanol	29940	<1	$n \rightarrow \sigma^*$
Aqueous phase of V(V)	33113	2590	Charge transfer
Organic phase of V(V)	33557	246	Charge transfer
3-methyl-1-butanol	30674	< 1	$n \rightarrow \sigma^*$
Aqueous phase of V(V)	33333	3429	Charge transfer
Organic phase of V(V)	33333	805	Charge transfer

3.8 Infrared Spectra

The infrared spectra data for the extractant and organic phases of 1-pentanol and 3-methyl -1-butanol are presented in Table 4. With 1-pentanol extractant, infrared absorption bands are observed at 3338 and 2958 cm^{-1} ascribable to Ʋ (O-H) and Ʋ(C-H) vibrations, respectively.

In the organic phases, the Ʋ (O-H) vibration shifted downwards by 72 cm^{-1} indicating the involvement of the oxygen atom of the hydroxyl group in bonding to the VO_2^+ ion.

A new band observed at 790 cm^{-1} assigned to υ (V = O) is indicative of coordination of the pervanadyl ion (VO_2^+) to the oxygen atom of the hydroxyl group.

As regards 3-methyl-1-butanol extractant, the observed bands are 3341 and 2958 cm^{-1} assigned to υ (O-H) and υ (C-H) vibrations. In the organic phase, the υ (O-H) shifted downwards by 72 cm^{-1} indicating that the hydroxyl group was involved in bonding to the VO_2^+ ion. In addition, a new band observed at 790 cm^{-1} which was absent in the exctractant has been assigned to υ (V=O) is a confirmation of the coordination of the pervanadyl ion to the alkanolic hydroxyl group (Ojo, 2010; Ojo, & Oyegoke, 2013; Kurbatova, Kurbatov, & Medvedeva, 2004; Kurbatova & Kurbatov, 2006; Lamber, Sharvell, Lightner, & Cooks, 1987).

Table 4. Infrared spectra data for the extractant and organic phase of V(V) in 1-pentanol and 3-methyl-1-butanol

Compound	Band (cm^{-1})	Assignment
1-pentanol	3338	υ (O-H)
	2958	υ (C-H)
Organic phase of V(V)	3266	υ (O-H)
	2958	υ (C-H)
	790	υ (V=O)
3-methyl-1-butanol	3341	υ (O-H)
	2958	υ (C-H)
Organic phase of V(V)	3265	υ (O-H)
	2960	υ (C-H)
	790	υ (V=O)

4. Conclusion

From the current study, the extraction efficiency of 1-pentanol and its isomer, 3-methyl-1-butanol are at par, with the presence of branched methyl chain exhibiting an insignificant effect on the overall efficiency of the extractant. However, at lower temperatures, 3-methyl-1-butanol has a slight edge in terms of extraction efficiency and recycling coefficient over the straight-chain analog. But on the whole, either could be selected for the extraction of V(V) from HCl solutions with high prospects.

References

Greenwood, N. N., & Earshaw, N. (1997). *Chemistry of the Elements* (2^{nd}ed). England: Pergamon Press Plc.

Kurbatova, L. D., & Kurbatov, D. I. (2006). Extraction recovery of vanadium (V) from sulphuric acid solutions. *Russian Journal of Applied Chemistry, 79*, 850-852. http://dx.doi.org/10.1134/S1070427206050302

Kurbatova, L. D., Kurbatov, D. I., & Medvedeva, N. I. (2004). Mechanism of the extraction of vanadium (V) from sulphuric acid solutions with di (2 –ethylhexyl) phosphoric acid. *Russian Journal of Applied Chemistry, 77*, 676-679. http://dx.doi.org/10.1123/B:RJAC.0000038692.79239.19

Lambert, J. B., Sharvell, H. F., Lightner, D. A., & Cooks, R. G. (1987). *Introduction to organic spectroscopy*. New York: Macmillian Publishing Company.

Morrison, R. T., & Boyd, R. N. (2009). *Organic Chemistry* (6^{th}ed). New Delhi, Dorling Kindersley Pvt. Ltd.

Ojo, J. O. (2007). Solvent Extraction of Vanadium (V) and Molybdenum(VI) from hydrochloric and nitric acids with tri-n-butyl phosphate and di (2 –ethylhexyl) phosphoric acid. *Ph.D Thesis*. Federal University of Technology, Akure, Nigeria.

Ojo, J. O. (2010). Solvent extraction of vanadium (V) from nitric acid solutions with di (2-ethylhexyl) phosphoric acid and tri-n-butyl phosphate. *Malaysian Journal of Chemistry, 12*(1), 175-183.

Ojo, J. O. (2013). Separation of simulated mixed Mo(VI) and V(V) from HNO_3 and HCl solutions by selective extraction and stripping with tri-n-butyl phosphate as extractant. *Separation Science and Technology, 48*(10), 1577-1584. http://dx.doi.org/10.1080/01496395.2012.745.154

Ojo, J. O., & Adeeyinwo, C. E. (2011). Dependence of vanadium recovery on oxidation state in its solvent extraction from hydrochloric acid solutions with tri-n-butyl phosphate. *Ife Journal of Science, 31*(1), 175-184.

Ojo, J. O., & Oyegoke, D. A. (2013). Alkanols for the extraction of vanadium (V) from hydrochloric acid solutions.

International Journal of Chemistry, *5*(3), 123-134. http://dx.doi.org/10.5539/ijc.v5n3p123.

Ojo, J. O., Adebayo, A. O., & Ojo I. O. (2012). Comparative study of the extraction of V(V) and V(IV) from hydrochloric acid solutions with di (2-ethylhexyl) phosphate acid. *Ife Journal of Science, 14*(1), 15-22.

Ojo, J. O., Ipinmoroti, K. O., & Adeeyinwo, C. E. (2007). Kinetics and mechanisms of V(V) extraction from hydrochloric acid solutions with tri-n-butyl phosphate. *Journal of Chemical Society of Nigeria, 32*(2), 166-173.

Ojo, J. O., Ipinmoroti, K. O., & Adeeyinwo, C. E. (2008). Solvent extraction of molybdenum (VI) from diluted and concentrated hydrochloric acid solutions with tri-n-butyl phosphate. *Global Journal of Pure and Applied Sciences, 14*(3), 289-294. http://dx.doi.org/10.4314/gjpas.v14i3.16810

Talla, B. G., Gaikwad, S. U., & Pawar, S. D. (2010). Solvent extraction separation of Mo(VI) and W(VI) from hydrochloric acid solutions using cyanex-923 as extractant. *Indian of Chemical Technology, 17*, 436-440.

Thomas, J., Surender, G. D., & Reddy, M. L. P. (2003).Solvent extraction separation of vanadium (V) from multimetal chloride solutions using tributylphosphate. *Separation Science and Technology, 38*(15), 3761-3774. http://dx.doi.org/10.1081/SS-120024228

Vogel, A. I. (1998). *Vogel's Textbook of Quantitative Inorganic Analysis* (5th.ed). Essex: Longman Group Limited.

Yamada H., Hayashi, H., & Yasui, T. (2006). Utility of 1-octanol/octane mixed solvents for the solvent extraction of Aluminium (III), and Gallium(III), and Indium(III) with 8-Quinolinol. *Analytical Sciences, 22*, 371-376. http://dx.doi.org/10.2116/analsci.22.371

5

One Step Synthesis of Polythiophenes from the Partially Purified Crude Extract of the Roots of *Tagetes Erecta*

Udari Kodithuwakku[1], Chathura de Alwis[1], M.A.B Prashantha[1], Dilru R. Ratnaweera[1]

[1]Department of Chemistry, University of Sri Jayewardenepura, Nugegoda, Sri Lanka

Correspondence: Dilru R. Ratnaweera, Department of Chemistry, University of Sri Jayewardenepura, Nugegoda, Sri Lanka, E-mail: dilru@sci.sjp.ac.lk.

Abstract

Polythiophene and its derivatives are valuable conjugated polymers due to their physical properties and applications in various fields. Precursors of polythiophenes were conventionally obtained from petrochemical byproducts, which are non-renewable. Thiophene and its derivatives had been found naturally in *Tagetes* species such as *Tagetes erecta, Tagetes tenuifolia*, etc. *Tagetes* species produce 2, 2':5', 2''-terthienyl (Alpha-T), 5-(3-buten-1-ynyl)-2, 20-bithienyl (BBT), 5-(4-hydroxy-1-butynyl)-2, 20-bithienyl (BBTOH), and 5-(4-acetoxy-1-butynyl)-2, 20-bithienyl (BBTOAc), which are accumulated in different parts, specifically in the roots of the plants. These thiophene derivatives were extracted and identified by GC-MS, Uv-vis, FTIR and TLC analysis. The partially purified plant extract is polymerized through one step oxidative free radical polymerization to obtain polythiophenes. Product characterization has revealed that the selective polymerization of thiophene derivatives results in the phase-separated end product from the rest of the materials in the plant-extract.

Keywords: conjugated polymer, free radical oxidative polymerization, polythiophene, thiophene derivative

1. Introduction

Thiophene and its derivatives have versatile applications in pharmaceuticals, agrochemicals, dye manufacturing, bio diagnostics, organic electronics and optoelectronic devices.(Barbarella & Sotgiu, 2005),(Abdou, 2013),(Mabkhot et al, 2015) Polythiophene has become a very valuable material for altering the photostabilization, brittleness, hardness, and other properties of plastic, and to alter the coloring of dye.(Abdou, 2013),(Balakit, Ahmed, El-Hiti, Smith, & Yousif, 2015). In the pharmaceutical industry, thiophene and its derivatives are used to impart pharmacological properties such as the anti-inflammatory, the antimicrobial, and the antianxiety.(Souza et al, 2012) Oligo and polythiophene are utilized as an indicator to recognize DNA or protein.(Barbarella & Sotgiu, 2005) Azo thiophene is used in the dye manufacturing industry, where its heterocyclic nature allows excellent sublimation rapidity on the dyed fiber.(Abdou, 2013) Polymer based solar devises (also known as organic solar cells) have been recently introduced as a competitive system for inorganic Si based solar cells.(Tang, 2010) Coatings of such materials have the ability to convert many physical objects into active light harvesting solar cell devices. Poly 3-hexylthiophene and poly(3,4-ethylenedioxythiophene) are commercially available conducting polymers, which are famous for designing organic solar devices, field effect transistors and thermoelectric devices.(Yao, Wen, Xu, Zhang, & Duan, 2013)

Precursors of polythiophenes are mainly obtained from non – renewable petrochemical byproducts, and they have competitive pricing. The non-renewability of petroleum resources and their high consumption rates will cause the depletion of petroleum in the near future as predicted by the Hubbert peak theory.(Hubbert, 1956) Therefore, identifying natural sources for green chemicals and developing green manufacturing processes have become the trend as an alternative for petrochemicals.

Precursors of polythiophenes are also available naturally in some plants of the Asteraceae family, such as, *Tagetes tenuifolia*, *Tagetes erecta*, *Tagetes patula*, *Porophyllum ruderale* and *Bidens radiata species*. (Takahashi et al, 2011),(Silva et al, 2011) Out of such plants *Tagetes species* are commonly available in tropical countries as an ornamental flowering plant, which are commonly known as Marigold. Thiophenes and thiophene derivatives in *Tagetes* plants are biologically active compounds in the defense system of the different parts of the plant.(Marotti et al, 2010) The concentration of thiophene in plant tissues increases as the plants get older, and reaches the maximum during the reproductive stages.

Several studies have been carried out to identify the different thiophene derivatives that exist in *Tagetes* plants. Natural thiophene derivatives identified in *Tagetes* species consist of two or three thiophene rings connected at α- position of the ring. These compounds include 2, 2':5', 2''-terthienyl (Alpha-T, αT), 5-(3-buten-1-ynyl)-2, 20-bithienyl (BBT), 5-(4-hydroxy-1-butynyl)-2, 20-bithienyl (BBTOH), and 5-(4-acetoxy-1-butynyl)-2, 20-bithienyl (BBTOAc).(Munhoz, 2014) Alpha-T consists of three connected thiophene rings, whereas BBT, BBTOH and BBTOAc consist only two thiophene rings with one alkyl side chain attached to the α- position of the ring. Therefore, alpha T can be considered a natural trimer that can be used to synthesize polythiophene. The structural characterization of the above thiophene derivatives were studied by many research groups and are summarized in Table 1. (Marotti et al, 2010),(Takahashi et al, 2011),(Silva et al, 2011),(Saha, Walia, Kundu, Kumar, & Joshi, 2012),(Vijay, Laxman, Balasaheb, Yuvraj & Janardhan, 2013),(Deineka, Tret'yakov, Lapshova, & Deineka, 2014)

Table 1. Different type of thiophene derivatives identified in roots of *Tagetes* species

IUPAC Name of the compound	Chemical structure	Abbreviation
2,2',5',2''-terthiophene	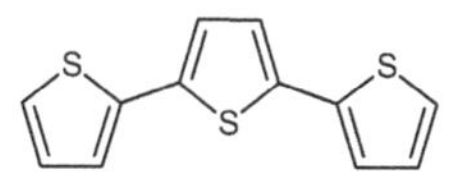	α – T
5-(3-butan-1-ynyl)-2,2'-bithienyl	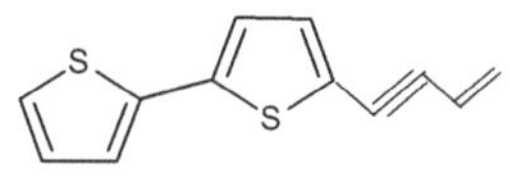	BBT
5-(4-hydroxyl-1-butynyl-2,2'-bithienyl	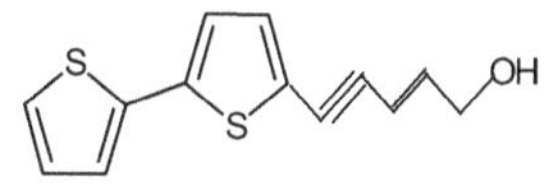	BBTOH
5-(4-acetoxyl-1-butynyl)-2,2'-bithienyl	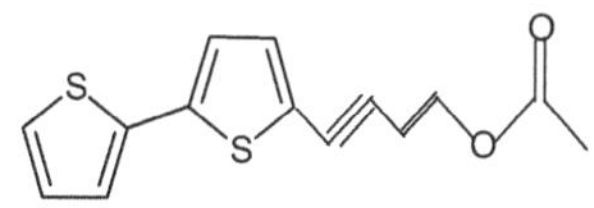	BBTOAc

The aim of this study is to extract thiophene derivatives from *Tagetes erecta* and partially purify the thiophene derivatives from the crude extract. The potential of synthesizing polythiophenes from the partially purified crude extracts were studied and the resulting polymers were compared with polythiophenes made from commercially available thiophene monomers. One of the major objectives of this study was to establish a single step synthesis of polythiophene derivatives from the plant extract, and establish the purification of thiophenes during the polymerization.

2. Materials and Methodology

The roots of *Tagetes erecta* plant were harvested after their full flowering stage (about three months from planting) from the Gampaha botanical garden (Longitudes = 79.9857°, Latitudes = 7.0993°), Sri Lanka. Analytical grade solvents and chemicals were used throughout the research.

2.1 Extraction and Partially Purification of Thiophenes

Lateral (hairy) and primary roots were separated from collected *Tagetes erecta* plants. Roots were washed several times with distilled water to remove soil particles and other water-soluble impurities. Thereafter, roots were dried under a shade at room temperature until obtaining a constant weight. Dried roots were ground into a fine powder using a lab scale mechanical grinder. Defatting of root samples was done using petroleum ether to remove fat components present in the roots. GC-MS analysis was carried out to check whether thiophene components were also removed during defatting.

Fine powders of air-dried roots were refluxed with methanol: water (3:1) at 78 °C. The extract was filtered through cotton wool to remove insoluble impurities. The filtered extract was concentrated *in vacuo* (Laborota 4000 Efficient) at 50 °C. The concentrated extract was sequentially partitioned into hexane: chloroform (2:1) solvent system. The extract was treated with anhydrous sodium sulfate (1 g for 400 ml of the extract) to remove water. Then the partitioned fraction in the hexane: chloroform system was subjected to rotary evaporation at 35 °C under reduced pressure to further dry the extract to get a dark brown residue. The dried product was kept in a desiccator for 24 hours to achieve a completely dried solid crude extract.

Separation of compounds for the partial purification was carried out for the crude extract by using hexane and acetone.

A 16.60 g of the dried crude extract was successively extracted three times with 30 ml of hexane at each extraction. The hexane fractions were collected to a 100 ml round-bottomed flask and desolvatization was carried out using the rotary evaporation under reduced pressure at 45 ºC to near dryness and completely dried by purging gas until it obtained a dark yellow solid residue. This crude residue was further dried for 24 hours using a desiccator. The weight of the partially purified hexane soluble fraction (HSF) was obtained and the mass yield was calculated. The hexane insoluble solid residue was extracted into 30 ml of acetone. Acetone fraction was dried by the rotary evaporation and further dried by keeping it in a desiccator for 24 hours. The weight of the acetone soluble fraction (ASF) was recorded and the mass yield was calculated.(Saha, Walia, Kundu, Kumar, & Joshi, 2012)

2.2 Analysis of the Partially Purified Product

HSF and ASF were analyzed using GC-MS, UV-Visible, FTIR and TLC in order to confirm the availability of thiophene derivatives in each solvent fraction.

2.2.1 Gas Chromatography-mass Spectroscopy (GC-MS)

The GC-MS analysis of plant extracts was carried out on a Agilent 5975C chromatograph coupled with a quadrupole mass detector and a HP-5MS 5% Phenylmethylsiloxane capillary column (30.00 m × 250 μm, film thickness 0.25 μm). A small aliquot (2 μl) of the sample was injected into the column. High pure Helium gas was used as the inert carrier gas at a flow rate of 1 mL min^{-1}. The initial temperature of the column was 70 ºC which was maintained for 3 minutes and then the temperature was increased at a constant rate of 10 ºC min^{-1} for 21 minutes. The final temperature of the column was 280 ºC and it was maintained for 6 minutes. The total retention time was 30 minutes. The chromatograms were analyzed using the NIST08 mass spectral database.

2.2.2 Ultraviolet-visible Absorption Spectroscopy (UV-Vis)

A small amount of the partially purified product (0.050 g) was dissolved in hexane (10 ml) and it was diluted to 100 times to get a colorless solution. The absorption spectra of these diluted samples were obtained using UV-Visible spectrophotometer (Perkin-Elmer, Lambda 35 spectrophotometer) within the spectral range of 200-400 nm. Hexane was used as a blank.

2.2.3 Fourier Transform Infrared Spectroscopy (FTIR)

A small amount of the extract (0.01 g) was dissolved in hexane (5 ml). One drop of that solution was taken to the analysis. The FTIR spectra were recorded in the range 600–4000 cm^{-1} with a 4 cm^{-1} resolution from ATR on a FT-IR system (Thermo Scientific Nicolot IS10). FT-IR bands were analyzed using OMNIC software (version Windows XP professional).

2.2.4 Thin Layer Chromatography (TLC)

A small quantity of (10 mg) of the partially purified extract having thiophene derivatives was dissolved in hexane (5 ml) and a drop of the solution was spotted on commercially available TLC plates (pre-coated TLC- sheets ALUGRAM® Xtra SIL G/UV254) using capillary tubes and solvent systems were used as the mobile phase as shown in Table 2. The best solvent system for separation of components in the partially purified extract was selected by observing the maximum separation of spots.

Table 2. TLC solvent systems used to separate the plant extract

System no.	Solvent system	Composition
1	Hexane : Acetone	10 : 0.1
2	Hexane : Acetone	9 : 1
3	Hexane : Acetone	8 : 2
4	Hexane : Acetone	7 : 3
5	Hexane : Acetone	6 : 4

Spots were visualized either by putting plates in an iodine bath or using UV lamp having wavelengths of 254 and 366 nm. The distribution of spots was recorded and R_f values were calculated with respect to the distance traveled by the solvent front.

The best solvent system was used to separate alpha T from the extract having thiophene derivatives. The extract having thiophene derivatives was purified to separate alpha T by using preparative TLC (silica gel 60 F254, layer thickness 0.25 mm, 10×20 cm). Each separated band was scratched out from TLC plates, dissolved into the relevant solvent, filtered and subjected to GC-MS analysis to confirm the availability of thiophene derivatives.

2.3 Synthesis of Polymers from Thiophene Monomers

2.3.1 Polymerization of Pure Thiophene

Thiophene (0.48 ml) was mixed with 70 ml of chloroform in a round-bottomed flask. Anhydrous $FeCl_3$ (1.0000 g) was dissolved in 180 ml of chloroform and added into the monomer solution in drop wise manner while stirring. The weight ratios of pure thiophene monomer to the oxidant were changed as 1:2, 1:5 and 1:7. The polymerization was carried out for 4 hours at room temperature. The dark-brown precipitate was collected by filtration and washed with anhydrous $CHCl_3$ to remove excess $FeCl_3$. The precipitated products were further washed with anhydrous methanol to remove the residual oxidant. The powder products were dried in a vacuum drier at 55 °C for 24 hours.(Gok, Omastova, & Yavuz, 2007)

2.3.2 Polymerization of the Plant Based Thiophene Monomers

The composition of the plant extract was determined based on the area under the curve of the calibrated Gas Chromatogram. The plant extract (0.9823 g) was mixed with 70 ml of chloroform in a round-bottomed flask containing a magnetic stir bar. Anhydrous $FeCl_3$ (4.9115 g) was dissolved in 180 ml of chloroform and it was added drop wise into the monomer solution while stirring. The weight ratios of plant extract and the oxidant were varied as 1:2, 1:5 and 1:7. Polymerization was carried out for 4 hours at room temperature. The dark-brown precipitate of the product was collected by filtration and washed with anhydrous $CHCl_3$ to remove excess $FeCl_3$.

The product was further washed with anhydrous methanol to remove the residual oxidant. During this procedure, the colour changed from black to dark brown. The powder product was dried in a vacuum drier at 55 °C for 24 hours.

2.4 Characterization of the Polymerized Products

2.4.1 Fourier Transform Infrared Spectroscopy (FTIR)

About 2 mg samples of the synthesized polymers were taken to the analysis. The FTIR spectra were taken using Thermo Scientific Nicolot IS10 within the wavenumber ranging from 600 to 4000 cm^{-1}. Instrument was used at the ATR mode with resolution of 4 cm^{-1}. FTIR spectral bands were analyzed using OMNIC software. The resulted spectra were compared with the reference FTIR spectra in literature which were obtained for polythiophene from pure monomers. (Marani, & Entezami, 1993), (Camurlu, Cirpan, & Toppare, 2005), (Gnanakan, Rajasekhar, & Subramania, 2009), (Karim, 2012)

2.4.2 X–Ray Diffraction (XRD)

Wide angle X-ray diffraction patterns of powder form of polymerized products were recorded using "Rigaku-Ultima IV" X-ray diffractometer with Cu – K_α radiation (λ= 1.54 Å). Samples were scanned at the rate of 2° min^{-1} in reflection mode over the 2Θ range of 3° - 45°. The resulted XRD spectra were compared with the literature available data.(Sakthivel, & Boopathi, 2014) Peak deconvolution was carried out using the Peak Fit 4.12 software. Degree of crystallinity of polymer was calculated using peak area measurements, which is shown in equation

$$\textit{Degree of crystallinity}=(\textit{Peak area of crystalline peaks/Total area under the curve}) \quad (1)$$

2.4.3 Ultraviolet-Visible Absorption Spectroscopy (UV-Vis)

A small amount of powder of the product (0.050 g) was dissolved in mono chlorobenzene (5 ml). Monochlorobenzene was selected as the blank. UV – Visible spectra of polymer solutions were recorded with respect to the blank using UV-Visible spectrophotometer (Perkin-Elmer, Lambda 35 spectrophotometer) within the spectral range of 200-400 nm.

3. Results and Discussions

3.1 Extraction Process

Few extraction methods are mainly reported in literatures such as steam distillation, soxhlet extraction, room temperature solvent extraction (maceration) and reflux extraction. In high thermal treatment thiophene derivatives are susceptible to degrade. Therefore, the steam distillation method was not used for the extraction. The soxhlet extraction method is suitable for a single solvent and is not possible to use for a solvent mixture. Therefore, the soxhlet technique was not used for this extraction. Moreover, room temperature solvent extraction may not be the most effective method because of its low temperature conditions. Furthermore, this extraction process consumes time because the plant material (roots) should be dipped for several days to obtain a considerable amount of the extract. Compared with other natural products present in the extract thiophene content is very low. Accordingly, the maximum attainable amount of thiophene derivatives should be extracted into the solvent. Therefore, the reflux extraction method was selected at 78 °C as the most appropriate extraction method.

GC-MS analysis of defatted samples had revealed that some thiophene derivatives (especially alpha – T) was extracted with the fat components. Hence, it caused to reduce the yield of thiophene derivatives. Therefore, the extraction process was optimized without defattening the roots. Methanol-water solvent system was directly used to extract thiophene derivatives from the roots of the plant. The methanol – water solvent system is a polar solvent system, which can break down both cell walls and cell membranes thereby thiophene and its derivatives can easily be extracted into the solvent system.

Thiophene derivatives can be dissolved well in non-polar organic solvents. Hence, hexane was used to partially purify the crude plant extract. The hexane soluble fraction (HSF) was yellow in color. The remaining hexane insoluble residue was soluble in acetone and it has a dark red color. The mass yield of the dried HSF was obtained with respect to the mass of dried roots using equation 2 and the results are tabulated in Table 3.

Percentage mass yield of HSF= (Weight of HSF (g) /Weight of the dried roots (g)) x 100 (2)

Table 3. Percentage yield of HSF and ASF with respect to the weight of the dried roots

Total weight of dried roots/g	Percentage yield of HSF	Percentage yield of ASF
2500	0.288	0.376

3.2 GC-MS Analysis of the Partially Purified Extract

Zoomed GC-MS spectra obtained for both the hexane and the acetone soluble fractions are shown in Figure 1(a) and (b) respectively. According to the NIST08 mass spectral data base, the peak at the retention time of 22.46 minutes in Figure 1(a) corresponds to the elution band of alpha – T and 22.23 minutes to the BBTOH.

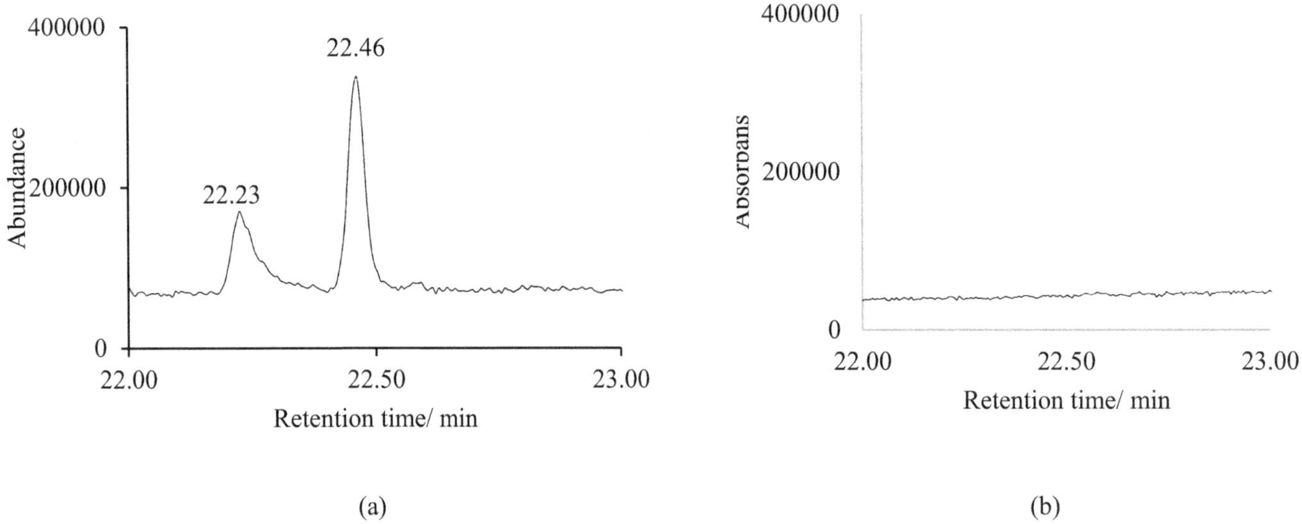

Figure 1. Gas chromatograms (GCs) of (a) HSF, (b) ASF of the plant extract of *Tagetes erecta*

Table 4. Major compounds observed from the Gas Chromatogram of HSF

Retention time (R_t)/min	Chemical Name	Chemical Structure
16.536	Benzeneacetic acid, alpha, 3,4 tris[(trimethylsilyl)oxy]-, trimethylsilyl ester	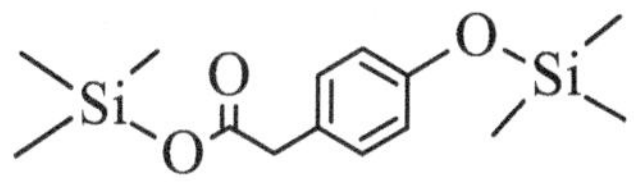
19.336	Pentadecanoic acid, 14-methyl, methyl ester	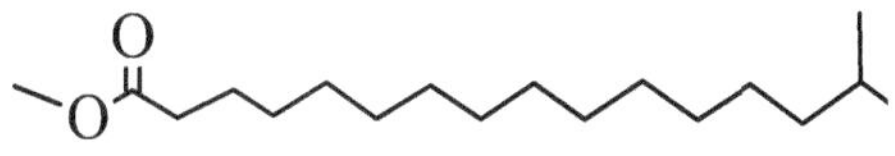
20.994	Octadecanoic acid, methyl ester	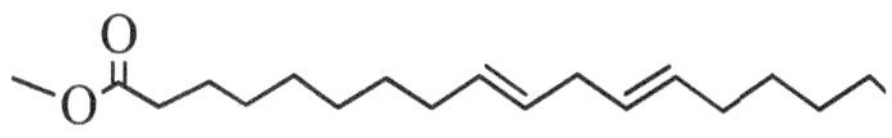
22.231	5-(4-hydroxy-1-butenyl) 2, 2-bithienyl, 5-(4-acetoxy-1-butenyl)	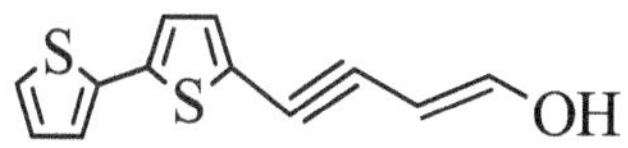
22.400	2, 2':5', 2"-Terthiophene	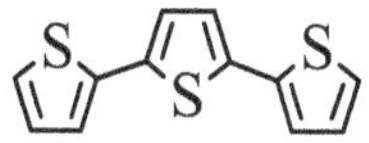
24.909	1,2- Benzenedicarboxylic acid, mono(2-ethylhexyl) ester	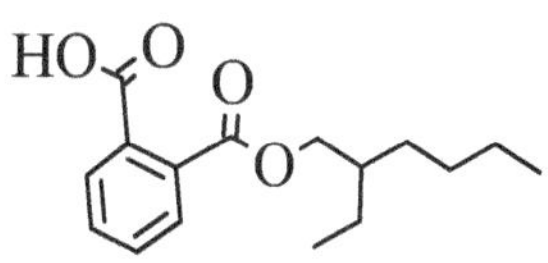

Mass spectra of the GC – MS for both alpha –T and BBTOH in the HSF are shown in Figure 2 (a) and (b). These MS spectra for alpha –T and BBTOH were further confirmed by comparing molar mass of molecule ions and other major fragments. The summary of possible molecular fragments for highly abundance peaks are summarized in Table 5 and 6.

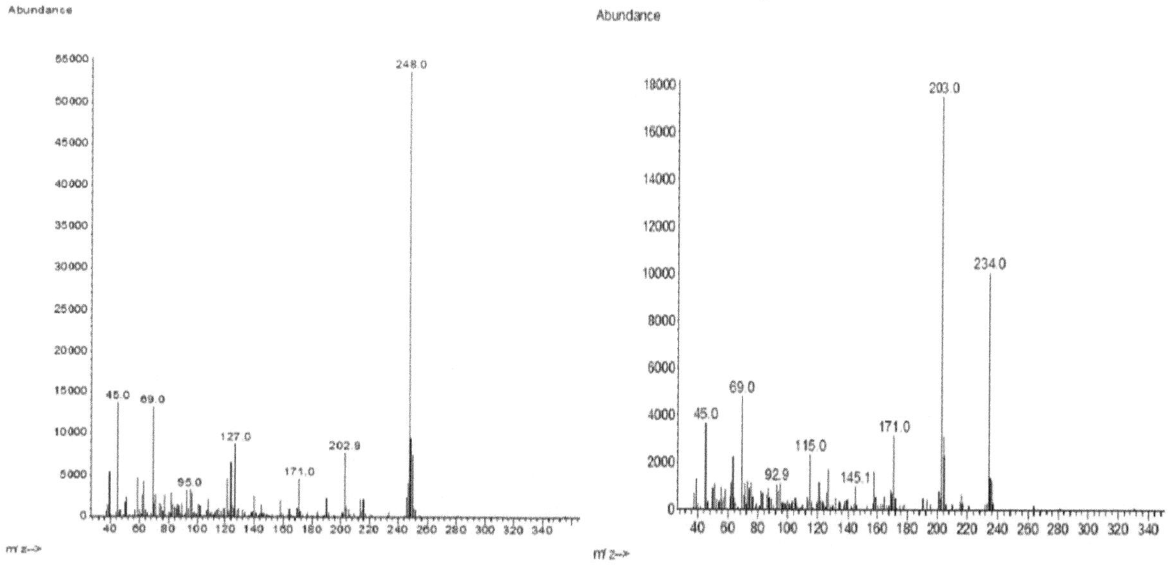

(a) (b)

Figure 2. Mass spectra of GC – MS peaks of (a) alpha – T and (b) BBTOH

Table 5. Possible molecular ion fragments of Alpha – T

m/z value	Possible molecular ion fragments
248	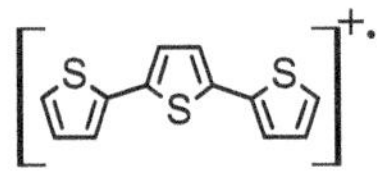
203	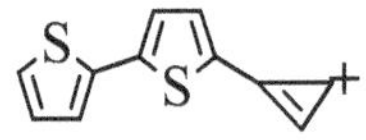
171	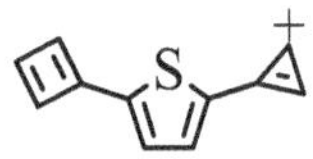
127	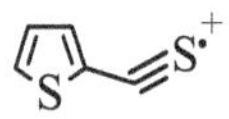
69	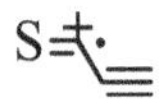
45	$HC\overset{+}{=}S$

Table 6. Possible molecular ion fragments of BBTOH

m/z value	Possible molecular fragments
234	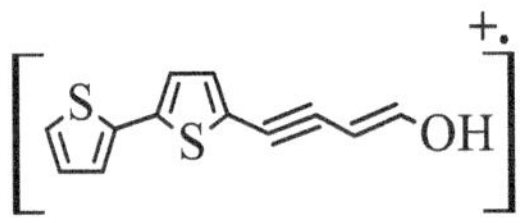
203	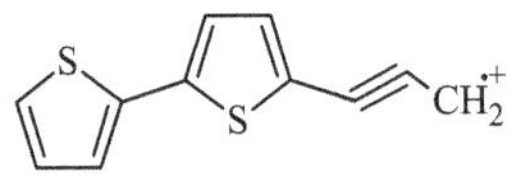
171	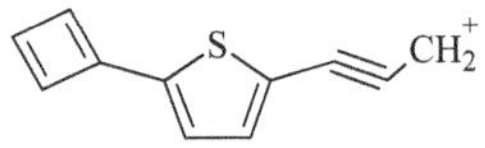
69	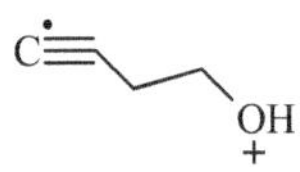
45	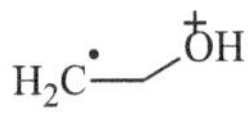

3.3 UV-Vis Analysis of the HSF

UV-Vis spectra of pure thiophene monomers and HSF are shown in Figure 3 (a) and (b) respectively.

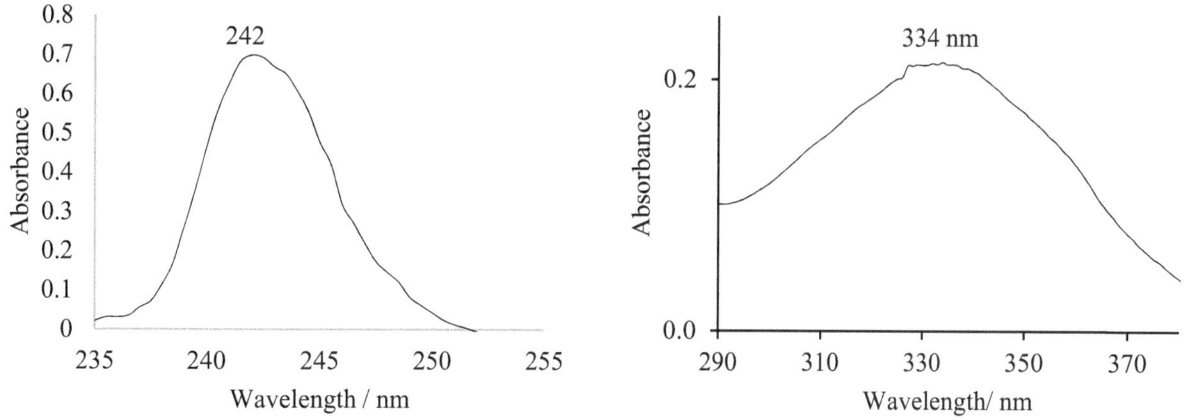

Figure 3. Uv-vis absorption spectrum of (a) thiophene monomer and (b) HSF of the extract of *Tagetes erecta*

The absorption maximum (λ_{max}) of pure thiophene monomers was observed at 242 nm. However, the λ_{max} of HSF was observed around 330 nm. This difference in λ_{max} values might be due to the presence of other components in the HSF and due to the conjugation of thiophene monomers. Due to the conjugation, thiophene derivatives in HSF show a bathochromic shift with respect to pure thiophene.

3.4 FTIR Analysis of the HSF

FTIR spectroscopic analysis has been carried out to identify the possible functional groups available in HSF of the extract. Relevant FTIR spectrum of HSF is shown in Figure 4.

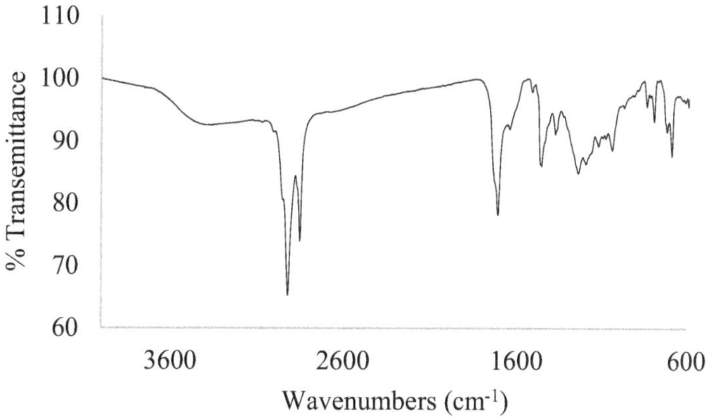

Figure 4. FT-IR spectrum of HSF of the plant extract

A detailed summary of the FTIR peak analysis is shown in Table 7. The tabulated results confirmed the presence of functional groups available in both alpha – T and BBTOH in the HSF of the plant extract. These results are summarized in Table 7.

Table 7. Summary of the band analysis of FTIR spectrum of the hexane soluble fraction

Wavenumber/cm^{-1}	Bond	Possible origin of functional groups
3417	O-H stretch, H-bonded	Alcohols
2954	ν_{as}–CH_3 stretching	Alkanes
2921-2853	aliphatic –CH_2 stretching	Alkanes
1710	C=O stretch	α, β- unsaturated esters
1642	C-C stretch (in- ring)	Aromatics
1457	The ring vibration of 2,5-disubstituted thiophene	Terthiophene and BBTOH
1377	C-C ring stretch	Aromatics
1241	The ring vibration of 2,5-disubstituted thiophene	Terthiophene and BBTOH
1041	-CH_2-CH_2–OH group	BBTOH
841	Angular deformation out of the plane of thiophene ring	Terthiophene and BBTOH
796	C-H deformation vibration of 2,5-disubstituted thiophene	Terthiophene and BBTOH
721	C-S bend	Thiophene

The availability of thiophene rings in HSF can be confirmed by FTIR absorption bands observed at 721 cm^{-1} for C-S bending and aromatic C-C ring stretching bands at 1642 cm^{-1} and 1377 cm^{-1}. FTIR absorption bands observed at 1457 cm^{-1}, 1241 cm^{-1} and 796 cm^{-1} show the connectivity of thiophene rings through 2,5 coupling. (Marani, & Entezami, 1993), (Camurlu, Cirpan, & Toppare, 2005), (Gnanakan, Rajasekhar, & Subramania, 2009), (Karim, 2012), (Sakil, Singh, & Roy, 2013) Other absorption bands may correspond to the side chains attached to thiophene and the impurities present in HSF.

3.5 TLC of the HSF

Several solvent systems with different polarities were used to separate the components available in HSF using TLCs. Those solvent systems are summarized in Table 2. The gas chromatogram obtained for the component extracts from the upper band of TLC is shown in Figure 5. However, BBTOH was not able to observe from GC. This might be due to the very low concentration of BBTOH in the extract.

Table 8. Results of the TLC analysis for the HSF

R_f of the separated spot	Under illumination of 254 nm	Under illumination of 366 nm
0.83	Brown	Yellow
0.78	Brown	Blue
0.70	Brown	Absent
0.48	Brown	Absent
0.33	Brown	Pale yellow
0.12	Brown	Blue

Literatures have been reported that n-hexane: acetone 10:0.1 solvent system was used to develop TLCs for the plant extract, which was extracted from the flowers of *Tagetes* species such as *Tagetes tenuifolia*, *Tagetes patula* and *Tagetes erecta*. Reported R_f value for thiophene derivatives was around 0.9.(Deineka, Tret'yakov, Lapshova, & Deineka, 2014)

In this study, only two separated spots for the plant extract were observed on the TLC plate for *n*-hexane: acetone 10:0.1 solvent system. Out of those solvent systems hexane: acetone (9:1) had shown the maximum number of spots with better separation among the spots. The polar stationary phase retards the movement of polar compounds due to better intermolecular interaction between compounds and the stationary phase. Alpha – T (terthiophene) is a non- polar compound and has less preference to the stationary phase. Therefore, it travels easily with the mobile phase to give high R_f value (R_f= 0.83). Thiophenes isolated from *Tagetes patula* had given brown color spots under UV radiation at 254 nm. Further visual observation of thiophene compounds can be obtained by illuminating those with UV radiation at 366 nm, which gives a characteristic yellow color for alpha - T.(Adekunle, 2007) The summary of this TLC analysis is shown in Table 8. Through UV illumination it has been confirmed that the thiophene derivatives can only be available in the spot with the highest R_f value.

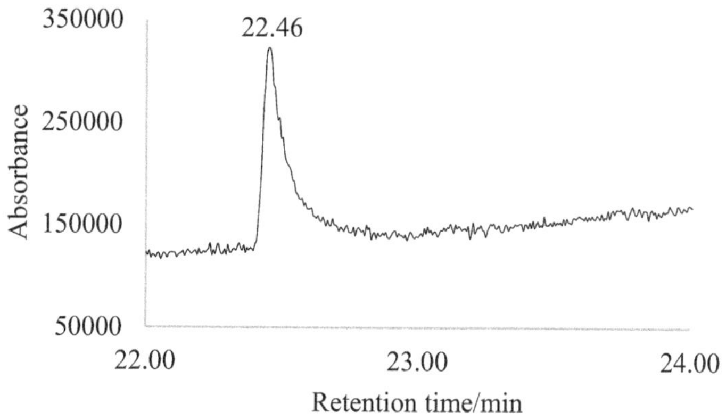

Figure 5. GC of the upper band in preparative TLC of HSF

These bands observed in preparative TLC plates were separately dissolved in hexane and subjected to GC-MS analysis. It was confirmed that the presence of Alpha – T in the upper band of the TLC.

3.6 Polymerization of Pure Thiophene Monomers and Plant Based Thiophene Derivatives

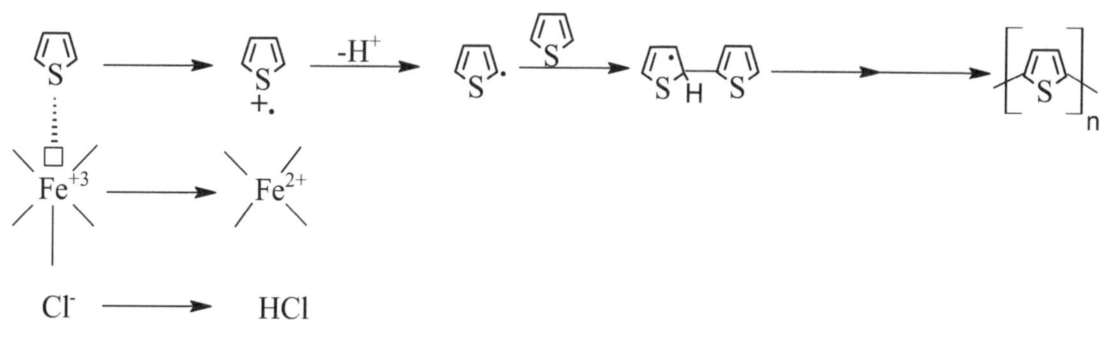

Figure 6. The possible polymerization mechanism of thiophene in the presence of $FeCl_3$ catalyst

Chemical oxidative polymerization was followed to obtain the polymerized product from the plant based thiophene derivatives and pure thiophene monomer separately. These thiophene polymerization reactions were catalyzed by anhydrous $FeCl_3$. In this process, polymerization starts from solid $FeCl_3$, due to the surface crystal Fe^{3+} ions being the active sites, which has one unshared chloride and one vacant orbital that has the Lewis acidity to accept electrons from sulphur atom of thiophene. This ability of Fe^{3+} ions is enough to oxidize the thiophene molecule into a radical cation. In parallel, Fe^{3+} reduces to Fe^{2+}. Therefore, ferric chloride acts as an oxidant and it initiates the radical oxidative cationic polymerization of thiophenes.(Niemi, Knuuttila, & Osterholm, 1992)

One of the possible mechanisms of thiophene polymerization is shown in Fig. 6. Chloroform is used as the medium of polymerization. The excess amount of $FeCl_3$ was dissolved in $CHCl_3$ prior to adding thiophene monomer or plant based thiophene derivatives. After polymerization, the related mixture was filtered, and the remaining crude was washed with the excess volume of chloroform to remove residual unreacted alpha – T and other organic impurities. Thereafter, the product was washed with excess methanol in order to remove the remaining ferric chloride catalyst.

The best average percentage yield of 89% for the pure polythiophenes was obtained for 1:7 weight ratio of pure thiophene monomer to oxidant. The weight ratios of HSF to oxidant for the polymerization was altered to obtain the highest average percentage yield, which was 21.8 %, received for weight ratio of HSF to oxidant 1:7. When compared with the percentage yield of pure polythiophene, the plant based product is less. The yield can reduce due to the other side products of ferric chloride with aromatic compounds and the effects of iron chelating.

3.7 FTIR Analysis for the Polymerized Product

The FTIR spectra for the pure polythiophene and the plant based polythiophene are shown in Figure 7 (a) and (b) respectively.

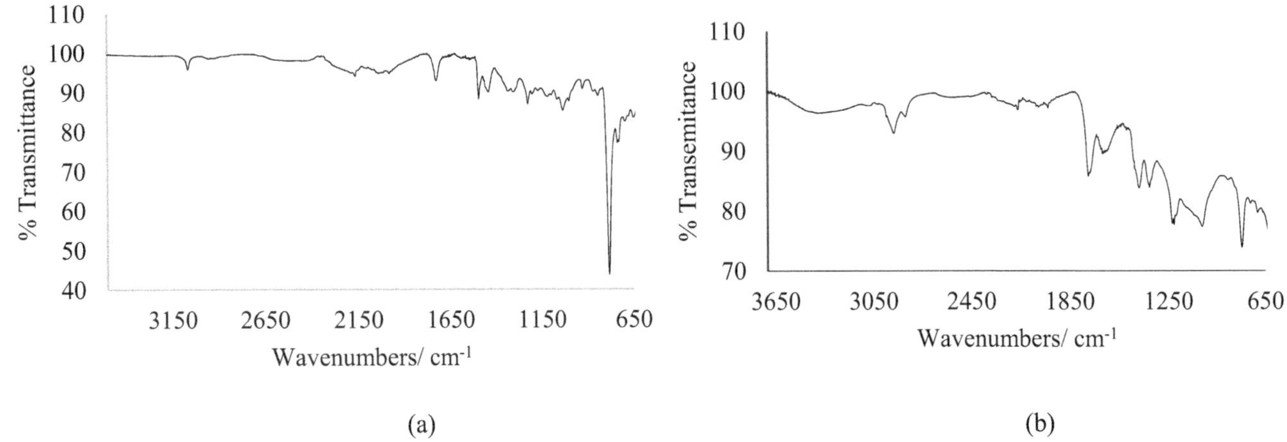

Figure 7. FTIR spectrum of (a) pure polythiophene, (b) plant based polythiophene

The common FTIR bands for both polythiophenes are summarized in Table 9.

Table 9. Summary of the common FTIR band analysis for polythiophenes made from pure thiophene monomer and plant based thiophene derivatives

Wavenumber/cm^{-1}	Bond	Possible origin of functional groups
2950-3050	ν_{as} –CH_3 stretch	Aromatic CH_3
1029-1086	C-S stretch	Thiophene
781-801	α, α' – coupling of poly-2, 5-thiophene	Thiophene

The most significant band for identifying polythiophene is the band between 781 cm^{-1} – 801 cm^{-1} due to 2, 5-disubstitution of thiophenes. According to the polymerization mechanism, polymerization prefers to occur via 2, 5 coupling, which is also known as α, α' coupling of monomers.(Kelkar, & Chourasia, 2011) This is a very strong and narrow band. The FTIR absorption bands at 1029 and 1086 cm^{-1}, show C – S stretching. The asymmetric vibration of the -CH_3 group shows in between 2950 – 3050 cm^{-1}. When the FTIR bands of the plant-based product was compared with the pure polythiophene, it confirms the presence of similar functional groups in the plant-based product.

Table 10. Summary of FTIR band analysis of the plant based polythiophene

Wavenumber/cm^{-1}	Bond	Possible origin of functional groups
3390	-OH stretch, H-bonded	Alcohols from BBTOH
2924 – 2854	aliphatic –CH_2 stretch	Alkane from BBTOH
1456	–CH_2 bend	Alkane from BBTOH
1376	-CH_3 bend	Alkane
833	ω/τ thiophene ring	Terthiophene from BBTOH

In addition to common FTIR bands of polythiophenes, other bands present in the plant based polythiophene are tabulated in Table 10. The band at 3390 cm^{-1} shows the availability of hydroxyl groups in the plant based polythiophene.

These hydroxyl groups might come from the polymerization of BBTOH in the plant extract. Other bands for aliphatic hydrocarbon chains may correspond to side chains attached to thiophenes.

3.8 Uv-vis Analysis of the Polymerized Products

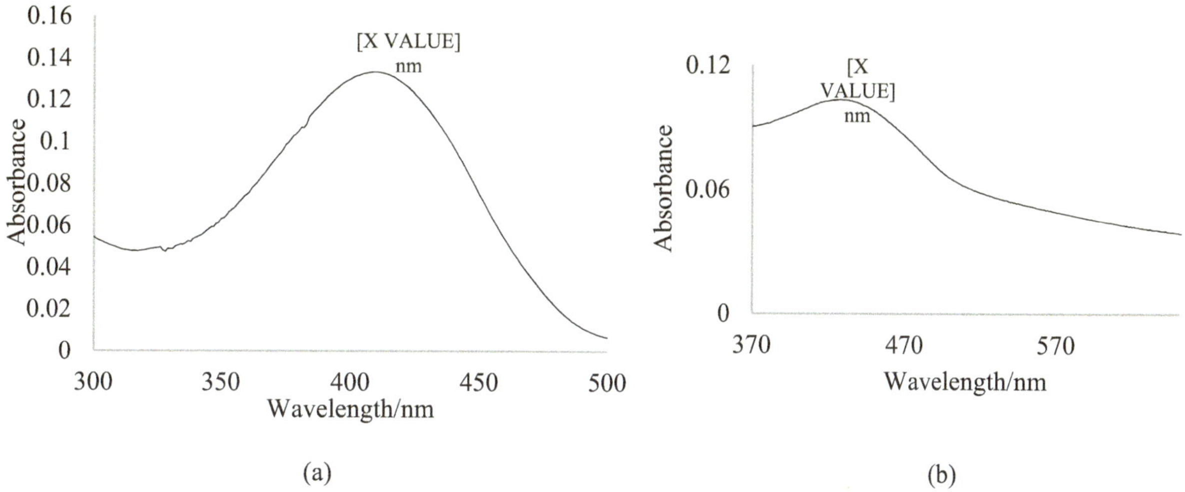

Figure 8. Uv-vis spectrum for (a) the pure polythiophene, (b) plant based polythiophene

UV spectroscopy looks at double bonds and lone pairs in compounds. Electrons can be promoted from the HOMO level to the LUMO level by absorbing radiation. The energy gap of bonding to antibonding (π-π^*) transition is higher than that of non-bonding to antibonding (n- π^*). Therefore, π-π^* transition requires short wavelengths compared to n-π^*transition. When the length of conjugation increases the gap for π-π^*, the transition narrows down gradually. The Uv-vis spectrum for the pure polythiophene and the plant based polythiophene are shown in Fig.8 (a) and (b) respectively.

The pure polythiophene obtained the absorption band at 413 nm and also the plant based polythiophene associated the absorption band at 432 nm. Therefore, the conjugation length of the plant based polythiophene is a little higher than that of pure polythiophene. The literature reported that these λmax values may correspond to π-π^* transitions due to the main charge carriers such as polarons and bipolarons in polythiophenes.(Gnanakan, Rajasekhar, & Subramania, 2009)

3.9 XRD Analysis of the Polymerized Product

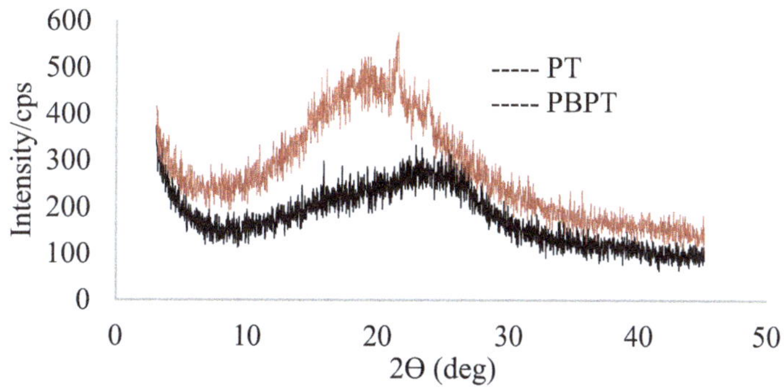

Figure 9. XRD spectrum for both polythiophenes

The XRD spectra showed for both pure polythiophene (PT) and plant based polythiophene (PBPT), which revealed three peaks near 2Θ values of 19.61°, 21.34° and 23.79°. Among them the first large broad peak (at 2Θ = 19.61°) relates to amorphous peak and other two at 2Θ = 21.34 ° and 23.79° small peaks indicates two crystalline peaks. These three peaks are overlapped together and the overall peak is given on Figure 9. But the pure polythiophene has shown only the amorphous peak. Polymer chain packing of polythiophenes made from pure thiophene monomers and the thiophene derivatives of the plant extract are different. Plant based polythiophenes may contain other side chains due to the polymerization of BBTOH with Alpha-T. Therefore, peak center of the amorphous diffraction may slightly shifter to lower theta (higher d values).

The peak deconvolution and degree of crystallinity of plant based polythiophene was calculated and summarized in Table 11.

Table 11. XRD peak characteristics for plant based polythiophene

Peak position at 2Θ/degree	Nature of peaks	Inter planar distance/ Å	Area of peaks	Degree of crystallinity
19.61	Amorphous	2.29	2690.1	Amorphous
21.34	Crystalline	2.10	88.6	3.1
23.79	Crystalline	1.91	103.6	3.6
	Total		2882.3	6.7

4. Conclusions

The study had confirmed the possibility of polymerizing plant based thiophene derivatives in their crude extract into polythiophenes. This study was the first green polymerization reported in literature to direct polymerization of the crude extract of plant based thiophene derivatives to get the final polymerized product. Polythiophene formation from HSF confirms the minimal effect from other impurities present in the plant extract on polymerization of thiophene derivatives.

Thiophene derivatives in the roots of *Tagetes erecta* were extracted from methanol-water solvent system and partitioned into the hexane fraction. By analyzing the HSF through GC-MS, it was confirmed that two thiophene derivatives (alpha – T and BBTOH) were available with many other impurities like hydrocarbons, fatty acids, esters and carbonyl compounds in the root extract of *Tagetes erecta*. The FTIR and Uv-Vis spectra recognized the existence of connected thiophene rings while TLC results confirmed that the presence of thiophene derivatives in the HSF.

Polythiophene was obtained through direct cationic oxidative polymerization of thiophene derivatives in the HSF in the presence of anhydrous $FeCl_3$ catalyst. The percentage yields for plant based polythiophenes and pure polythiophenes concluded that the monomer to the oxidant weight ratios affect the overall yield. Synthesized plant based polymer and pure polythiophene were characterized by using FTIR, XRD and Uv-Vis techniques. The FTIR spectra also confirmed that certain other groups have been attached to the polythiophene chain apart from α-T, which were mainly aliphatic side chains, and hydroxyl groups that come from BBTOH and organic impurities present in the partially purified extract. XRD studies have shown that the resulting plant based polythiophenes were assembled into a semi-crystalline form. Through the analysis of XRD spectrum by Peak fit software, it was calculated that the degree of crystallinity of the prepared polymer was about 6.7%.

Acknowledgement

This work was financially supported by the Research Grant (ASP/06/RE/SCI/2013/07), University of Sri Jayewardenepura, Sri Lanka. UV-Visible studies were carried out with the instrument provided by the NSF, Sri Lanka (NSF Grant RG/2013/EQ/07)

References

Abdou, M. M. (2013). Thiophene-Based Azo Dyes and Their Applications in Dyes Chemistry. *American Journal of Chemistry, 3*(5), 126-135. http://dx.doi.org/10.5923/j.chemistry.20130305.02

Adekunle, O. I. (2007). Biotransforation of Acetylenic Thiophenes Isolated from *Tagetes patula* (African Marigold). *Research Journal of Biological Sciences, 2*(4), 424-426.

Balakit, A. A., Ahmed, A., El-Hiti, G. A., Smith, K., & Yousif, E. (2015). Synthesis of New Thiophene Derivatives and Their Use as Photostabilizers for Rigid Poly(vinyl chloride). *International Journal of Polymer Science,* 1-10. http://dx.doi.org/10.1155/2015/510390

Barbarella, G. M. M., & Sotgiu, G. (2005). The Versatile Thiophene: An Overview of Recent Research on Thiophene-Based Materials. *Advanced Materials, 17*, 1581-1593. http://dx.doi.org/10.1002/adma.200402020

Camurlu, P., Cirpan, A., & Toppare, L. (2005). Conducting polymers of octanoic acid 2-thiophen-3-yl-ethyl ester and their electrochromic properties. *Journal of Materials Chemistry and Physics, 92*, 413-418. http://dx.doi.org/10.1016/j.matchemphys.2005.01.040

Deineka, V. I., Tret'yakov, M. Y., Lapshova, M. S., & Deineka, L. A. (2014). Thiophenes of *Tagetes* Flowers and Partial Purification of Xanthophyll Esters. *Universal Journal of Agricultural Research, 2*(3). http://dx.doi.org/10.13189/ujar.2014.020304.

Gnanakan, S. R. P., Rajasekhar, M., & Subramania, A. (2009). Synthesis of Polythiophene Nanoparticles by Surfactant - Assisted Dilute Polymerization Method for High Performance Redox Supercapacitors. *Int. J. Electrochem. Sci., 4*, 1289 - 1301.

Gok, A., Omastova, M., & Yavuz, A. G. (2007). Synthesis and characterization of polythiophenes in the precense of surfactants. *The Journal of Electronic Polymers and Electronic Molecular Materials, 157*(1), 23-29. http://dx.doi.org/10.1016/j.synthmet.2006.11.012

Hubbert, M. K. (1956). Nuclear Enaergy and the Fossil Fuels *Drilling and Production Practice*: American Petroleum Institute

Karim, M. R. (2012). Synthesis and Characterizations of Poly(3-hexylthiophene) and Modified CarbonNanotube Composites. *Journal of Nanomaterials*. http://dx.doi.org/10.1155/2012/174353

Kelkar, D., & Chourasia, A. (2011). Structural Properties of Polythiophene Doped with $FeCl_3$. *Chemistry and Chemical Technology, 5*(3).

Mabkhot, Y. N., Kaal, N. A., Alterary, S., Al-Showiman, S. S., Barakat, A., Ghabbour, H. A., & Frey, W. (2015). Synthesis, In-vitro Antibacterial, Antifungal, and Molecular Modeling of Potent Anti-microbial Agents with a Combined Pyrazole and Thiophene Pharmacophore. *Molecules, 20*(5), 8712-8729. http://dx.doi.org/10.3390/molecules20058712

Marani, A. D., & Entezami, A. (1993). New Synthesis Method of Polythiophenes. *Iranian Journal of Polymer Scimcr and Technology, 3*(1).

Marotti, I., Marotti, M., Piccaglia, R., Nastri, A., Grandi, S., & Dinelli, G. (2010). Thiophene occurrence in different Tagetes species: agricultural biomasses as sources of biocidal substances. *J Sci Food Agric, 90*. http://dx.doi.org/10.1002/jsfa.3950

Munhoz, V. M., Longhini, R., Souza, J. R. P., Zequi, J. A. C., Mello, E. V. S. L., Lopes, G. C., & Mello, J. C. P. (2014). Extraction of flavonoids from Tagetes patula: process optimization and screening for biological activity. *Brazilian Journal of Pharmacognosy, 24*, 576-583. http://dx.doi.org/10.1016/j.bjp.2014.10.001

Niemi, V. M., Knuuttila, P., & Osterholm, J. E. (1992). Polymerization of 3-alkylthiophenes with $FeCl_3$. *Polymer, 33*(7).

Saha, S., Walia, S., Kundu, A., Kumar, B., & Joshi, D. (2012). Antifungal Acetylinic Thiophenes from Tagetes minuta: Potential Biopesticide. *Journal of Applied Botany and Food Quality, 85*, 207-211.

Sakil, M., Singh, A. K., & Roy, G. S. (2013). Study of the Properties of Nanocomposite Cadmium Sulphide (CdS)\Polythiophene(PTh) By TGA/DTA, XRD, UV-VIS Spectroscopy, SEM-EDXA and FTIR. *Researcher, 5*(1).

Sakthivel, S., & Boopathi, A. (2014). Synthesis and Preparation of Polythiophene Thin Film by Spin Coating Method. *International Journal of Science and Research.*

Silva, F. L., Fischer, D. C. H., Tavares, J. F., Silva, M. S., De Athayde-Filho, P. F., & Barbosa-Filho, J. M. (2011). Compilation of secondary metabolites from Bidens pilosa L. *Molecules, 16*(2), 1070-1102. http://dx.doi.org/10.3390/molecules16021070

Souza B. C.C., De Oliveira T. B., Aquino T. M., De Lima M. C.A., Pitta I. R., Galdino S. L., Lima, E. O., Goncalves-Silva, T., Militao, G.C.G., Scotti, L., Scotti, M. T., & Mendonca F. J.B. (2012). Preliminary antifungal and cytotoxic evaluation of synthetic cycloalkyl[b]thiophene derivatives with PLS-DA analysis. *Acta Pharm, 62*, 221–236. http://dx.doi.org/10.2478/v10007-012-0017-y

Takahashi, H. T., Novello, C. R., Ueda-Nakamura, T., Filho, B. P. D., De Mello, J. C. P., & Nakamura, C. V. (2011). Thiophene Derivatives with Antileishmanial Activity Isolated from Aerial Parts of *Porophyllum ruderale* (Jacq.) Cass. *Molecules, 16*(5), 3469-3478. http://dx.doi.org/10.3390/molecules16053469

Tang, Z. (2010). *Studies of Inverted Organic Solar Cells Fabricated by Doctor Blading Technique.*

Vijay, K. P., Laxman, B. C., Balasaheb, S. R., Yuvraj, N. R., & Janardhan, P. M. (2013). Pharmacognostic, Physicochemical and Phytochemical Investigation of *Tagetes erecta* Flowers (Asteraceae). *Journal of Biological & Scientific Opinion, 1*(1). http://dx.doi.org/10.7897/2321-6328.01124

Yao, Y., Wen, Y., Xu, J., Zhang, L., & Duan, X. (2013). Application of commercial poly(3,4-ethylenedioxy-thiophene):poly(styrene sulfonate) for electrochemical sensing of dopamine. *Journal of the Serbian Chemical Society, 78*(9), 1397-1411. http://dx.doi.org/10.2298/jsc120927036y

6

Correlating the Layer Properties of Ni-alumina Composite Coatings and the Mechanism of Codeposition

Amir Sadeghi[1], Maximilian Sieber[1], Hosein Hasannejad[2], Ingolf Scharf[1], Thomas Lampke[1]

[1]Chemnitz University of Technology, Institute of Materials Science, Chemnitz, Germany

[2]Arak University, Department of of Materials Engineering, Iran

Correspondence: Amir Sadeghi, Chemnitz University of Technology, Institute of Materials Science and Engineering, Chemnitz, Germany. email: amir.sadeghi@mb.tu-chemnitz.de

Abstract

Ni-Al_2O_3 composite coatings electrodeposited from Watt's electrolyte include Al_2O_3 nano-and submicron particles. The effect of particle size and concentration of the particles in the electrolyte were investigated on the morphology and incorporation value of particles into the deposits. The influence of alumina particles on the electrodeposition behavior of Ni was also studied by means of electrochemical impedance spectroscopy. The results achieved from the impedance measurements and the correlation with the layer characterization could help to better understand the codeposition mechanism derived from the impact of different particle characteristics including size and concentration on the nature of the double-layer. The increase in the concentration of particles from 1 to 20 g/l resulted in an increase of the double-layer capacity and decrease of the charge transfer resistance, while the addition of submicron particles had a higher influence on the characteristics of the double-layer compared to the nano particles. Although the alumina particles with submicron size could stimulate the incorporation of particles faster than those with nano size, the strengthening performance of the layers not only depended on the incorporation value of the particles, but also on the microstructure of the deposits.

Keywords: Ni-Al_2O_3 electrodeposited coating, particle size, codeposition mechanism, electrochemical impedance spectroscopy, layer properties

1. Introduction

Composite coating has been well developed for the last decades. The simultaneous deposition of the non-metal inclusions and metal phases to form composite layers entails a significant improvement on several physical and mechanical properties of the coating. Among the non-metal inclusions, ceramic particles like Al_2O_3, SiC, TiO_2, WC, etc. have been frequently used in the literature for synthesizing composite coatings particularly with better mechanical properties (Guglielmi, 1972; Viswanathan, 1979; Sun, 1996; Ghouse, 1984; Periene, 1993; Zahavi 1982; Stankovic, 1996; Bakhit, 2013).

The final performance in the composite coatings is highly dependent on the characteristics of the particles used in the electrodeposition process including type, size and concentration (Zahavi 1982; Stankovic, 1996; Bakhit, 2013; Ehrhardt, 1985; Bazzard, 1972; Verelest,1991; Chen, 1971; Roos, 1978; Stojak, 1999; Stojack, 2001). The characteristics of Al_2O_3 as a typical reinforcement used in the electrodeposition process have already been investigated by others (Stojak, 1999; Stojack, 2001; Osborne, 2006; Lee, 1988; Shao, 2002). It is stated that the size of alumina particles can strongly influence the codeposition value of particles in the deposited metal. As it was reported by Shao and his coworkers (Shao, 2002), the volume percentage of Al_2O_3 300 nm is much higher than Al_2O_3 50 nm, although the number of codeposited nano particles is higher compared to submicron particles. Meanwhile, except for the effect of the particle size, the influence of the particle type cannot be neglected. It seems that there are some characteristics between alumina alpha and gamma like different zeta-potential or magnetic properties, which can eventually change their codeposition behavior in different solutions (Chen, 1971; Roos, 1978; Stojak, 1999; Stojack, 2001).

Up to now, different models have been reported for the codeposition mechanism of particles in the electrolyte under various working conditions by researchers like Guglielmi (Guglielmi, 1972), Celis (Roos, 1978), and other researchers. Generally, the early suggested models for the codeposition mechanism of solid particles in the metallic matrix initiated from three main phenomena: electrophoresis, mechanical entrapment and adsorption. More details about each of these phenomena are broadly explained in ref (Low, 2006). These theoretical models can provide us with a good understanding of the codeposition phenomena. The validity of the various theoretical models underlying particle incorporation in the metal matrix, still requires more empirical laboratory trials and attention to the interactive variables such as the particle characteristics like particle size and particle concentration.

In recent years, it has been established that electrochemical impedance spectroscopy (EIS) can be a reliable tool in order to better understand the effect of particles in the codeposition process (Benea, 2001; Hu, 2005; Hu, 2007; Sohrabi, 2012). For instance, Benea et al. (Benea, 2001) studied the influence of silicon carbide nano particles on nickel electroplating. They stated that particles affected the nickel reduction, which became obvious by the lower charge-transfer resistance. They also stated that the surface morphology of nanostructured composite layers is different compared to a pure nickel coating. The purpose of the SiC particles was to increase the number of nucleation sites with a reduction in crystal growth resulting in a smaller grain size of the nickel matrix.

The aim of this work is to investigate the influence of nano- and submicron Al_2O_3 particles with different concentrations in the electrolyte on the codeposition behavior of particles in a Ni matrix by the aid of impedance spectroscopy and also by relying on the effect of particle size and concentration on the layer properties including morphology, microstructure and hardness.

2. Experimental Setup

In this study, a typical Watt's electrolyte including 250 g/l $NiSO_4 .7H_2O$ (Sigma Aldrich), 30 g/l $NiCl_2.6H2O$ (Merck), 30 g/l boric acid (Merck) and 0.3 g/l sodium dodecyl sulfate (Merck) was used for the fabrication of Ni and Ni-alumina composite coatings. One liter of each electrolyte was prepared with distilled water in a 2000 cc glass container. The pH and temperature were held at 3.5 and 55 °C, respectively. Suspensions were prepared by separately adding alumina (Alfa) 13 nm and (Gamma) 150-200 nm to the solution. The mix of suspension and solution stirred for at least 6 h before the deposition process. The coatings electrodeosited from electrolyte with three different particle concentration of 0, 5, 20 g/l. It should be pointed that higher concentration of particles, due to the agglomeration challenge and decrease in incorporation value of particles into the deposit was not investigated. An IM6-Zahner working station in a galvanostatic mode was simultaneously used for the production of the deposits and also for the impedance measurements in a three-electrode cell as it can be seen in figure 1. Ultrasound vibration (25 Hz and power of 55 W) was applied during the deposition while the electrolyte was agitating with a rate of 400 rpm. Carbon steel with the polishing surface area of 20 cm^2 and pure nickel with 45 cm^2 were used as working electrode and counter electrode, respectively. A distance of 5 cm was maintained between working electrode and counter electrode, and the Ag/AgCl reference electrode was placed in the middle. The applied current was -800 mA, which is equal to 4 A/dm^2, according to the active area of the working electrode. The frequency range for impedance measurements was between 100 KHz and 10 Hz with a 10 mV amplitude. Before performing the first impedance test, in order to reach a constant and stable condition, each sample was deposited in the electrolyte for 10 minutes and after that the impedance measurements were recorded in three more cycles each three minutes with adjusting in Polysequencer configuration available on the Zahner-software. Each impedance measurement took 81 seconds. In order to ensure the reproducibility of the results, except performing four cycle tests on each sample, some of the tests were repeated randomly with new samples, and the average values were finally reported for the impedance measurements.

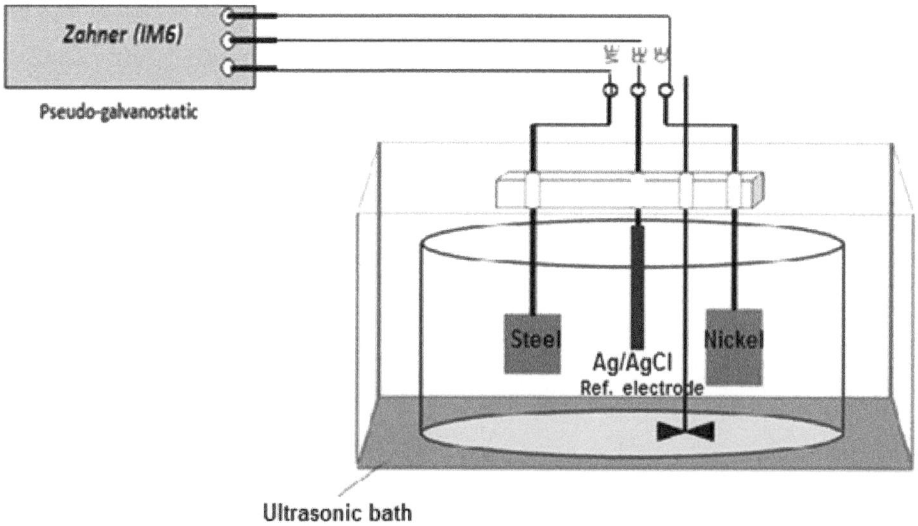

Figure 1. Experimental setup

The thickness of coatings was measured at 10 different points by means of optical microscopy (Olympus GX 51). The morphology of the deposits and also the value of the incorporation of the particles were examined by the help of SEM (NEON 40EsB, Zeiss) and EDX (Genesis).The EDX measurements were performed around the cross-section. Image analyzer software (a4i-Analysis) was used for the estimation of the volume percentage of the particles in a certain surface area. The nickel grain size was estimated by the X-ray diffraction (XRD) method of Scherrer's equation (Patterson, 1939) using the Co–Kα line (D8 Discover, Bruker AXS).

It should be mentioned that the obtained values for the grain size represent the average size in the analyzed volume and not allow for a statement about the distribution of the grain size. Moreover, the evaluation of the grain size achieved by this method is only true for the grain size under about few hundreds of microns; over almost 300 nm the real average of grain size can be even larger. For detecting the Martens hardness value, instrumented indentation (HM2000XYm, HELMUT FISCHER GmbH) was applied on the cross-sections of the deposits by applying a 25 mN force for 30 s loading time, 30 s holding time and another 30 s unloading time on a Vickers indenter. The final value quoted for a deposit represents the average of 5 measurements.

3. Results and Discussion

3.1 Effect of Alumina Particles on the Morphology, Microstructure and Mechanical Properties

As illustrated in figure 2, the morphology of Ni-Al_2O_3 composite coatings is different compared to a pure nickel coating. With the addition of Al_2O_3 particles in the electrolytic, the boarders of the Ni grains become fuzzy and more compact in comparison to pure Ni with no particles. In the case of adding nano-Al_2O_3, the morphology seems to be even more compact and homogenous. Moreover, for the coating electrodeposited with submicron particles, it is obvious that the particles were mainly indulged with a certain gap from the Ni matrix and growth of nickel around the particle becoming rough. This behavior can be related to the non-inductive nature of alumina particles as it was also mentioned by (Erler, 2002). Furthermore, according to the cross-section SEM images in figure 3, it is observed that the dark spherical Al_2O_3 particles are uniformly distributed in the white nickel matrix, thanks to the use of ultrasound during the deposition process. Nevertheless, due to the higher surface energy of nano particles, agglomeration at higher particle concentrations is still unavoidable despite of the application of ultrasound.

The EDS analysis in figure 4 displays the atomic percentage of Al for Ni-Al_2O_3 composite coatings electrodeposited from Watt's bath including nano- and submicron Al_2O_3 particles with different particle concentrations in the electrolyte. As can be seen, under the same conditions, the atomic percentage of Al for submicron particles is higher than the coatings prepared with nanoparticles. In addition, in both cases, the incorporation value of particles increases with increasing the loading of particles in the electrolyte.

However, as plotted in figure 5, the measurement of the volume percentage of particles in a certain surface by means of the image analyzer shows that, under the same working conditions, the number of incorporated particles, in the case of adding nanoparticles, is slightly higher compared to the addition of submicron particles, namely 5.1 to 4.6 vol%. The same behavior for Al_2O_3 50 nm and 300 nm was also detected by Vereecken et al. Shao, 2002). In fact, by the reduction of the particle size, the number of the particles codeposited into the matrix is higher; but, due to the smaller size, the atomic percentage would be lower.

Figure 2. SEM images of the surface morphology:
(a) Ni, (b) Ni-submicron Al_2O_3 20 g/l (c) Ni-nano Al_2O_3 20 g/l electrodeposited at $i=4A/dm^2$ for 40 min

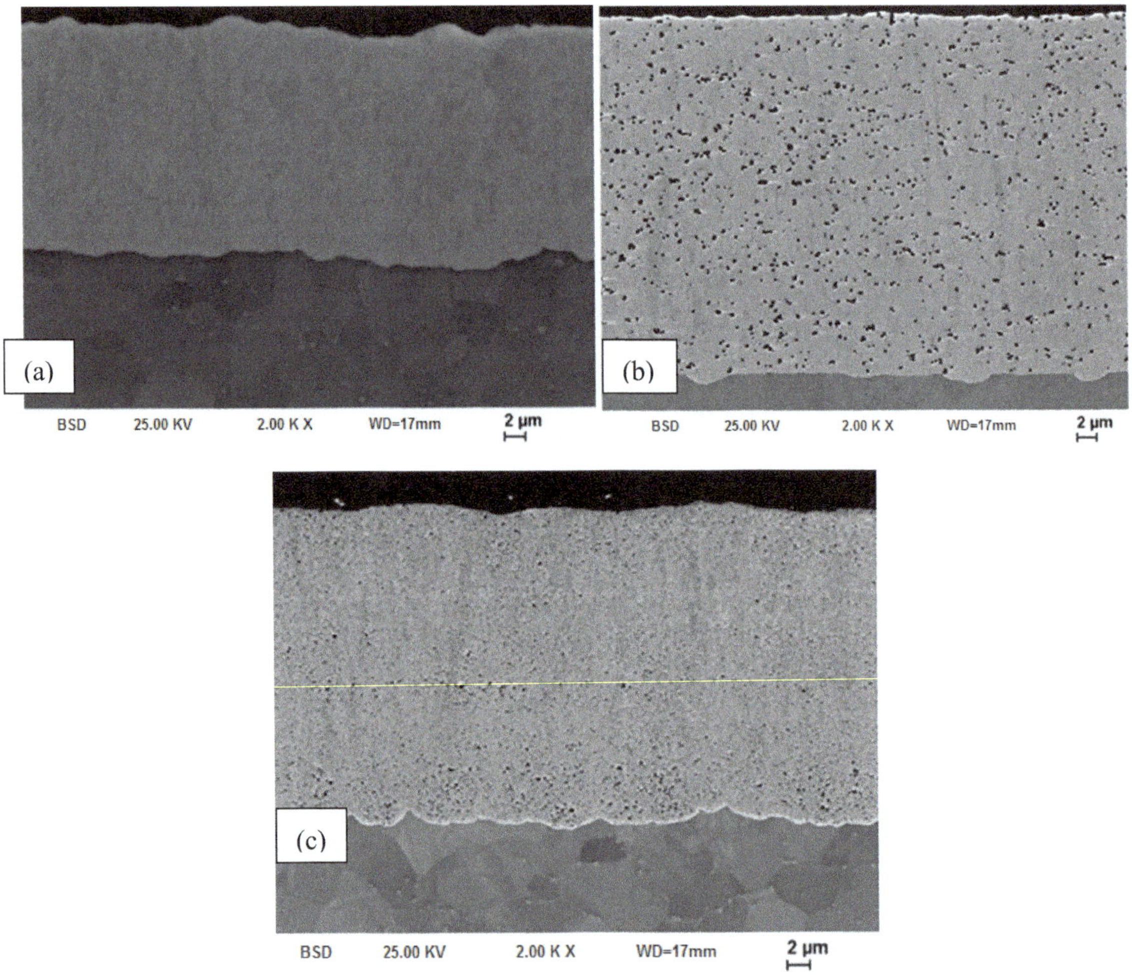

Figure 3. SEM images of cross-section:

(a) Ni, (b) Ni-submicron Al_2O_3 20 g/l (c) Ni-nano Al_2O_3 20 g/l electrodeposited at i=4A/dm^2 for 40 min

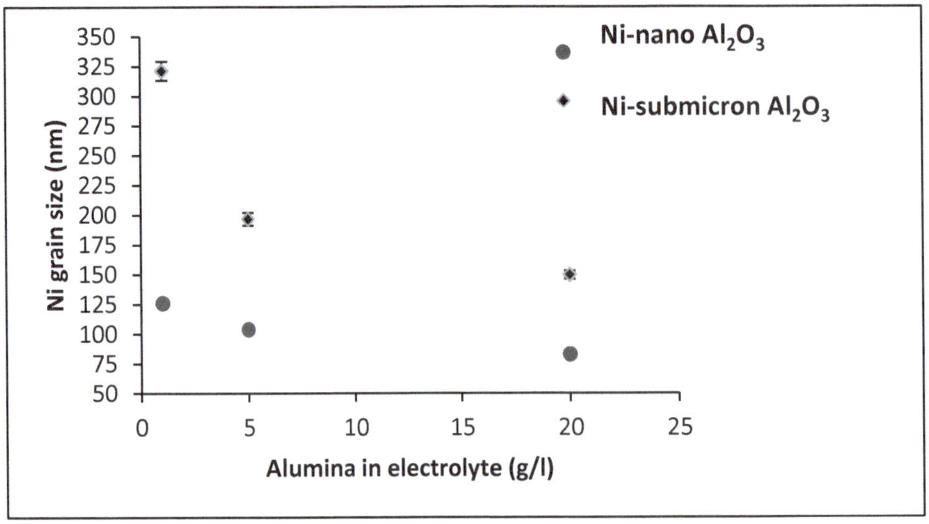

Figure 4. Effect of particle size and concentration on the incorporation value of particles achieved from EDX mapping analysis from the thickness of the deposits

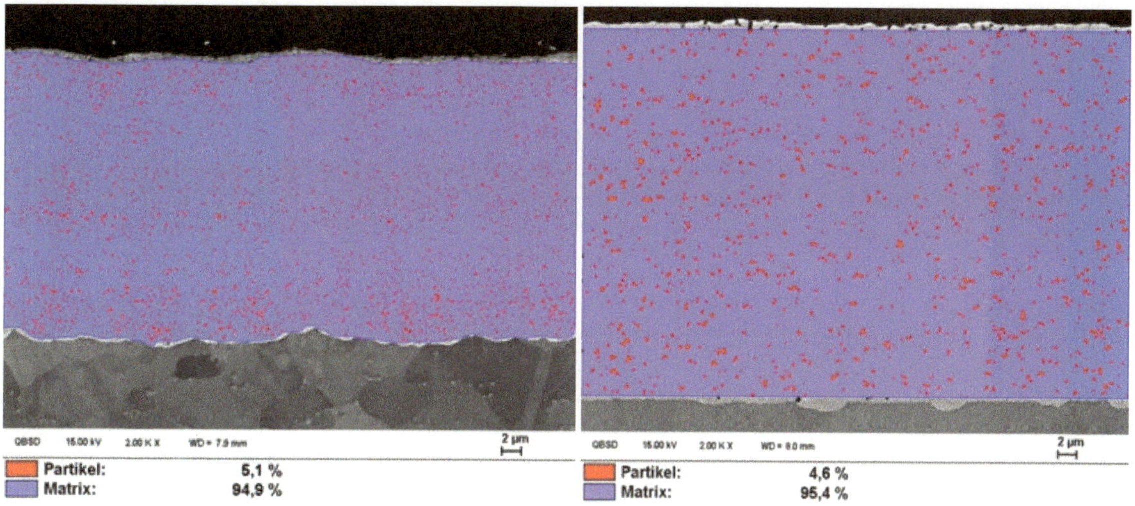

(a) (b)

Figure 5. Volume percentages (vol%) of alumina particles estimated by image analyzer software: (a) Ni-nano Al_2O_3 20 g/l (b) Ni-submicron AL_2O_3 20 g/l electrodeposited under the same working conditions

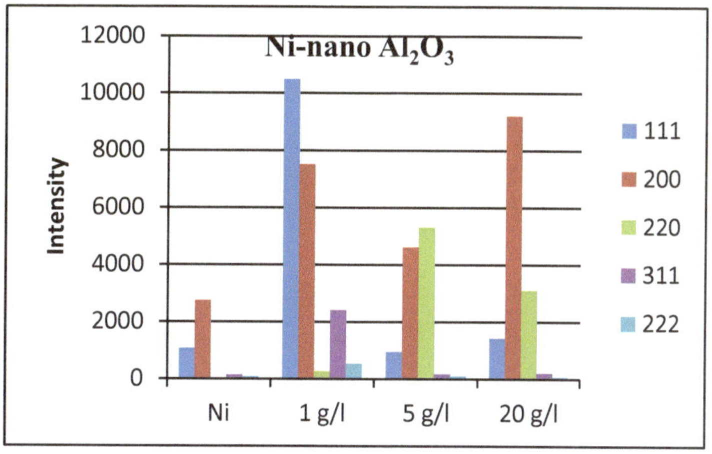

(a)

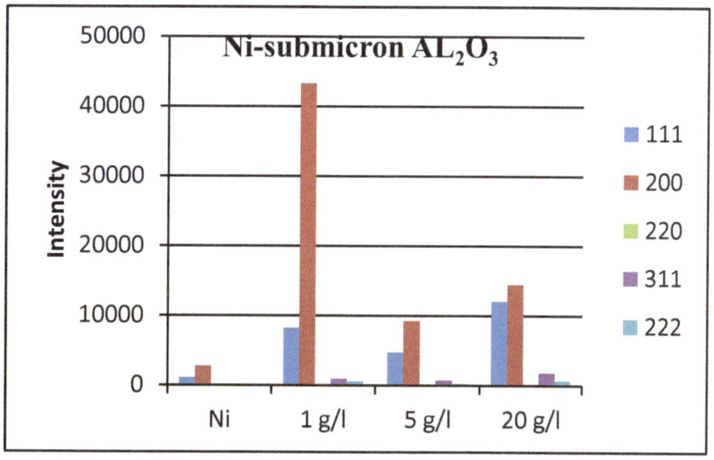

(b)

Figure 6. Texture values in (a) Ni-nano Al_2O_3 (b) Ni-submicron Al_2O_3 electrodeposited at current density of $4A/dm^2$ with 20 g/l particle concentration into Watts's electrolyte

According to the crystal mapping which has been done specifically for the coatings with nano and submicron particle for the concentration of 20 g/l, as shown in Figure 6, the co-deposition of Al_2O_3 obviously affects the relative intensity corresponding to different crystal faces. The relative intensity corresponding to crystal face (1 1 1) increases, but the relative intensity of crystal face (2 0 0) decreases. For composite coatings with nanoparticles, the peak of crystal face (2 2 0) is more obvious. It seems that also the peak of (311) becomes stronger with increasing of Al_2O_3 content. Erler et al. (Erler, 2002) demonstrated that the XRD patterns of nickel nano-composite coatings indicate changes in texture of such coatings which are dependent on the particle content in electrolyte. Chang et al. (Chang, 2003) was also presented similar results.

Figure 7 exhibits the grain size of the Ni matrix for the samples estimated from Scherrer's equation using the XRD peak width. In fact, the reduction in the grain size of the Ni crystal by adding particles approves the compactness of the coatings as also confirmed by the results of the morphology. It can be seen that, by increasing the concentration of particles in the electrolyte, on the one hand, the incorporation value for the case submicron particles will be higher than nano particles; but, on the other hand, the grain size value shows a higher reduction compared to when nano particles are added. The reason here is that smaller particles can offer more sites for nucleation, which is detrimental for the growth of Ni crystal while on larger particles; Ni crystal finds more opportunity to grow (Zhang, 2014).

One possible mechanism to explain the codeposition behavior of particles here is the phenomenon proposed by Guglielmi (Guglielmi, 1972). In the first step, the particles in the electrolyte clouded by nickel ions and due to the agitation in the electrolyte, clouded particles approach the cathodic surface and there, a certain amount of particles can be trapped and buried by reducing new ions over them. According to the thickness measurements in figure 8, it is understandable that in lower particle concentrations till 5 g/l, the possibility of transferring Ni ions through the particles is high such as the value of the thickness at the given concentration is large, namely around 65 μm. At the higher concentration of 20 g/l, however, the thickness values for both particle types showed a declining trend. It is possible that the addition of more particles into the solution could block the cathodic surface and therefore, a more active area for the reduction of Ni ions.

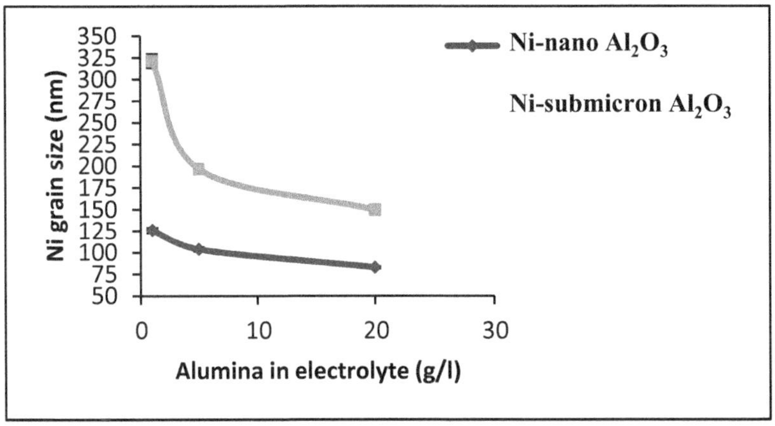

Figure 7. Effect of particle size and concentration on the Ni grain size

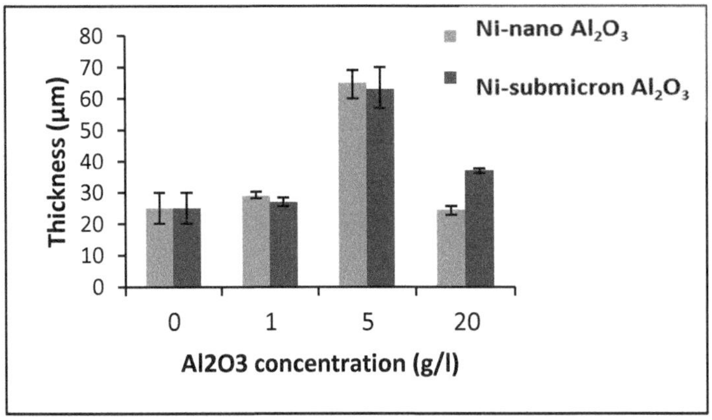

Figure 8. Effect of particle size and concentration on the thickness value of the deposits electrodeposited under the same conditions

Nevertheless, at the higher concentration of 20 g/l, the thickness of the layer deposited with submicron particles is higher than the layer deposited with nano particles, 37±0.7 and 25±1.2 µm respectively. The reason here, as explained above, can be the better opportunity of Ni crystals to grow on larger particles rather than on smaller ones. Since the particle hinders the current flow, the local overvoltage should rise respectively, thus the number of crystallization sites would be increased.

The Martens hardness of Ni-Al_2O_3 composite coatings was also carried out on the cross section of the layers by means of the Martens test as it was already explained in the experimental section (Chang, 2003)

According to the results of the Martens hardness presented in figure 9, it can be discerned that the Martens hardness of the Ni-Al_2O_3 composite coatings electrodeposited with both nano- and submicron particles follow an increasing trend by increasing the particle concentration in the electrolyte. It is obvious that the addition of nano particles leads to a higher Martens hardness at each single concentration in contrast to the addition of submicron particles. According to the results of the Martens hardness and the suggested codeposition mechanism for alumina particles with nano and submicron size, it can be concluded that, even though the codeposition can arise easier for the larger particles, which can be related to easier transfer of the larger particles from the double layer to the surface, this does not mean that in the case of larger particles, the nucleation rate of nickel and subsequently, the hardness value should also necessarily be higher.

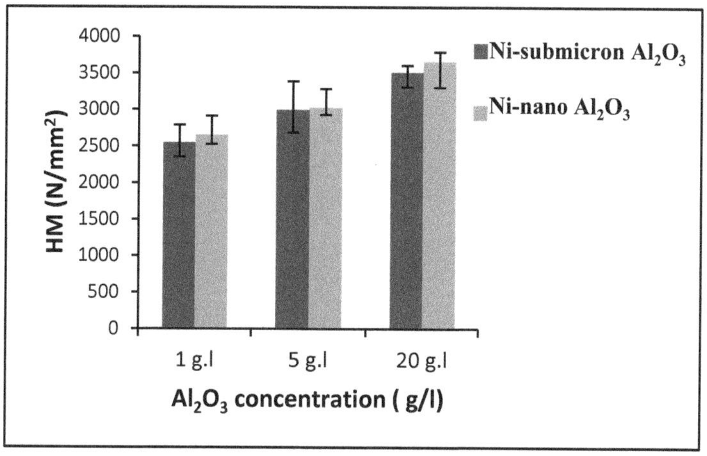

Figure 9. concentrations including the effect of grain size and particle content

The Martens hardness can be controlled by dispersion hardening or the amount of dispersed particles embedded in the layer based on the Orowan mechanism (Epelboin, 1971). As it could be seen, by increasing the concentration of particles in the electrolyte, the incorporation value for submicron particles will be higher than that of nano particles but on the other hand the number of incorporated particles in the case of applying nano particles is higher. This means that the possibility of hindering the dislocations in the coating deposited by nano particles is higher. Moreover, as it was already seen, the grain size of the nickel matrix is smaller in the case of nano particles. This means that the grain size of the matrix, based on the Hall-Petch effect (Orowan, 1984), plays an important role in the hardening mechanism of coatings.

In order to shed more light on our assumptions about the codeposition behavior of alumina particles with different sizes and concentrations, impedance spectroscopy was used to explain the effect of particle features on the behavior of the charge transfer and double layer capacity for the Ni electrodeposition.

3.2 In-situ EIS Study of Ni-Al_2O_3 Electrodeposition

Figures 10 (a-c) show the Nyquist plots from electrochemical impedance spectroscopy results during the electrodeposition process for Ni and Ni-Al_2O_3 with particle nano- and submicron particles and different concentrations of 1, 5 and 20 g/l under the same deposition conditions. The impendence data comprises of a single capacitance loop in high and medium frequencies and an inductive loop in lower frequencies.

According to the mechanism of Ni electrodeposition proposed by Epelboin and Wiart (Wiart, 1990; Epelboin, 1971), the simple circuit model excluding inductive in very low frequencies was suggested in this study. Therefore, the impedance measurements could be achieved much faster within a shorter range of frequency and provide us with information about charge transfer and double layer capacity, which can be directly related to the Ni reduction. Nevertheless, the trace of inductive loops is indeed visible in rather lower frequencies. For simplicity we have just considered the effect of particles on higher frequencies which can achieve faster without variations in working parameters during electrodeposition process such as pH. In this model, C_{dl} is the double layer capacitance of the interface of the electrode/electrolyte, and R_{dl} is a resistance against the charge transfer through the double layer at the electrode/electrolyte interface. As it was seen in figure 10, the loop size for both particle sizes was found to decrease with increasing particle concentration till 20 g/l. Moreover, as it is seen in the figure 10c, at the concentration of 20 g/l, the usage of submicron particles caused higher shrinkage in the semicircle of Nyquist spectrums compared to the addition of nanoparticles to the electrolyte.

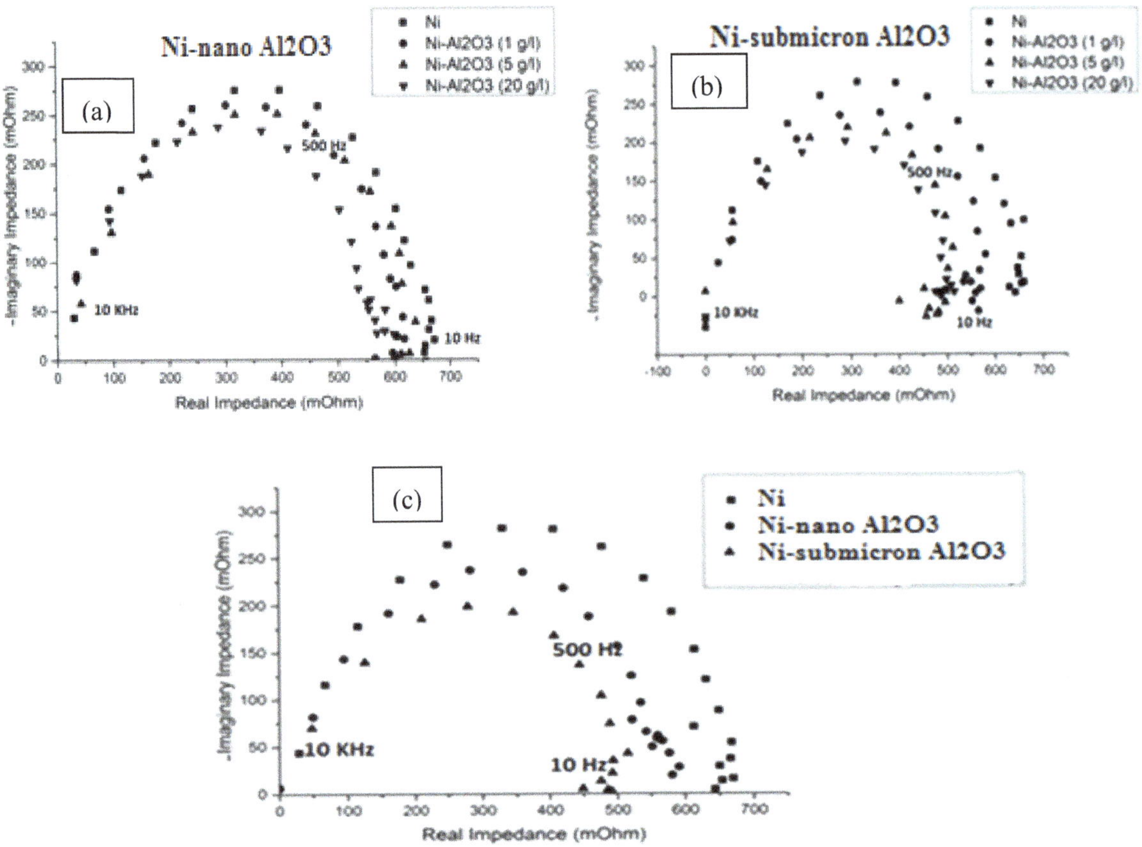

Figure 10. Effect of particle size and concentration on the Nyquist spectrums:

(a) Ni-nano Al_2O_3 (b) Ni-submicron Al_2O_3 (c) comparison at the concentration of 20 g/l

Increasing the concentration of the particles resulted in the decrease of the charge transfer resistance or the increase of the double layer capacity as it was also shown by Benea or Dolati for SiC particles (Sohrabi, 2012; Benea, 2001). As it can be deduced from the impedance results in figure 10, a descending trend in the impedance of the charge transfer as well as an ascending trend in the doubled-layer capacity was observed by increasing the Al_2O_3 particle concentration till 20 g/l. According to the impedance results, by the increase in the concentration of the particles in the electrolyte, the possibility that nickel ions are transferred by means of the particles towards the cathodic surface is higher. In the other words, the declines in the thickness of the double layer can finally inffluence the nickel reduction. Moreover, according to the results of figure 12, the value of the charge transfer resistance for submicron particles is lower than that of the nano particles, and the double layer capacity for larger particles is moderately higher compared to smaller particle sizes.

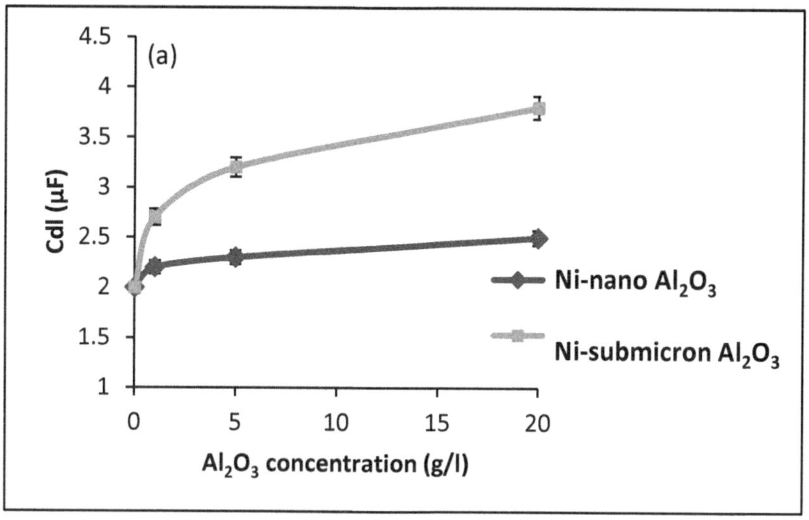

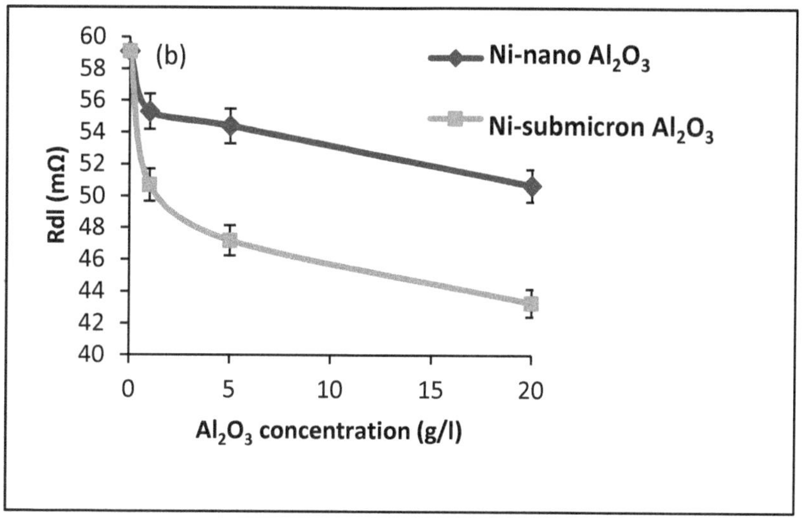

Figure 12. Results of C_{dl} (a) and R_{dl} (b) calculated by fitting the impedance results with the suggested circuit model by means of Matlab

This means that the thickness of the double layer on the cathodic surface is less than that of nano particles in the case of adding submicron particles. Thus, as it is illustrated in the schematic of figure 13, submicron particles can be easier incorporated into the deposit than nanoparticles due to having a larger active surface for free ions and a higher mass transfer. Nevertheless the deposition of nickel at the higher concentration of 20 g/l arises faster due to the higher charge transfer and double layer capacity. The reason for the decline in the thickness of coatings or for the reduction of the thickness of deposits at the higher particle concentration of 20 g/l, as it is seen in figure 13b, can be partly related to the blocking of the active surface of the cathode due to the accumulation of particles on the surface.

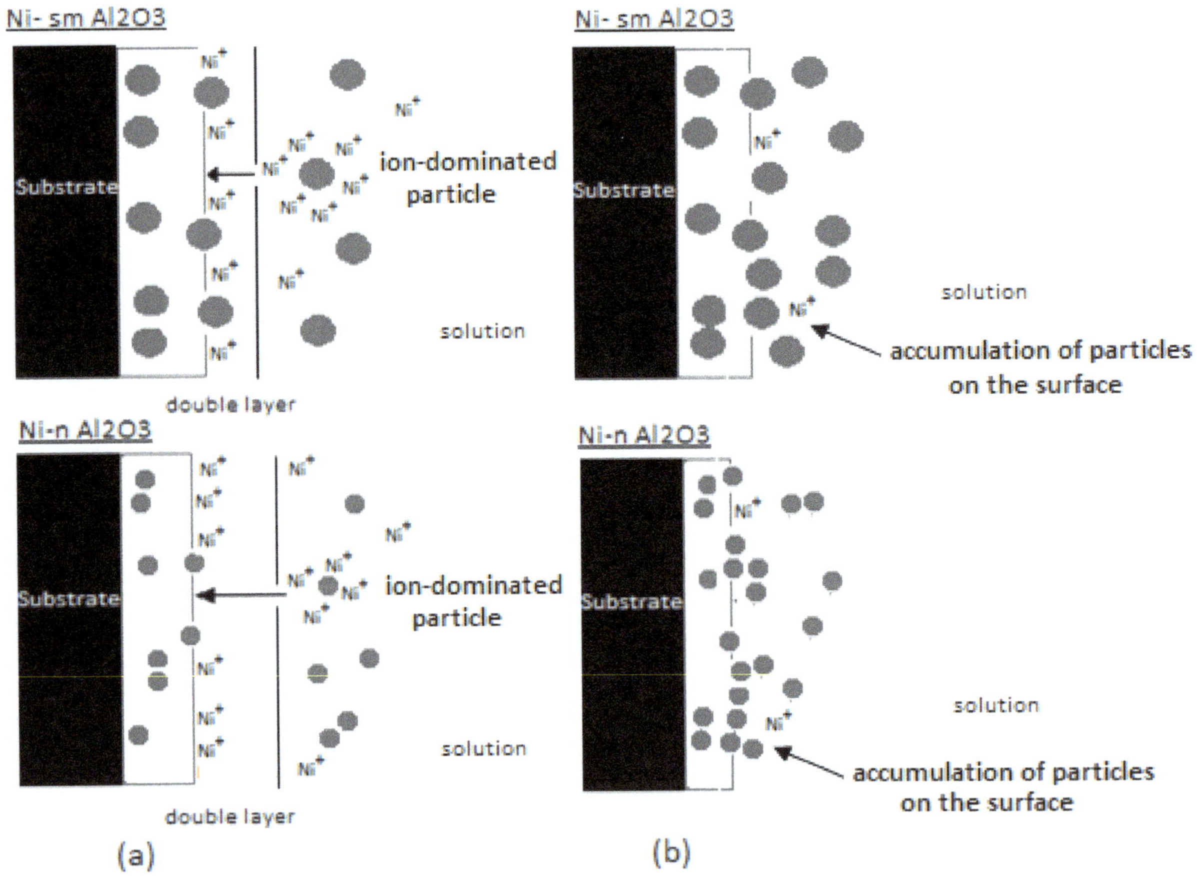

Figure 12. Schematic of the particle incorporation mechanism based on particle size and concentration: (a) at low concentration of particles (b) at high concentration of particles

4. Conclusion

A descending trend in the impedance of the charge transfer as well as an ascending trend in the doubled-layer capacity was observed by increasing the Al_2O_3 particle concentration till 20 g/l. The value of the charge transfer resistance for submicron particles is also lower than that of nano particles, and the double layer capacity for larger particles is moderately higher compared to smaller particle sizes. The results of the impedance measurements considering the layer characteristics could correspond well with the mechanism proposed by Guglielmi. The Martens results showed that for both nano- and submicron alumina particles, increasing the particle concentration to 20 g/l leads to highest hardness values. Based on the results of the Ni grain size, despite of the lower incorporation value of nano particles compared to submicron particles, the hardness ability of coating deposited with nano particles is rather higher than of the coatings prepared with submicron particles even with higher incorporation value. For both coatings produced with submicron and nano particles, at the middle concentration of 5 g/l, the current efficiency is much higher and both of the coatings show an almost similar strengthening performance, which relates to the simultaneous effect of Hall-Petch and the Orowan mechanisms.

Acknowledgements

This research has been supported by the institute of materials science and engineering of Technical university of Chemnitz.

References

Bakhit, B., & Akbari, A. (2013). Synthesis and characterization of Ni–Co/SiC nanocomposite coatings using sediment co-deposition technique J.Alloys and Compounds, 560, 92-104, http://doi.org/10.1016/j.jallcom.2013.01.122

Bazzard, R., & Boden, P. J. (1972). Nickel-Chromium Alloys by Codeposrtion: Part l-Oodepositton of Chromium, *Transactions of the Institute of Metal Finishing, 50*. http://doi.org/10.4139/sfj1950

Benea, L. (2009).Electrodeposition and tribocorrosion behaviour of ZrO_2–Ni composite coatings, *J. Appl. Electrochem., 39*. http://doi.org/10.1007/s10800-009-9859-5

Benea, L., Bonora, P. L., & Borello, A. (2001). Wear corrosion properties of nano-structured SiC-nickel composite coatingsobtained by electroplating, *Journal of the Electrochemical Society, 148*. http://doi.org/10.1016/S1003-6326(13)62688-6

Chen, S., Lakshiminaryanan, G. R., & Sautter, F. K. (1971). The codeposition of alumina and titania with copper, *Metallurgical Transactions, 937*. http://doi.org/10.1007/BF02664222

Cheng, S., Spencer, J. A., & Milligan, W. W. (2003). Strength and tension/compression asymmetry in nanostructured and ultrafine-grain metals, *Acta Mater., 51*. http://doi.org/10.1016/S1359-6454(03)00286-6

Ehrhardt, J. (1985). Composite electroplating by using brush technique Product Research and Development, *24*. http://doi.org/10.1021/i300020a016

Epelboin, I., & Wiart, R. (1971). Mechanism of the Electrocrystallization of Nickel and Cobalt in Acidic Solution, *J. Electrochem. Soc. 118*, 1577-1582. http://doi.org/10.1149/1.2407788

Erler, F., Jakob, C., Romanus, H., Spiess, L., Wielage, B., Lampke, T., & Steinhauser, S. (2002). Interface problems in nickel composite coatings with oxide ceramic nano-particles, Oral presentation, 53rd Annual Meeting of the International Society of Electrochemistry, *Düsseldorf*, 15-20. http://doi.org/10.1016/S0013-4686(03)00380-3

Foster, J., & Cameron, B. (1976). A Study of the Mechanism of Formation of Electrodeposited Composite Coatings Trans. of the Inst. of Metal Finishing, *54*.

Ghouse, M. (1984). Wear Characteristics of Sediment Codeposited Ni-SiC Composite, *Met. Finish. 82*, 33–37. http://doi.org/10.1016/0376-4583(84)90162-6

Guglielmi, N. (1972). Kinetics of the deposition of inert particles from electrolytic baths, *J. Electrochem. Soc., 119*, 1009-1012. http://doi.org/10.1149/1.2404383J

Hu, F., & Chan, K. C. (2005). Deposition behaviour and morphology of Ni–SiC electro-composites under triangular waveform, *Applied Surface Science, 243*. http://doi.org/10.1016/j.apsusc.2004.09.068

Hu, F., Chan, K. C., & Qu, N. S. (2007). Effect of magnetic field on electrocodeposition behavior of Ni–SiC composites *J. Solid State Electrochem, 11*. http://doi.org/10.1007/s10008-006-0103-5

Lee, C. C., & Wan, C. C. (1988). A Study of the Composite Electrodeposition of Copper with Alumina Powder, *J. Electrochem. Soc., 135*. http://doi.org/10.1149/1.2096182

Low, C. T. J., Wills, R. G. A., & Walsh, F. C. (2006). Electrodeposition of composite coatings containing nanoparticles in a metal deposit, *Surface & Coatings Technology 20*, 371–383. http://doi.org/10.1016/j.surfcoat.2005.11.123

Orowan, E. (1984). Symposium on internal stresses in metals and alloys. *Institute of Metals, London, 451*.

Osborne, S. J. (2006). Electrocodeposition of nanoparticle composite films using an impinging jet electrode, Ph.D Dissertation., UC San Diego. http://escholarship.org/uc/item/2q71k3qk

Palumbo, G., Erb, U., & Aust, K. T. (1990). Triple line disclination effects on the mechanical behaviour of materials, *Scripta Metall. Mater. 24*. http://dx.doi.org/10.1051/jphys:019840045060102500

Patterson, A. (1939).The Scherrer Formula for X-Ray Particle Size Determination. *Phys. Rev. 56*, 978–982, http://dx.doi.org/10.1103/PhysRev.56.978

Periene, N., & Taicas, L. (1993). Nickel electrodeposits with improved hot oxidation and corrosion resistance by co-deposition of submicron powder, *Plat. Surf. Finish. 80*, 73-76. http://doi.org/10.1080/0144341930130306

Robert, P., & Socha et al, (2004). Influence of adsorption on the charge transfer reactions at the pyrite surface, *Colloids and Surfaces A: Physicochem. Eng. Aspects, 235*. http://doi.org/10.2138/am-2000-5-628

Roos, J. R., & Celis, J. P. (1978).Dispersion-hardened electrolytic copper-alumina coatings, *Thin solid films, 54(2)*, 16 October, 173-182. http://doi.org/10.1016/0040-6090(78)90196-7

Shao, I., Vereecken, P. M, Cammarata, R. C., & Searson, P. C. (2002). Kinetics of Particle Codeposition of Nanocomposites, *J. Electrochem. Soc., 149*. http://doi.org/10.1149/1.1514672

Sheng, C. W., Wen, C. J., & Wie, J. (2003). Characterization of electroplated Ni/SiC and Ni/Al2O3 composite coatings bearing nanoparticles, *Mater. Res., 18*(7). http://doi.org/10.1557/JMR.2003.0216

Sohrabi, A., & Dolati, A. (2012). A Study on the Mechanism of Electrodeposition of Ni/SiC Composite Coatings using, *Impedance Technique ECS Trans. 41*. http://doi.org/10.1149/1.4718390

Stankovic, D. V., & Gojo, M. (1996). Electrodeposited composite coatings of copper with inert, semiconductive and conductive particles, *Surf. Coat. Technol. 81*, 225-232. http://dx.doi.org/10.1016/0257-8972(95)02486-7

Stojack, J. L., & Talbot, B. (2001). Effect of particles on Polarization during Electrocodeposition using a Rotating Cylinder Electrode, *Journal of Applied Electrochemistry, 31.*

Stojak, J. L., & Talbot, J. B. (1999). Investigation of Electrocodeposition Using a Rotating Cylinder Electrode, *Journal of the Electrochemical Society, 146*. http://doi.org/10.1149/1.1392665

Sun, K. N., Hu, X. N., Zhang, J. H., & Wang, J. R. (1996). Electrodeposited Cr-Al2O3 Composite Coating for Wear Resistance, *Wear 196*, 295-297. http://doi.org/10.1016/0043-1648(95)06860-0.

Verelest, M., & Bonino, J. P. (1991). Electroforming of Metal Matrix Composite:Dispersoid Grain Size Dependence of Thermostructural and Mechanical Properties, *Materials Science and Engineering,135.* http://doi.org/10.1016/0921-5093(91)90536-V

Viswanathan, M. (1979). Occlusion plating of nickel–graphite composite, *Met. Finish, 77*, 67-69. http://doi.org/10.1016/j.surfcoat.2011.05.030.

Watson, S. W. (1993). Electrochemical Study of SiC Particle Occlusion during Nickel Electrodeposition, *J. Elechtrochem. Soc., 140*. http://doi.org/10.1149/1.2220801.

Watson, S., & Walters, P. (1991). The effect of chromium particles on nickel electrodeposition, *Journal of Electrochernical Society 138*. http://doi.org/10.1149/1.2085471

Wei, H., & Cent, J. (2010). Kinetics analysis of Ni-TiO2 composite system during initial stages of electro-crystallization, *South Univ. Techno. 17*. http://doi.org/10.1007/s11771-010-0507-3

Wiart, R. (1990). Elementary steps of electro-deposition analysed bymeans of impedance spectroscopy, *Elec.Acta 35*, 1587-1593. http://doi.org/10.1016/0013-4686(90)80014-F

Wu, G., et al, (2004). Effect of α-Al2O3 particles on the electrochemical codeposition of Co-Ni alloys from sulfamate electrolytes, *Materials Chemistry and Physics, 87*. http://doi.org/10.1016/j.matchemphys.2004.06.016

Zhang, Z., Wu, X., Jiang, C., & Ma, N. (2014). Electrodeposition of Ni matrix composite coatings containing ZrC particles, *Surface Engineering, 30*, 21-25. http://doi.org/10.1179/1743294413Y.0000000185

7

Scalable Synthesis of High-Purity (*R, R*)-Dexmethylphenidate Free Base

Guan Wang[1], Dongsheng Chen[1], Yuanyuan Sun[1], Qing Cao[1], Bonan Li[1], Jianqi Li[1]

[1]Novel Technology Center of Pharmaceutical Chemistry, Shanghai Institute of Pharmaceutical Industry, Shanghai , P. R. China

Correspondence: Bonan Li, Room 5C01, Xinghan Biulding, 285 Gebaini Road, Shanghai, 201203, P.R. China. E-mail: pkuthulbn@gmail.com

Abstract

A practical method for the preparation of high-purity (*R, R*)-dexmethylphenidate free base was developed. The method involves a substitution reaction of 2-chloropyridine and phenylacetonitrile via hydrolysis followed by hydrogenation, configuration inversion, chiral resolution, methyl esterification, and salification to give high-purity dexmethylphenidate hydrochloride. The hydrochloride salt was then neutralized by powder sodium hydroxide overnight to give dexmethylphenidate free base with over 99% purity. This method can be used for the industrial production of the dexmethylphenidate patch API, which could also be further applied for the preparation of other types of amino acid ester free bases.

Keywords: dexmethylphenidate, amino acid ester, free base, high purity

1. Introduction

The dexmethylphenidate hydrochloride slow-release capsule was developed by Novartis Inc. (Har et al., 2000), and it was approved by the US Food and Drug Administration (FDA) in May 2005 for the treatment of attention deficit hyperactivity disorder (ADHD) (Volkmar et al., 2003), which is the most commonly diagnosed behavioral disorder in children. In July of 2010, the FDA approved the methylphenidate patch developed by Shire Plc inc, Inc. It was the first non-oral drug for the treatment of ADHD and provided a more convenient treatment choice for children. The APIs used in the capsule or patch treatments have two major differences, dexmethylphenidate chloride is a chiral, optically pure molecule, where as methylphenidate is the racemic isomer. Furthermore, the hydrochloride salt of dexmethylphenidate is used in the capsule, but it is too hydrophilic to be used in the patch because of the hydrophobic property of skin, for which only the free base could be effective. The main functional groups on both dexmethylphenidate and methylphenidate are amino acid esters (Figure 1). Amino acid esters are not very stable and can be easily hydrolyzed when neutralizing the hydrochloride salt to free base, while there is no publication to report the synthesis method of high purity free base of this kind of compound. Thus, the development of high-purity dexmethylphenidate free base could be deemed as the key step for enabling the entire API development process to meet the safety and quality requirements currently accepted by the International Conference on Harmonization (ICH).

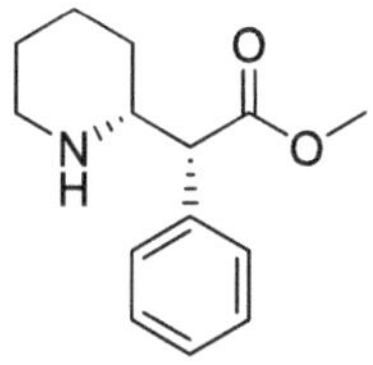

Figure 1. Structure of (*R,R*)-dexmethylphenidate free base (**1**)

As an effective therapeutic molecule, the synthesis of dexmethylphenidate has long attracted the attention (Prashad et al., 2001) of chemists around the world, and different synthetic methods have been invented and reported in the past decades. Of these, scientists from Novartis first invented a synthetic route (Prashad et al., 1999) in which the asymmetric aldol reaction was introduced as the key step; they later further improved (Prashad et al., 1999) upon this route. Following Novartis, additional scientists reported different methods (Gutman et al., 2007 & Davies et al., 1999), such as the use of Evans amide as the key intermediate for the synthetic route by Matsumura (Matsumua et al., 1999 & Matsumura et al., 2000) and coworkers. Moreover, Fox (Fox et al., 2000) and coworkers made an effective improvement on the basis of Matsumura's route by using methylbenzylamine as the key chiral auxiliary. For the purpose of future API industrialization, a concise method (Axten et al., 1998, Chan et al., 2010, Hu et al., 2000, Kumar et al., 2008, Prashad et al., 2000 & Zhang et al., 2016) was finally confirmed as the synthetic route to dexmethylphenidate. Herein, we report our work towards the convenient and high efficient synthesis of high-purity dexmethylphenidate free base.

2. Results and Discussion

The synthetic route began from the commercially available starting materials 2-chloropyridine (**2**) and phenylacetonitrile (**3**), which were treated with sodium amide to give 2-pyridyl-2-yl-phenylacetonitrile (**4**) in 78% yield. The cyano group on **4** was hydrolyzed by concentrated sulfuric acid to the amide group of intermediate **5** in 85% yield. Owing to the aromatic property of the pyridine ring on compound **5**, which had to be hydrogenated under pressure (above 1.0 mPa) to give 2-pepridyl-2-yl-phenacetamide (**6**) in 95% yield. Compound **6** had two chiral centers with both syn-form and anti-form, the anti-from was converted to the syn-form under the treatment of potassium tert-butoxide in 83% yield to give intermediate **7** with syn-form only. Then, **7** as a racemic mixture was resolved by anhydrous D-dibenzoyl tartaric acid in isopropanol to give the optically pure product **8** in 35% yield with over 99% ee values.

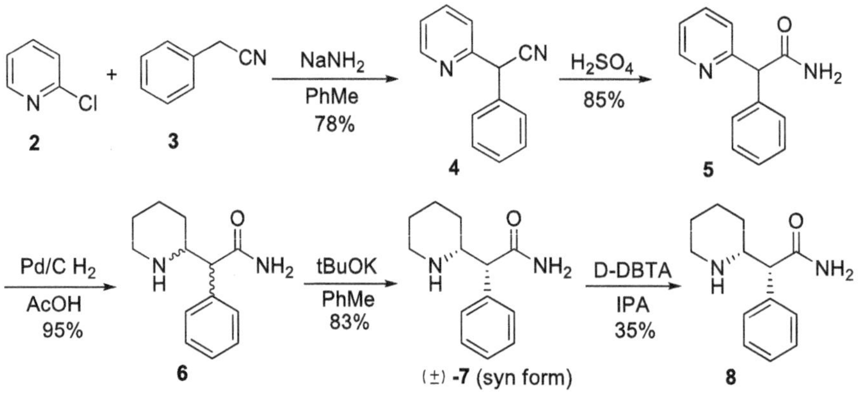

Scheme 1. Synthesis of (R,R)-2-piperidyl-2-yl-phenylacetamide (**8**)

With highly optically pure compound **8** in hand, the condition screening of the methyl esterification reaction was then processed. Concentrated sulfuric acid was used as catalyst, and the reaction was carried out in refluxing methanol for various times. The excess sulfuric acid was neutralized with sodium hydroxide aqueous solution after the reaction was completed to give compound **9** as light yellow oil, which was the crude dexmethylphenidate free base. Different reaction times were investigated to ensure that the conversion of the methyl esterification reaction was as high as possible. The results are listed in Table 1. Approximately half of reactant **8** was converted to **9** after 12 hours (Table 1, entry 1). After 1 d, the conversion increased to 78% (Table 1, entry 2). Two days later, 94.2% of product **9** had been synthesized (Table 1, entry 3). The reaction time was further extended to 3 d, and the conversion reached 96% (Table 1, entry 4). Unfortunately, reactant **8** did not completely disappear after the reaction had been processed for 7 d; at least 1.1% of reactant **8** could not be converted to **9** in this system (Table 1, entries 5–7).

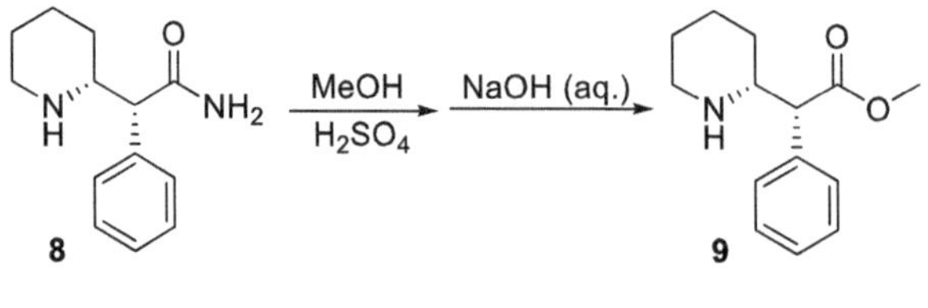

Table 1. Condition screening of the methyl esterification reaction

entry	Time (d)	Ratio of 8(%)	Ratio of 9(%)
1	0.5	52	47
2	1	21	78
3	2	5.1	94.2
4	3	3.0	96.0
5	4	1.9	98.0
6	5	1.1	98.4
7	6	1.2	97.6

The results shown in Table 1 indicated that the crude free base product was not qualified for API purity because of the residue of reactant **8**, so a purification procedure was needed. Consequently, concentrated hydrochloride acid was charged into the ethyl acetate solution of crude product **9**. The dexmethylphenidate hydrochloride salt was precipitated and filtered from the mother solution; then it was recrystallized from deionized water to give a white solid with up to 99.8% purity. Different neutralization methods were tried to obtain a high-purity free base form of dexmethylphenidate.

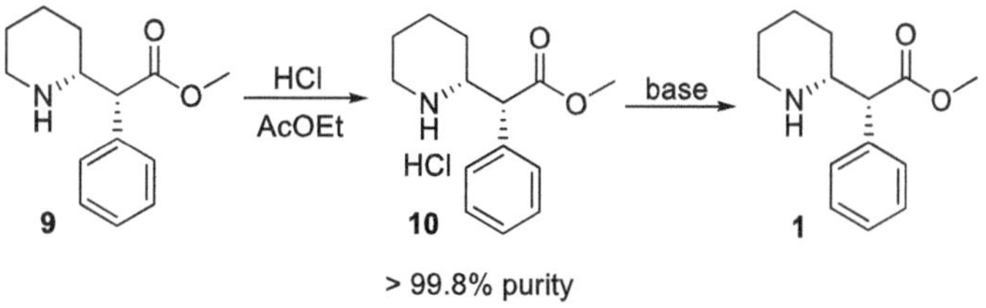

Table 2. Condition screening of the neutralization reaction to free base

entry	Forms of base	Solvent	time	Purity of **1** (%)	Yield(%)
1	10% NaOH (a.q.)	AcOEt/H_2O	10 min	90.7	88
2	10% LiOH (a.q.)	AcOEt/H_2O	10 min	94.2	90
3	$NaOCH_3$	MeOH	10 min	85.0	60
4	tBuOK	Toluene	10 min	76.2	53
5	NaOH (anhydrous solid)	CH_2Cl_2	12 h	99.7	86
6	LiOH (anhydrous solid)	CH_2Cl_2	12 h	99.8	91
7	KOH (anhydrous solid)	CH_2Cl_2	12 h	99.5	88

There are two requirements for high-purity free base, one is that no other impurity is formed or introduced and the second is that the hydrochloride acid must be neutralized completely in the neutralization process. As the piperidine on methylphenidate is a strong organic base, a stronger base is needed to neutralize hydrochloride acid completely from piperidine. Our attempts (Table 2) began with using 10% sodium hydroxide aqueous solution to neutralize hydrochloride acid, the result (Table 2, entry 1) indicated that only 90.7% purity of the product was obtained, and the main by-product (Figure 2) 2-piperdyl-2-yl-phenylacetatic acid (**11**) was identified in this process. Though the purity was improved to 94.2% by using 10% lithium hydroxide aqueous solution, it still remained unattractive (Table 2, entry 2) because the methyl ester group was still easily hydrolyzed in aqueous solution. Anhydrous conditions were introduced using sodium methoxide and potassium tert-butoxide, but the results (Table 2, entries 3 and 4) were less suitable because many unidentified impurities were observed. An alternative method was subsequently developed, anhydrous sodium hydroxide powder and dexmethylphenidate hydrochloride solids were mixed as slurry into dichloromethane under a nitrogen atmosphere and stirred vigorously overnight, followed by filtration of the insoluble solid and concentration of the solution in vacuo. The high-purity methylphenidate free base was successfully prepared using this method in 99.8% purity and 90% yield (Table 2, entry 5). The purity of the free base also up to 99.7% and 99.5% when using the anhydrous solids of lithium hydroxide and potassium hydroxide, respectively (Table 2, entries 6 and 7).

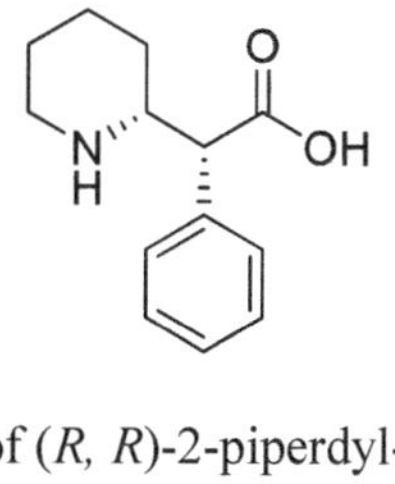

Figure 2. Structure of (*R, R*)-2-piperdyl-2-yl-phenylacetatic acid (**11**)

3. Experimental

Nuclear magnetic resonance (NMR) spectra were recorded on a Bruker Ultrashield-600 spectrometer with TMS as an internal standard. Chemical shifts (values) and coupling constants (J values) are given in ppm and Hz, respectively. ESI mass spectra were obtained using on an Agilent 6210 TOF spectrometer. Reaction progress and chemical purity were evaluated by HPLC analysis using a Waters symmetry C18 column (5 μm, 250 mm × 4.6 mm) with mobile phases A (MeOH + 0.05% TFA) and B (0.3% TEA), 88:12 v/v, and detection at 230 nm; flow rate: 1.0 mL/min, temperature: 25 °C. Palladium content was analyzed by Perkin Emler 800 AAS instrument. TLC analysis was conducted using prepared HSGF254 TLC plates. Solvents and reagents were used as received. All the intermediates and products were characterized by 1H & 13C NMR, high resolution ESI mass spectra, HPLC, melting point apparatus and optical rotation apparatus (if needed).

Synthesis of 2-pyridyl-2-yl-phenylacetontrile (4): Sodium amide (3.5 kg, 89.7 mol) was added to toluene (25.0 L) in a

100-L double layer reaction vessel under a nitrogen atmosphere at room temperature. Phenylacetonitrile (**3**, 6.5 kg, 55.6 mol) was added dropwise into the toluene solution over a period of 30 min; the temperature of the solutions was approximately 40 to 50 °C, then cooled to 15 °C. Afterwards, 2-chloropyridine (**2**, 6.0 kg, 53.1 mol) was added dropwise into the solution slowly; the reaction temperature temporarily increased to 70 °C and decreased to room temperature after 1 h. The reaction was kept at this temperature for 4 h, quenched by dripping water (5.0 L), and then washed by water (5.0 L, three times). Afterwards, approximately 20 L of toluene was distilled out, petroleum ester (5.0 L) was added in one portion, and the solution was cooled to 15 °C. The product was precipitated and filtered from the mother solution as a light yellow solid (8390.0 g) in 81.5% yield and 99.83% purity, Mp. 84.0 – 86.0 ^{0}C. ^{1}H NMR (600Hz, DMSO-d_6) δ 5.93 (s, 1H), 7.22 – 7.46 (m, 7H), 7.85 – 7.98 (m, 1H), 8.59 (s, 1H). ^{13}C NMR (150Hz, DMSO-d_6) δ 120.2, 122.9, 123.8, 128.6, 129.4, 129.6, 138.4, 150.2, 155.9. HRMS (ESI): Calcd for $[C_{13}H_{10}N_2+H]^+$ 195.0917, $[C_{13}H_{10}N_2+Na]^+$ 217.0736, found: 195.0921, 217.0739.

Synthesis of 2-pyridyl-2-yl-phenylacetyl amide (5): 2-pyridyl-2-yl-phenylacetontrile (**4,** 8000g, 41.2 mol) was added to concentrated sulfuric acid (10.0 L) in a 50-L round bottom reaction vessel and stirred at room temperature. After 24 h, the reaction was quenched by adding water dropwise (10.0 L) and neutralized by 30% sodium hydroxide solution until the pH was 13. Many solids were precipitated and filtered from the solution. The red-colored, crude product was dried by heating in an oven for 6 h and recrystallized from ethanol to give the purified product as a white solid (6720 g) in 77.0% yield and 99.70% purity, Mp. 133.8 – 134.6 ^{0}C. ^{1}H NMR (600Hz, DMSO-d_6) δ 5.12 (s, 1H), 7.23 – 7.46 (m, 8H), 7.75 – 7.95 (m, 2H), 8.59 (s, 1H). ^{13}C NMR (150Hz, DMSO-d_6) δ 59.8, 122.4, 123.5, 127.3, 128.7, 129.1, 137.0, 139.8, 149.2, 160.1, 172.7. HRMS (ESI): Calcd for $[C_{13}H_{12}N_2O+H]^+$, 213.1022, $[C_{13}H_{12}N_2O+Na]^+$ 235.0842 found: 213.1023, 235.0839.

Synthesis of *syn/anti*-2-piperidyl-2-yl-phenylacetyl amide (6): 2-pyridyl-2-yl-phenylacetyl amide (6000.0 g, 28.3 mol) and 10% wet Pd/C catalyst (1500.0 g, 40% dry base) were added into an acetic acid solution (25.0 L) and mixed well in one portion of a 50-L sealed hydrogenation vessel. The vessel was heated to 80 °C for 8 h under hydrogen pressure greater than 1.0 mPa. After the reaction was completed, the Pd/C catalyst was filtered, and the acetic acid was removed *in vacuo*. The residue was dissolved in water (10.0 L) and neutralized by 30% sodium hydroxide solution until the pH was 13. The product was precipitated and filtered from the solution and washed by water (10 L, three times) to give a white solid (5860.0 g) in 95% yield with 84.25% *anti*-form and 15.74% *syn*-form, the palladium content is 2 ppm, Mp. 156.0 – 157.5 ^{0}C. ^{1}H NMR (600Hz, DMSO-d_6) δ 0.82 – 0.96 (m, 1H), 0.98 – 1.10 (m, 4H), 1.14 – 1.31 (m, 4H), 1.38 – 1.48 (m, 3H), 1.57 – 1.65 (m, 1H), 1.72 (d, J = 9.0 Hz, 3H), 2.37 (t, J = 10.2 Hz, 2H), 2.47 – 2.54 (m, 2H), 2.66 (d, J = 11.4 Hz, 2H), 2.91 – 2.98 (m, 4H), 3.26 (t, J = 9.6 Hz, 4H), 6.82 – 6.86 (m, 2H), 7.16 – 7.37 (m, 14H), 7.51 – 7.56 (m, 3H). ^{13}C NMR (150Hz, DMSO-d_6) δ 24.6, 24.7, 25.5, 26.3, 26.4, 30.2, 31.0, 46.8, 46.9, 58.1, 58.6, 58.9, 127.1, 127.3, 128.6, 128.7, 128.8, 128.9, 138.9, 139.3, 174.1, 174.5. HRMS (ESI): Calcd for $[C_{13}H_{18}N_2O+H]^+$ 219.1492, found: 219.1493.

Synthesis of *syn*-2-piperidyl-2-yl-phenylacetyl amide (7): *Syn/anti*-2-piperidyl-2-yl-phenylacetyl amide (**6**, 2000.0 g, 9.2 mol) and potassium tert-butoxide (2000.0 g) were added into toluene solution (50.0 L) in a 100-L double layer reaction vessel and mixed well. The solution was heated to 70 °C for 16 h, quenched by water, and washed by 6 N hydrochloride acid solutions (1.0 L, twice). The aqueous phase was separated and combined. A 30% sodium hydroxide solution was used to neutralize the solution until the pH value was 13. The product was precipitated, filtered, and dried by heating in an oven for 6 h to give the purified product as a white solid (1710 g) in 85.5% yield and 97.76% purity. Mp. 172.9 – 173.9 ^{0}C. ^{1}H NMR (600Hz, DMSO-d_6) δ 0.82 – 0.88 (m, 1H), 1.09 – 1.16 (m, 2H), 1.22 – 1.28 (m, 1H), 1.45 (d, J = 12.0 Hz, 1H), 1.58 (d, J = 12.6 Hz, 1H), 1.72 – 2.11 (br, 1H), 2.47 – 2.56 (m, 1H), 2.84 – 2.95 (m, 2H), 3.88 (d, J = 10.2 Hz, 1H), 6.82 (s, 1H), 7.15 – 7.45 (m, 5H), 7.53 (s, 1H)..^{13}C NMR (150Hz, DMSO-d_6) δ 24.7, 26.4, 30.2, 46.8, 58.6, 58.9, 127.1, 128.4, 128.9, 139.3, 174.6. HRMS (ESI): Calcd for $[C_{13}H_{18}N_2O+H]^+$ 219.1492, found: 219.1494.

Synthesis of (*R,R*)-2-piperidyl-2-yl-phenylacetyl amide (8): To a 20-L double layer reaction vessel, *syn*-2-piperidyl-2-yl-phenylacetyl amide (7, 600.0 g, 2.8 mol) and anhydrous D-dibenzoyl tartaric acid (1002.0 g, 2.8 mol) were added into isopropanol solution (12.0 L) and heated to 60 °C until all of the solids were dissolved. Then, the reaction was cooled gradually to 30 °C. White solids were precipitated, filtered, and transferred into the 6 N hydrochloride acid solution and stirred for 1 h; the precipitates were eliminated, and the solution was neutralized using a 30% sodium hydroxide solution until the pH was 13. The product was precipitated, filtered, and dried by heating in an oven for 6 h to give the purified product as a yellow solid (213.0 g) in 35.5% yield, 99.30% purity, and 99.9% ee values, Mp. 173.1 – 174.0 ^{0}C. ^{1}H NMR (600Hz, DMSO-d_6) δ 0.81 – 0.87 (m, 1H), 1.04 – 1.16 (m, 2H), 1.21 – 1.27 (m, 1H), 1.34 (d, J = 7.2 Hz, 1H), 1.57 (d, J = 12.6 Hz, 1H), 1.83 (s, 1H), 2.39 – 2.51 (m, 2H), 2.93 – 2.96 (m, 2H), 3.33 (d, J = 10.2 Hz, 1H), 6.85 (s, 1H), 7.21 – 7.43 (m, 5H), 7.52 (s, 1H). ^{13}C NMR (150Hz, DMSO-d_6) δ 24.7, 26.0, 26.4, 30.2, 58.6, 58.9, 127.1, 128.6, 128.8, 139.9, 174.5. HRMS (ESI): Calcd for $[C_{13}H_{18}N_2O+H]^+$ 219.1492, found: 219.1495.

Synthesis of (*R,R*)-*O*-methyl-2-piperidyl-2-yl-phenylacetate hydrochloride (10):

(*R,R*)-2-piperidyl-2-yl-phenylacetyl amide (**8**, 200.0 g, 0.92 mol) was added to a methanol solution (500 mL) in a 2-L round bottom flask with four necks. Concentrated sulfuric acid (200 mL) was added dropwise, and the solution was refluxed for 3 d. afterwards, methanol was removed *in vacuo*. The residue was dissolved into a mixture of ethyl acetate (3.0 L) and water (1.0 L), which was neutralized with a 30% sodium hydroxide solution until the pH of the aqueous layer was 13. The top organic layer was separated and concentrated hydrochloride acid (50 mL) was added into the solution and mixed well. The white solid was precipitated, filtered, recrystallized from deionized water (150 mL), and finally dried by heating in an oven for 2 h to give the purified product as a white solid (186.0 g) in 75.5% yield and 99.82% purity. $[\alpha]_D^{20}$ +87.3°(c=1,MeOH), Mp. 222.0 – 223.0, ^{1}H NMR (600Hz, DMSO-d_6) δ 1.24 – 1.26 (m, 1H), 1.35 – 1.41 (m, 2H), 1.52 – 1.68 (m, 3H), 2.95 (s, 1H), 3.25 – 3.28 (m,1H), 3.39 – 3.47 (m,1H), 3.66 (s,3H), 3.78 – 3.84 (m,1H), 4.15(d, *J* = 9.6 Hz,1H), 7.27 – 7.49 (m,5H), 9.02 (br,1H), 9.74 (br,1H). ^{13}C NMR (150Hz, DMSO-d_6) δ 21.8, 21.9, 26.1, 44.9, 53.1, 53.7, 57.2, 128.7, 128.9, 134.7, 171.7. HRMS (ESI): Calcd for $[C_{14}H_{20}N_2OCl - Cl^-]^+$ 234.1489, found: 234.1492.

Synthesis of (*R,R*)-2-piperdyl-2-yl-phenylacetatic acid (11): To a solution of methanol (20.0 mL) (*R,R*)-O-methyl-2-piperidyl-2-yl-phenyl-acetate hydrochloride (**10**, 1.0 g) was added in one portion and 0.1 M sodium hydroxide aqueous solution (10.0 mL) was added dropwise, the reaction was crrried out at room temperature for 3 hours, then 1 N hydrochloride aqueous solution was dripping into until the pH values was up to 2. The white solids were precipitated from the solution and filtered, which was dried by oven at 60 ^{0}C overnight to afford the product as white solid (0.39g, 48% yield) with 99.25% purity, Mp. 220.0 – 221.0 ^{0}C, $[\alpha]_D^{20}$ +22.7°(c=1,H_2O), ^{1}H NMR (600Hz, DMSO-d_6) δ 1.25 – 1.29 (m, 1H), 1.38 – 1.42 (m, 2H), 1.60 – 1.67 (m, 3H), 2.94 – 2.99 (br, 1H), 3.26 (d, *J* = 12.6 Hz, 1H), 3.64 – 3.72 (br, 1H), 4.06 (d, *J* = 9.0 Hz, 1H), 7.21 – 7.46 (m, 5H), 8.56 (s, 1H), 9.69 (s, 1H), 12.75 – 13.65 (br, 1H). ^{13}C NMR (150Hz, DMSO-d_6) δ 21.9, 26.1, 34.4, 45.1, 53.8, 57.2, 128.4, 129.0, 129.4, 135.4, 172.9. HRMS (ESI): Calcd for $[C_{13}H_{17}NO_2+H]^+$ 220.1332, $[C_{13}H_{17}NO_2+Na]^+$ 242.1151, found: 220.1338, 242.1158.

Synthesis of (*R,R*)-*O*-methyl-2-piperidyl-2-yl-phenylacetate free base (1):

(*R,R*)-*O*-methyl-2-piperidyl-2-yl-phenylacetate hydrochloride (**10**, 30.0g, 112.0 mmol) and anhydrous sodium hydroxide solid (4.93 g, 123 mmol) were added into dichloromethane (500 mL) in a 1-L round bottom flask and stirred vigorously under a nitrogen atmosphere. After 12 h, the undissolved solids were filtered, and the organic solution was washed with water (200 mL, twice) and dried with anhydrous magnesium sulfate. The inorganic solid was filtered and the dichloromethane was removed *in vacuo* to give the final free base product as a colorless oil (22.3 g) in 86.1% yield and 99.69% purity, $[\alpha]_D^{20}$ +87.7°(c=1,MeOH), ^{1}H NMR (600Hz, DMSO-d_6) δ 0.82 – 0.88 (m, 1H), 1.07 – 1.14 (m, 1H), 1.15 – 1.27 (m, 2H), 1.38 – 1.43 (m, 1H), 1.59 – 1.61 (m, 1H), 2.46 – 2.51 (m, 1H), 2.92 – 2.99 (m, 1H), 3.01 – 3.03 (m, 1H), 3.34 (d, J = 9.6 Hz, 1H), 3.57 (s, 3H), 7.31 – 7.33 (m, 5H). ^{13}C NMR (150Hz, DMSO-d_6) δ 22.5, 24.6, 28.1, 44.6, 49.9, 52.1, 56.2, 125.7, 126.8, 127.1, 135.1, 171.7. HRMS (ESI): Calcd for $[C_{14}H_{19}N_2O+H]^+$ 234.1489, found: 234.1487.

4. Conclusion

In summary, a new, practical method towards high purity free base of methylphenidate was successfully developed via seven steps starting from simple commercially available materials. This innovative method could also be used for the synthesis of high-purity free base of other amino acid esters.

Acknowledgments

Project supported by the National Science and Technology Major Project (No. 2014ZX09507006-002) of the Ministry of Science and Technology of China and 2016 Shanghai Pujiang Talent Program (16PJ1432800).

References

Axten J. M., Krim, L., Kung, H. F., & Winkler, J. D. (1998). A Stereoselective Synthesis of dl-threo-Methylphenidate: Preparation and Biological Evaluation of Novel Analogues. *The Journal of Organic Chemistry, 63,* 9628-9629. http://dx.doi.org/10.1021/jo982214t

Chan, A. B., Gundecha S. S., Kadam, P. N., Maikap, G. C., & Gurjar, M. K. (2010). Dexmethylphenidate: An Efficient Process for the Racemization of Unwanted (2S, 2'S or l-threo)-α-Phenyl-α-(2-piperidyl)acetamide. *Organic Process Research & Development*, *14*, 1473-1475. http://dx.doi.org/ 10.1021/op100197g

Davies, H. M. L., Hansen, T., Hopper, D. W., & Panaro, S. A. (1999). Highly Regio-, Diastereo-, and Enantioselective C–H Insertions of Methyl Aryldiazoacetates into Cyclic N-Boc-Protected Amines. Asymmetric Synthesis of Novel C2-Symmetric Amines and threo-Methylphenidate. *Journal of The American Chemical Society*, *121*, 6509-6510. http://dx.doi.org/10.1021/ja9910715

Fox, M. E., & Paul, J. M. (2000). *7-amino-2-heptenoates and their use in the preparation of methylphenidate,* US Patent 6031124. https://worldwide.espacenet.com/publicationDetails/biblio?DB=EPODOC&II=0&ND=3&adjacent=true&locale=en_EP&FT=D&date=20000229&CC=US&NR=6031124A&KC=A

Gutman, A., Zaltsman, I., & Shalimov, A. (2007). *Process for the preparation of dexmethylphenidate hydrochloride*. US Patent 7247730. https://worldwide.espacenet.com/publicationDetails/biblio?II=0&ND=3&adjacent=true&locale=en_EP&FT=D&date=20040916&CC=US&NR=2004180928A1&KC=A1

Har, D., & Prashad, M. (2000). *Process for preparing the d-threo isomer of methylphenidate hydrochloride*. US Patent 6100401.https://worldwide.espacenet.com/publicationDetails/biblio?II=1&ND=3&adjacent=true&locale=en_EP&FT=D&date=20000808&CC=US&NR=6100401A&KC=A

Hu, B., & Prashad, M. (2000). *Process for preparing the d-threo isomer of methylphenidate hydrochloride*. US Patent 6162919.https://worldwide.espacenet.com/publicationDetails/biblio?II=0&ND=3&adjacent=true&locale=en_EP&FT=D&date=20001219&CC=US&NR=6162919A&KC=A

Kumar, A., Mahale, G. D., & Patil, S. H. (2008). *Manufacturing process for methyl phenidate and intermediates thereof.* US Patent 7459467. https://worldwide.espacenet.com/publicationDetails/biblio?II=0&ND=3&adjacent=true&locale=en_EP&FT=D&date=20051215&CC=US&NR=2005277667A1&KC=A1

Matsumura, Y., Kanda, Y., Shirai, K., Onomura, O., & Maki, T. (1999). A Convenient Method for Synthesis of Enantiomerically Enriched Methylphenidate from N-Methoxycarbonylpiperidine. *Organic Letters*, *1*, 175-178. http://dx.doi.org/10.1021/ol9905046

Matsumura, Y., Kanda, Y., Shirai, K., Onomura, O., Maki, T. (2000). A Convenient Method for Synthesis of Optically Active Methylphenidate from N-Methoxycarbonylpiperidine by Utilizing Electrochemical Oxidation and Evans Aldol-type Reaction. *Tetrahedron, 56,* 7411-7422. http://dx.doi.org/ 10.1016/S0040-4020(00)00653-0

Prashad, M. (2000). Approaches to the Preparation of Enantiomerically Pure (2R,2'R)-(+)-threo-Methylpheni-date Hydrochloride. *Advanced Synthesis & Catalysis, 343*, 379-392. http://dx.doi.org/10.1002/1615-4169(200107)343:5<379

Prashad, M., Hu, B., Repic, O., Blacklock, T. J., & Giannousis P. (2000). Resolution of (±)-threo-Methylphenidate with (R)-(−)-Binaphthyl-2,2'-diyl Hydrogen Phosphate: 0.5 M Equiv of Resolving Agent Is Better than 1 M Equiv. *Organic Process Research & Development*, *4*, 55-59. http://dx.doi.org/10.1021/op990195z

Prashad, M., Kim, H. Y., Lu, Y., Liu, Y., Har, D., Repic, O., Blacklock, T. J., & Giannousis, P. (1999). The First Enantioselective Synthesis of (2R,2'R)-threo-(+)-Methylphenidate Hydrochloride. *The Journal of Organic Chemistry, 64,* 1750-1753. http://dx.doi.org/10.1021/jo9821473

Prashad, M., Liu, Y., Kim, H. Y., Repic, O., & Blacklock, T. J. (1999). Enantioselective synthesis of (2S,2'R)-erythro-methylphenidate. *Tetrahedron: Asymmetry. 10*, 3479-3482. http://dx.doi.org/doi:10.1016/S0957-4166(99)00384-5

Zhang, F., Li, B., Sun, Y., & Li, J. (2016). Synthesis of Demethylphenidate hydrochloride and its related optical Isomers. Chinese Journal of Organic Chemistry, *36*. http://dx.doi.org/10.6023/cjoc201603001

Appendix A

NMR, HRMS and HPLC spectra of 2-pyridyl-2-yl-phenylacetontrile (4)

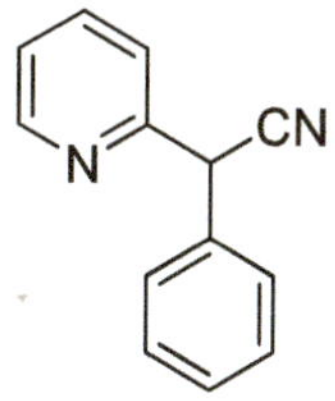

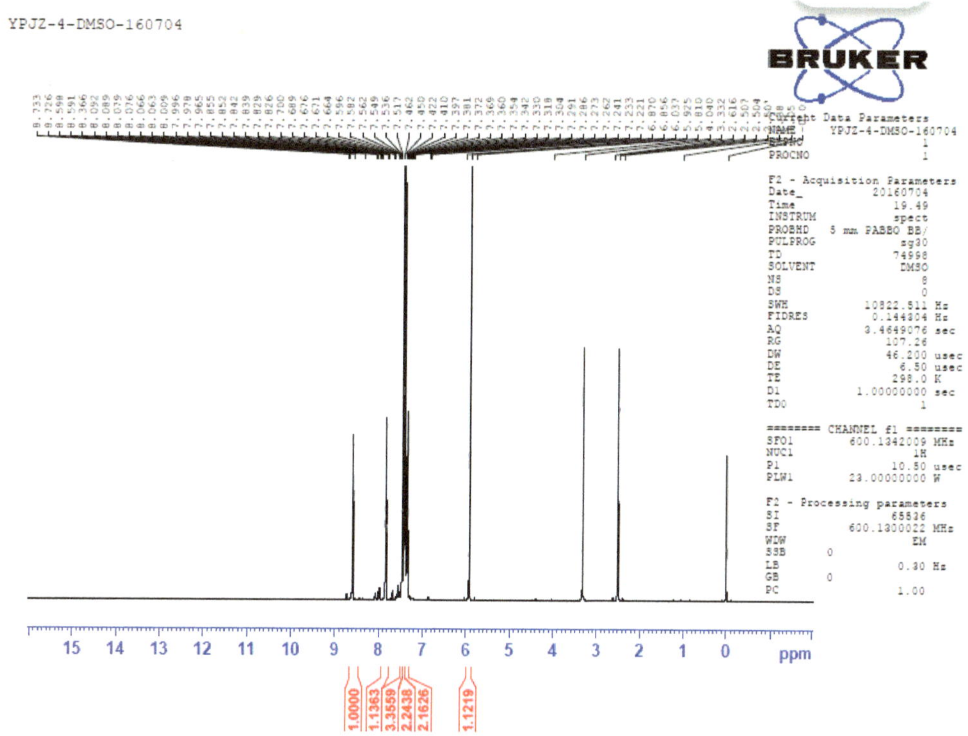

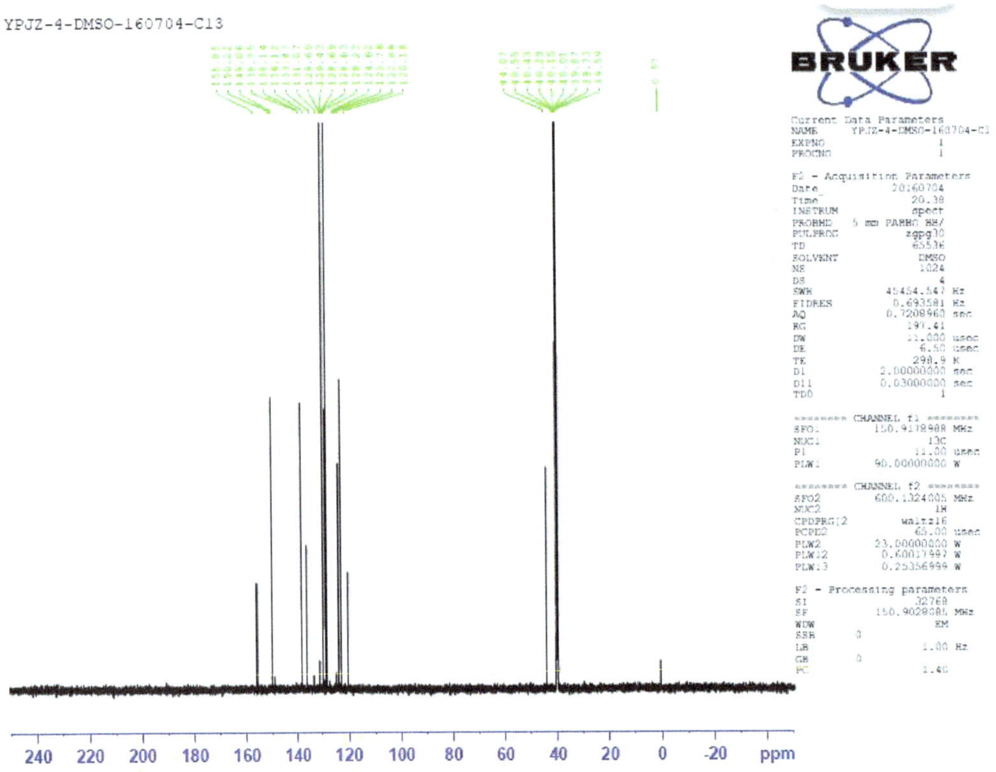

Qualitative Compound Report

Data File	20160704-005.d	**Sample Name**	YPJZ-4
Sample Type	Sample	**Position**	Vial 64
Instrument Name	Instrument 1	**User Name**	
Acq Method		**IRM Calibration Status**	Success
DA Method	MS.m	**Comment**	N20160704011

Compound Table

Compound Label	RT	Mass	Abund	Formula	Tgt Mass	Diff (ppm)
Cpd 1: C13 H10 N2	-0.25	194.0847	71407	C13 H10 N2	194.0844	1.69

Compound Label	RT	Algorithm	Mass
Cpd 1: C13 H10 N2	-0.25	Find By Formula	194.0847

MS Zoomed Spectrum

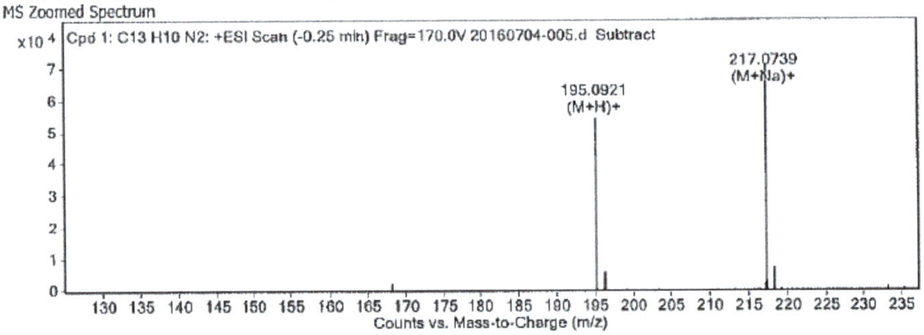

MS Spectrum Peak List

m/z	Calc m/z	Diff(ppm)	z	Abund	Formula	Ion
195.0921	195.0917	1.92	1	54458	C13 H11 N2	(M+H)+
217.0739	217.0736	1.41		71407	C13 H10 N2 Na	(M+Na)+

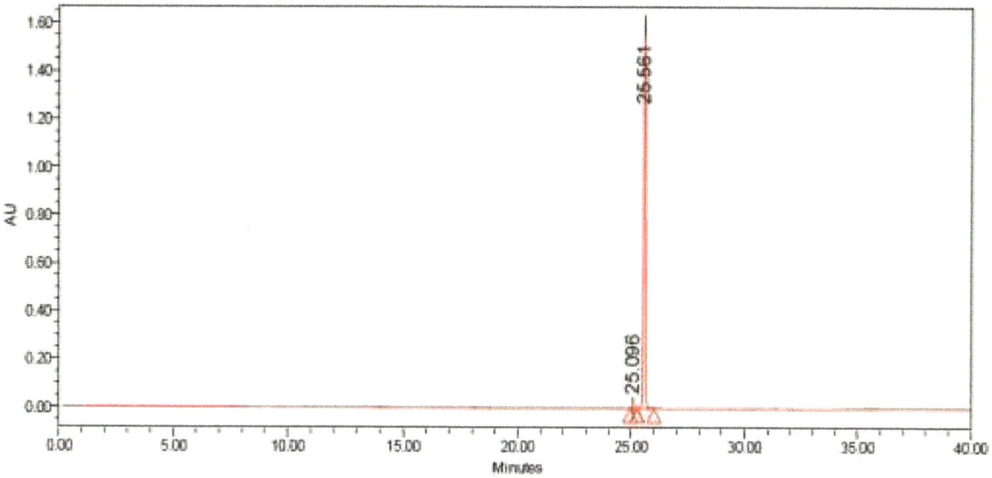

Peak Results

	Name	RT	Area	% Area	Height
1		25.096	14780	0.17	2549
2		25.561	8661812	99.83	1594325

NMR, HRMS and HPLC spectra of 2-pyridyl-2-yl-phenylacetyl amide (5)

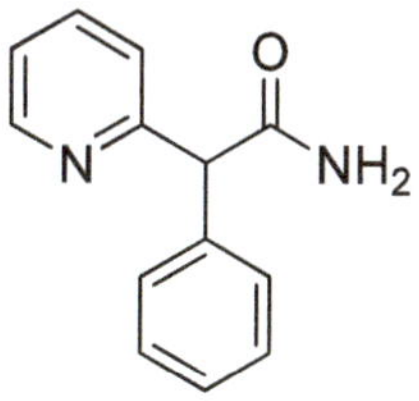

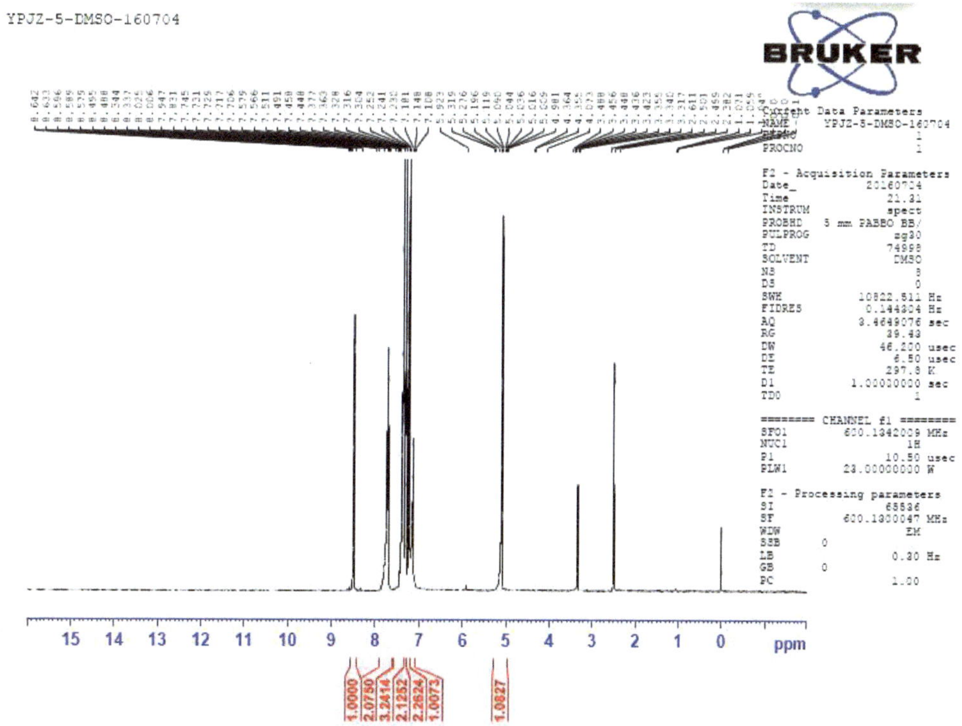

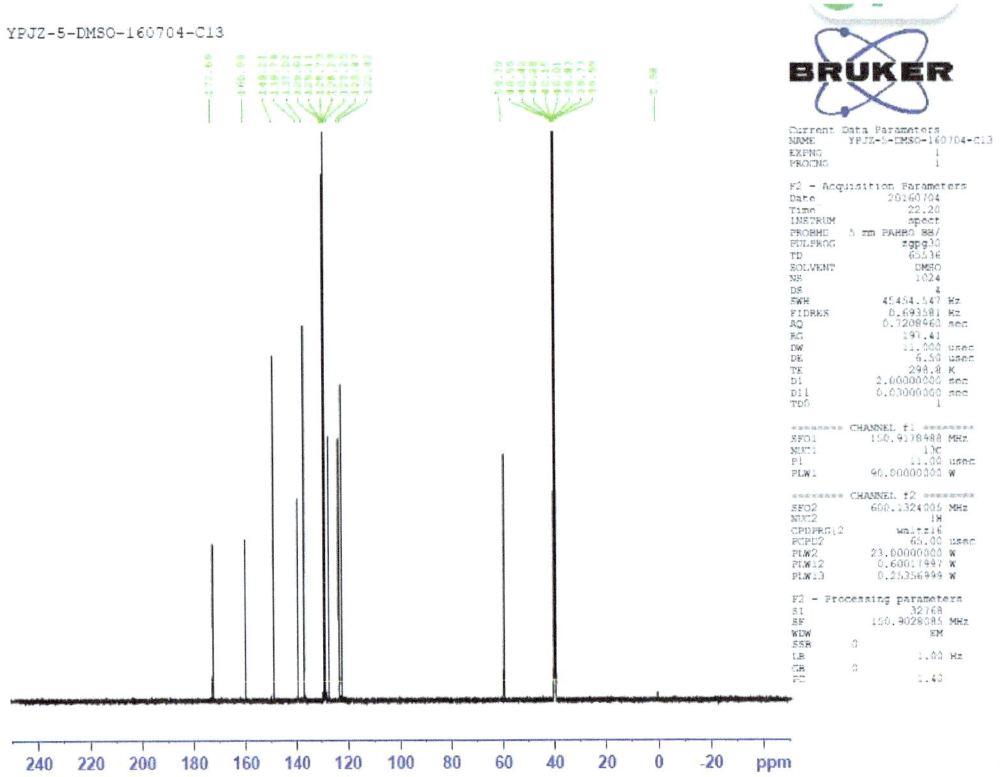

Qualitative Compound Report

Data File	20160704-006.d	**Sample Name**	YPJZ-5
Sample Type	Sample	**Position**	Vial 65
Instrument Name	Instrument 1	**User Name**	
Acq Method		**IRM Calibration Status**	Success
DA Method	MS.m	**Comment**	N20160704012

Compound Table

Compound Label	RT	Mass	Abund	Formula	Tgt Mass	Diff (ppm)
Cpd 1: C13 H12 N2 O	-0.28	212.0947	134184	C13 H12 N2 O	212.095	-1.06

Compound Label	RT	Algorithm	Mass
Cpd 1: C13 H12 N2 O	-0.28	Find By Formula	212.0947

MS Zoomed Spectrum

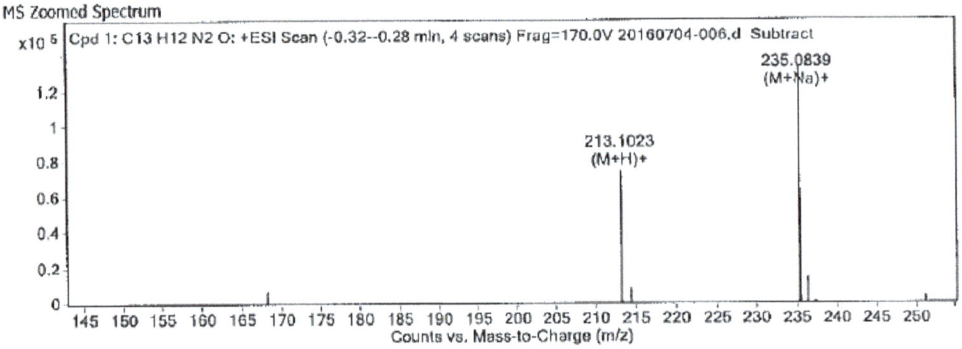

MS Spectrum Peak List

m/z	*Calc m/z*	Diff(ppm)	z	Abund	Formula	Ion
213.1023	213.1022	0.48	1	74546	C13 H13 N2 O	(M+H)+
235.0839	235.0842	-1.2		134184	C13 H12 N2 Na O	(M+Na)+

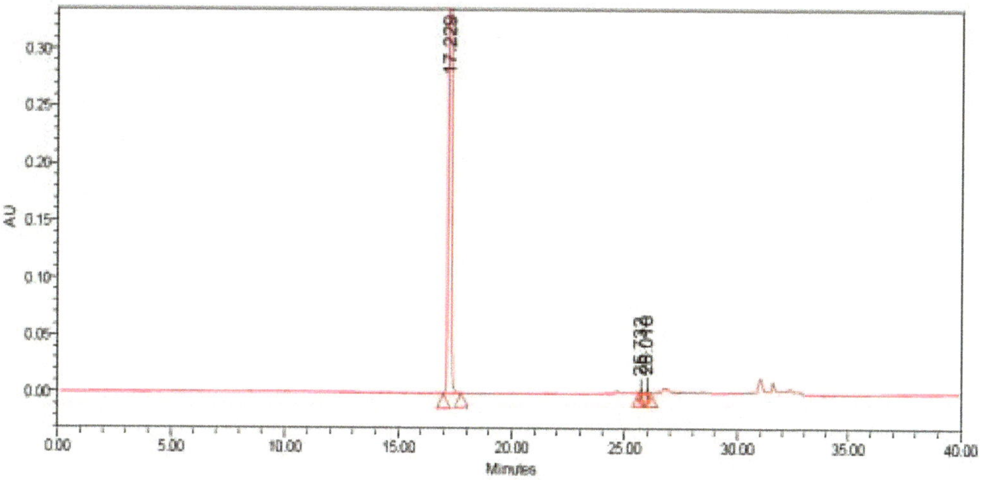

Peak Results

	Name	RT	Area	% Area	Height
1		17.229	9431958	99.70	1523750
2		25.732	9496	0.10	2107
3		26.010	18920	0.20	3546

NMR, HRMS and HPLC spectras *syn/anti*-2-piperidyl-2-yl-phenylacetyl amide (6)

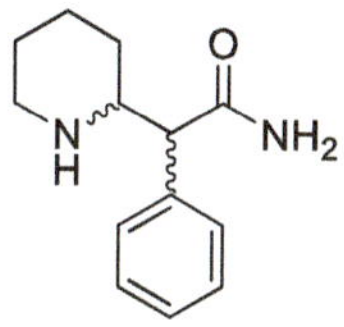

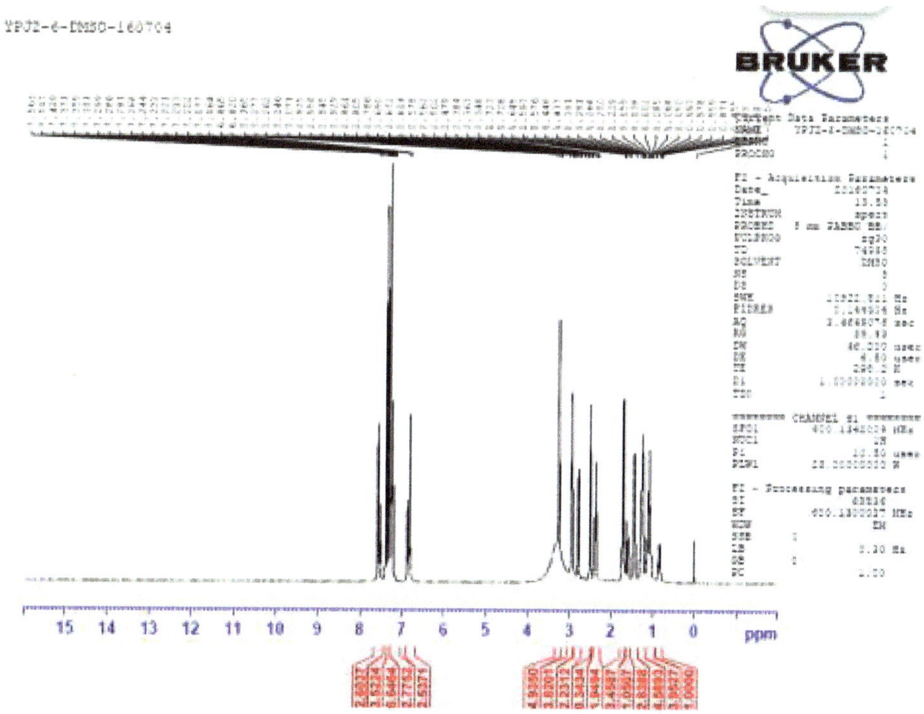

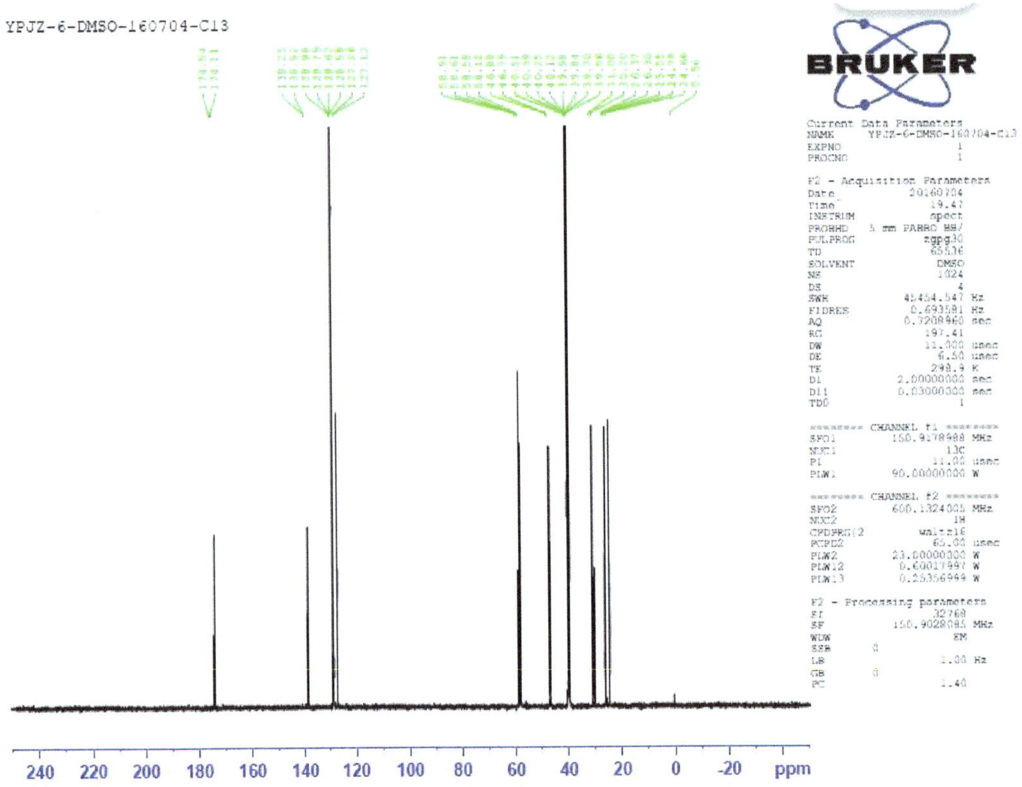

Qualitative Compound Report

Data File	20160704-007.d	**Sample Name**	YPJZ-6
Sample Type	Sample	**Position**	Vial 66
Instrument Name	Instrument 1	**User Name**	
Acq Method		**IRM Calibration Status**	Success
DA Method	MS.m	**Comment**	N20160704013

Compound Table

Compound Label	RT	Mass	Abund	Formula	Tgt Mass	Diff (ppm)
Cpd 1: C13 H18 N2 O	-0.25	218.142	530097	C13 H18 N2 O	218.1419	0.45

Compound Label	RT	Algorithm	Mass
Cpd 1: C13 H18 N2 O	-0.25	Find By Formula	218.142

MS Zoomed Spectrum

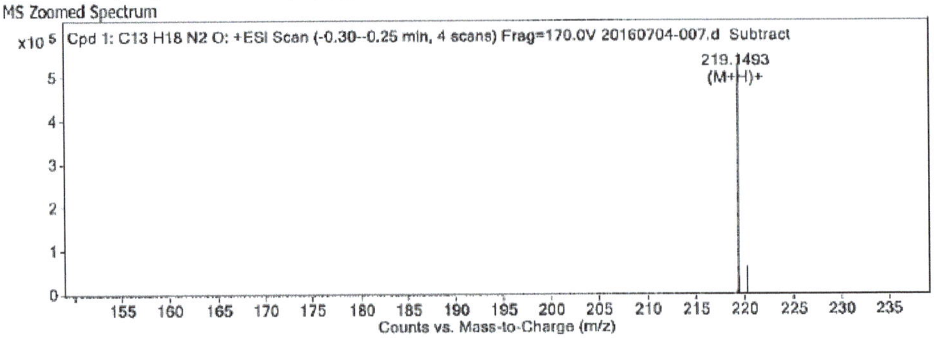

MS Spectrum Peak List

m/z	*Calc m/z*	Diff(ppm)	Abund	Formula	Ion
219.1493	219.1492	0.45	530097	C13 H19 N2 O	(M+H)+

--- End Of Report ---

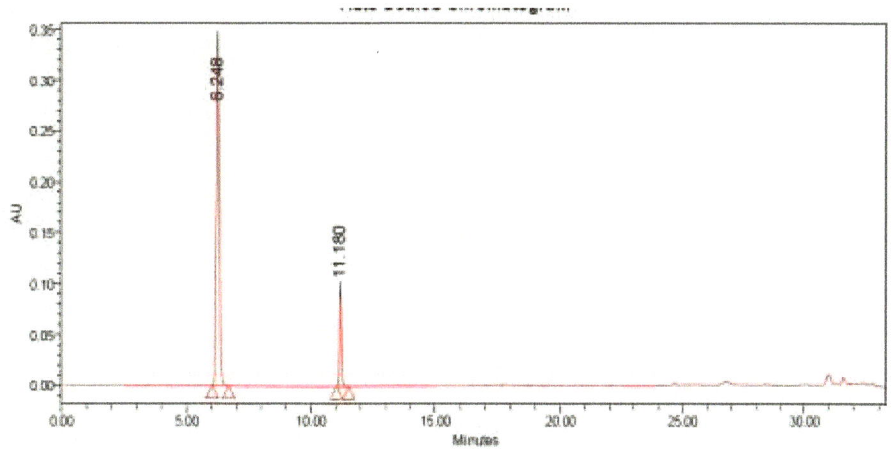

Peak Results

	Name	RT	Area	% Area	Height
1		6.248	2765140	84.26	338271
2		11.180	516352	15.74	93658

NMR, HRMS and HPLC spectra of syn-2-piperidyl-2-yl-phenylacetyl amide (7)

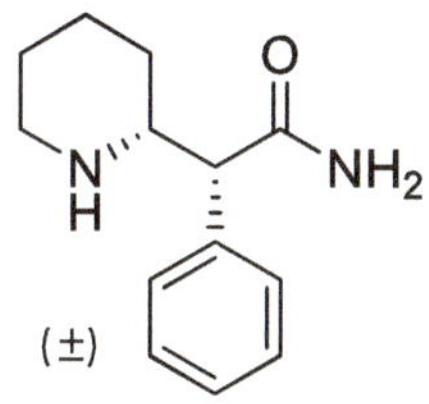

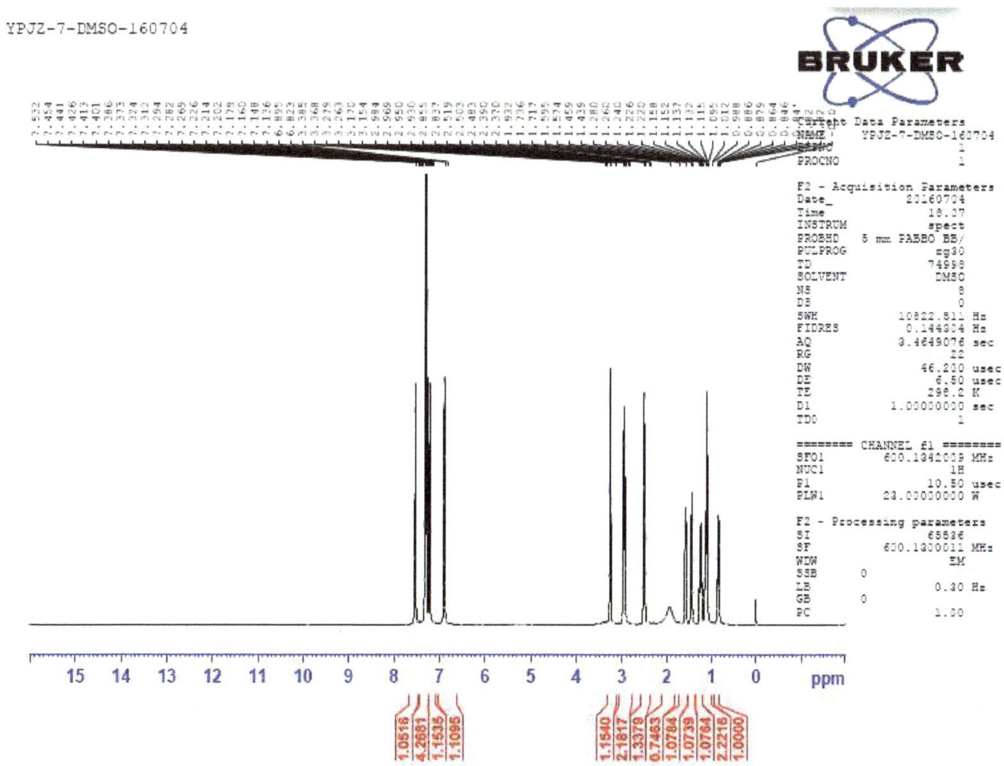

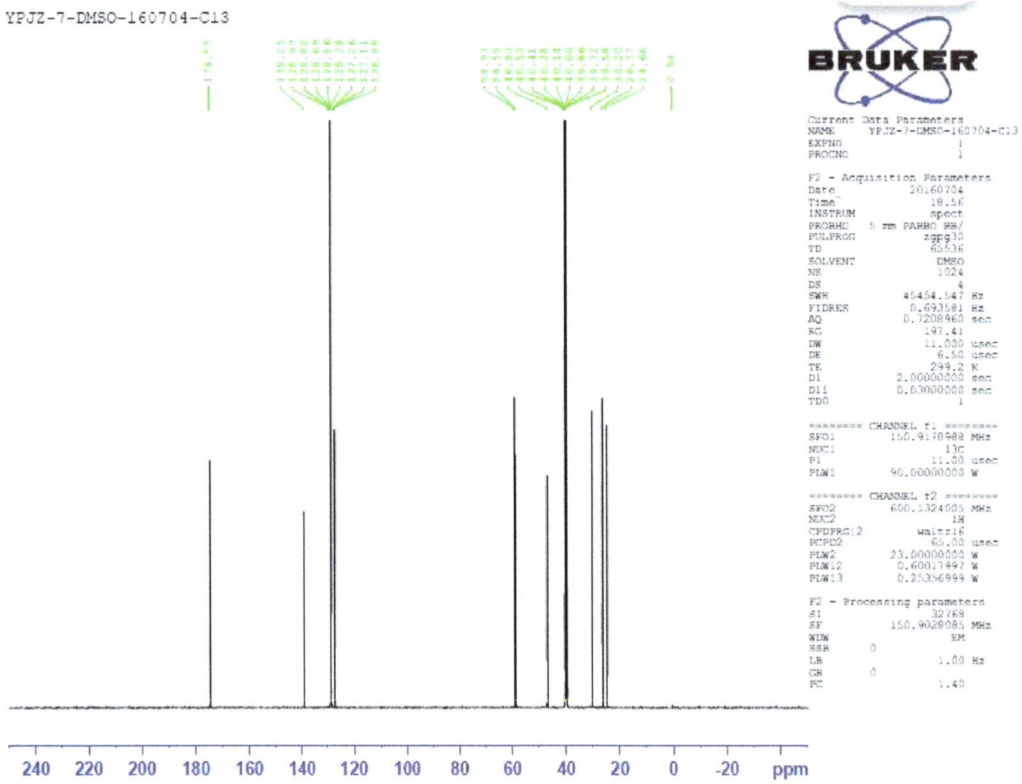

Qualitative Compound Report

Data File	20160704-008.d	**Sample Name**	YPJZ-7
Sample Type	Sample	**Position**	Vial 67
Instrument Name	Instrument 1	**User Name**	
Acq Method		**IRM Calibration Status**	Success
DA Method	MS.m	**Comment**	N20160704014

Compound Table

Compound Label	RT	Mass	Abund	Formula	Tgt Mass	Diff (ppm)
Cpd 1: C13 H18 N2 O	-0.28	218.1421	215881	C13 H18 N2 O	218.1419	0.95

Compound Label	RT	Algorithm	Mass
Cpd 1: C13 H18 N2 O	-0.28	Find By Formula	218.1421

MS Zoomed Spectrum

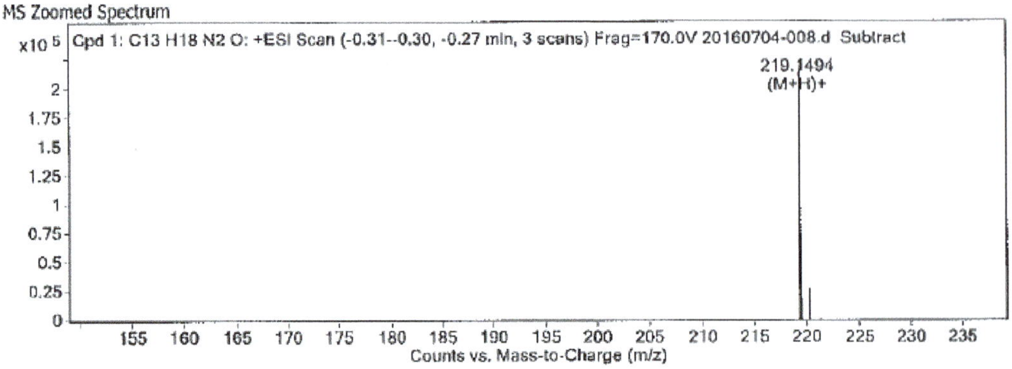

MS Spectrum Peak List

m/z	*Calc m/z*	Diff(ppm)	Abund	Formula	Ion
219.1494	219.1492	0.96	215881	C13 H19 N2 O	(M+H)+

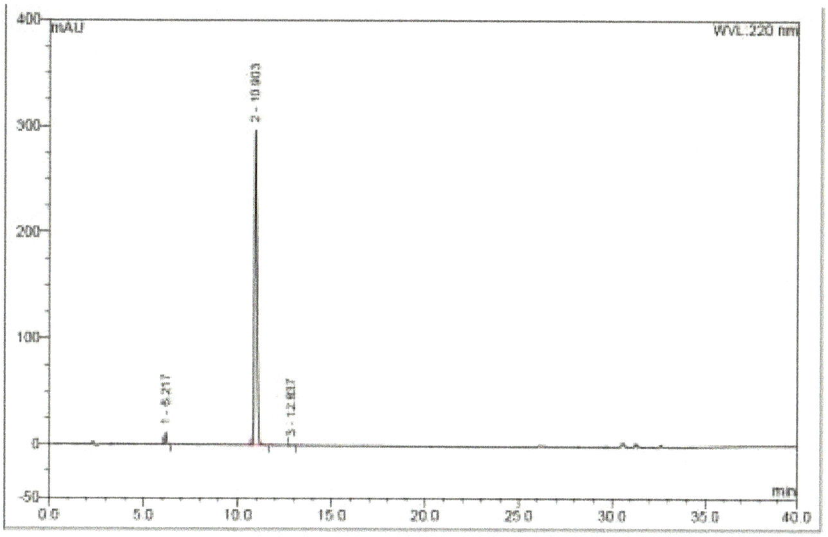

No.	Ret.Time min	Peak Name	Height mAU	Area mAU*min	Rel.Area %	Resolution	Plates
1	6.22	n.a.	10.8699	0.9576	2.09	24.45	32499
2	10.90	n.a.	296.5871	44.8663	97.76	9.78	31325
3	12.84	n.a.	0.7052	0.0699	0.15	n.a.	117437
Total:			308.162	45.894	100.00	34.236	

Chromatogram

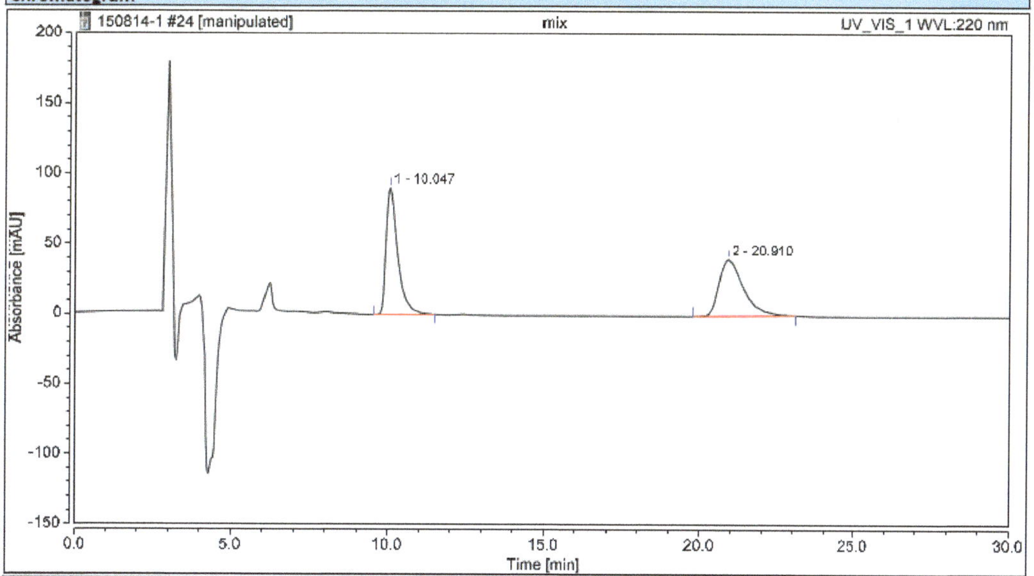

Integration Results

No.	Peak Name	Retention Time min	Area mAU*min	Height mAU	Relative Area %	Resolution (EP)	Plates (EP)
1		10.047	38.785	90.408	50.50	10.55	4023
2		20.910	38.016	40.669	49.50	n.a.	3415
Total:			76.801	131.077	100.00	10.55	

NMR, HRMS and HPLC spectra of (R,R)-2-piperidyl-2-yl-phenylacetyl amide (8)

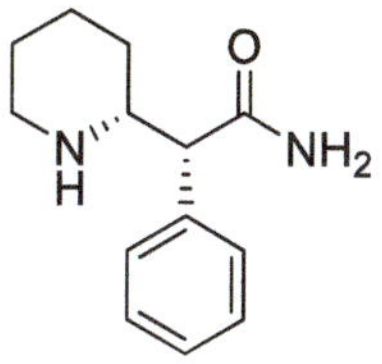

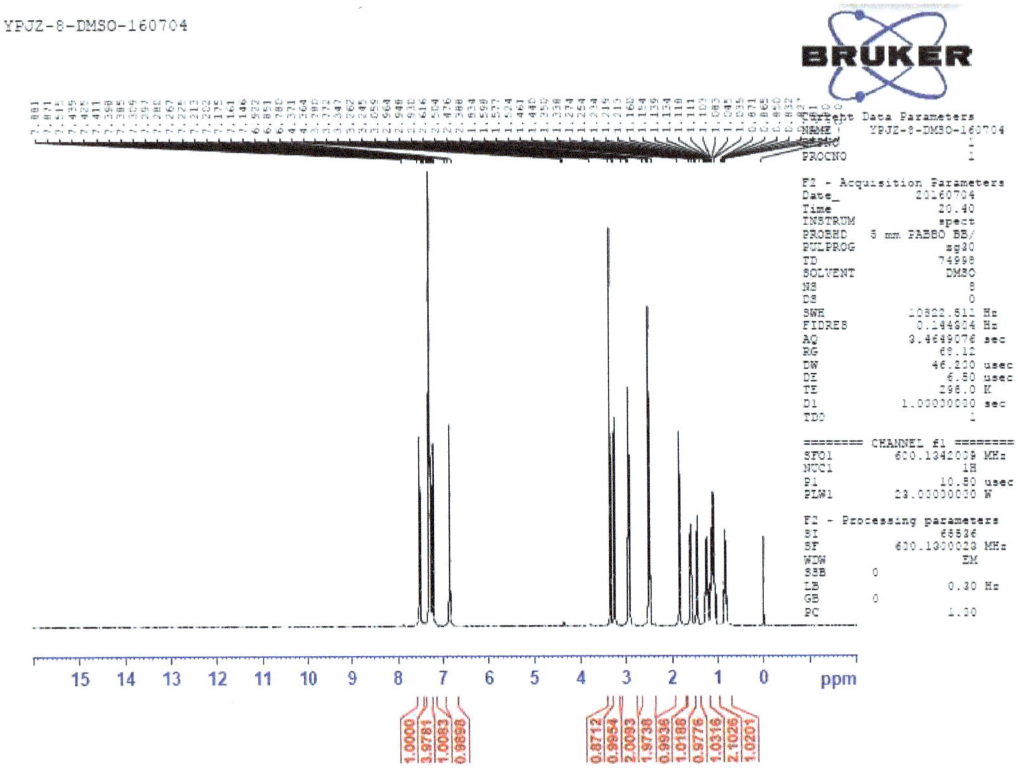
YPJZ-8-DMSO-160704
BRUKER
F2 - Acquisition Parameters
F2 - Processing parameters
15 14 13 12 11 10 9 8 7 6 5 4 3 2 1 0 ppm
1.0000 3.9781 1.0083 0.9898
0.8712 0.9954 2.0093 1.9738 1.0188 0.9936 0.9776 1.0316 2.1026 1.0201

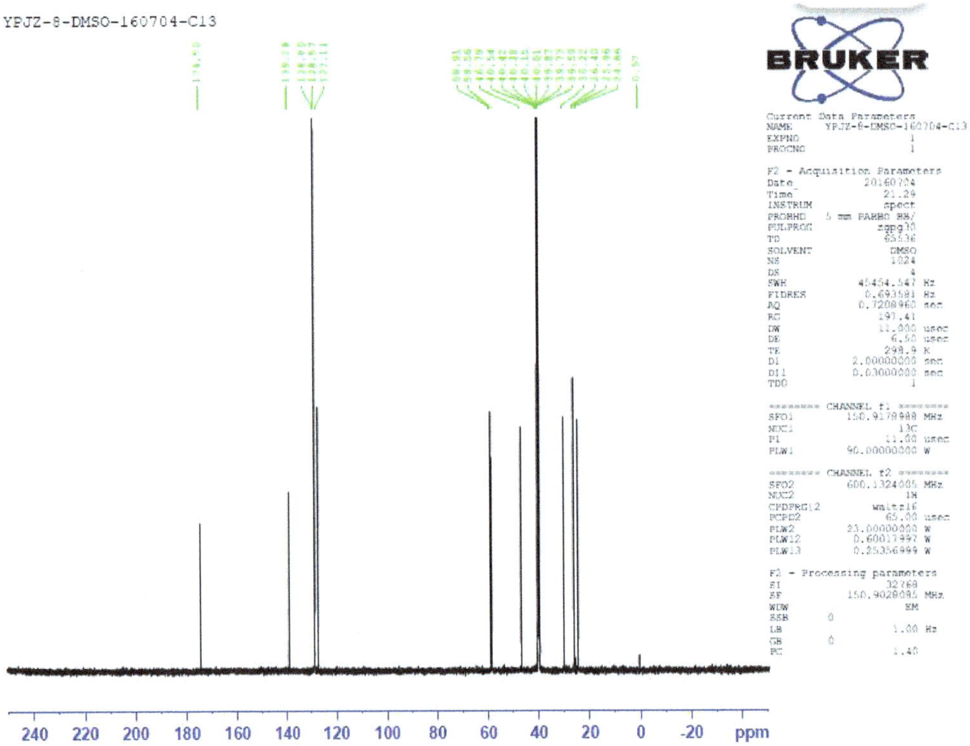
YPJZ-8-DMSO-160704-C13
BRUKER
240 220 200 180 160 140 120 100 80 60 40 20 0 -20 ppm

Qualitative Compound Report

Data File	20160704-009.d	**Sample Name**	YPJZ-8
Sample Type	Sample	**Position**	Vial 68
Instrument Name	Instrument 1	**User Name**	
Acq Method		**IRM Calibration Status**	Success
DA Method	MS.m	**Comment**	N20160704015

Compound Table

Compound Label	RT	Mass	Abund	Formula	Tgt Mass	Diff (ppm)
Cpd 1: C13 H18 N2 O	0.13	218.1423	243151	C13 H18 N2 O	218.1419	1.56

Compound Label	RT	Algorithm	Mass
Cpd 1: C13 H18 N2 O	0.13	Find By Formula	218.1423

MS Zoomed Spectrum

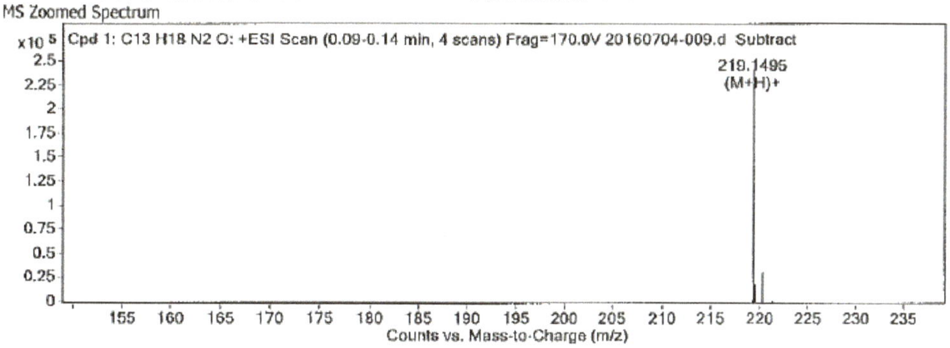

MS Spectrum Peak List

m/z	*Calc m/z*	Diff(ppm)	Abund	Formula	Ion
219.1495	219.1492	1.56	243151	C13 H19 N2 O	(M+H)+

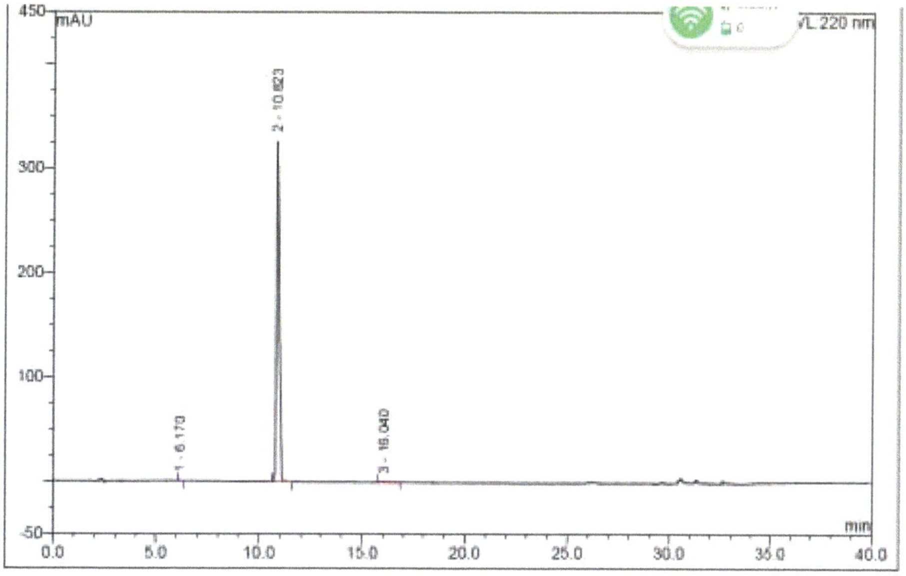

No.	Ret.Time min	Peak Name	Height mAU	Area mAU*min	Rel.Area %	Resolution	Plates
1	6.17	n.a.	0.2026	0.0192	0.04	23.23	28405
2	10.82	n.a.	326.4462	50.9630	99.30	9.15	28780
3	16.04	n.a.	0.6409	0.3419	0.67	n.a.	5224
Total:			327.290	51.324	100.00	32.387	

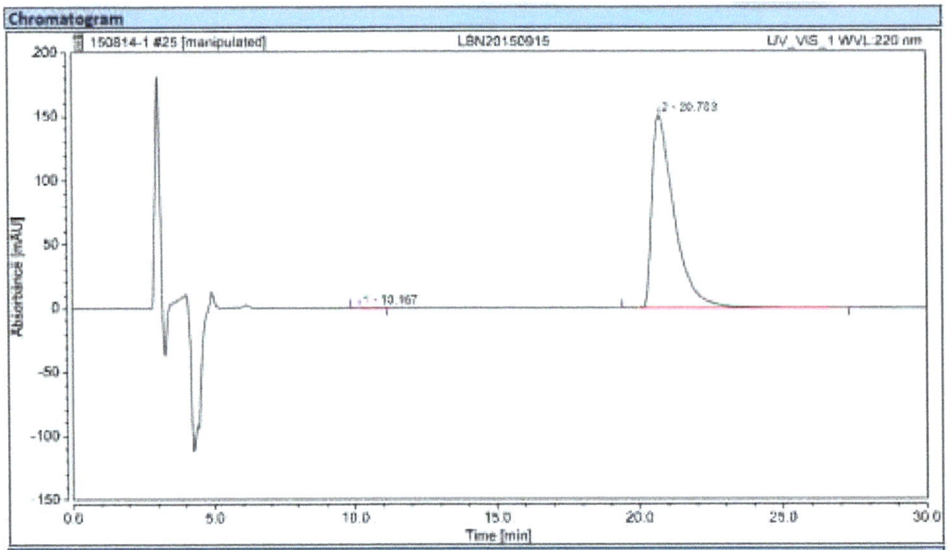

Integration Results

No.	Peak Name	Retention Time min	Area mAU*min	Height mAU	Relative Area %	Resolution (EP)	Plates (EP)
1		10.167	0.073	0.154	0.05	10.53	3483
2		20.703	133.680	150.529	99.95	n.a.	3954
Total:			133.753	150.683	100.00	10.53	

NMR, HRMS and HPLC spectra of (R,R)-O-methyl-2-piperidyl-2-yl-phenylacetate hydrochloride (10)

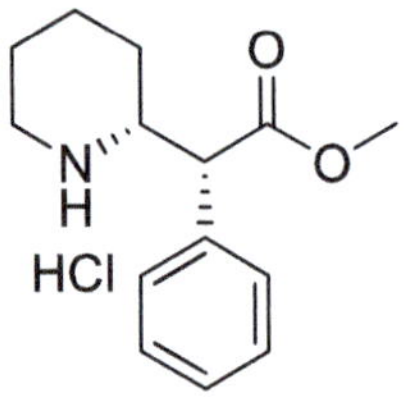

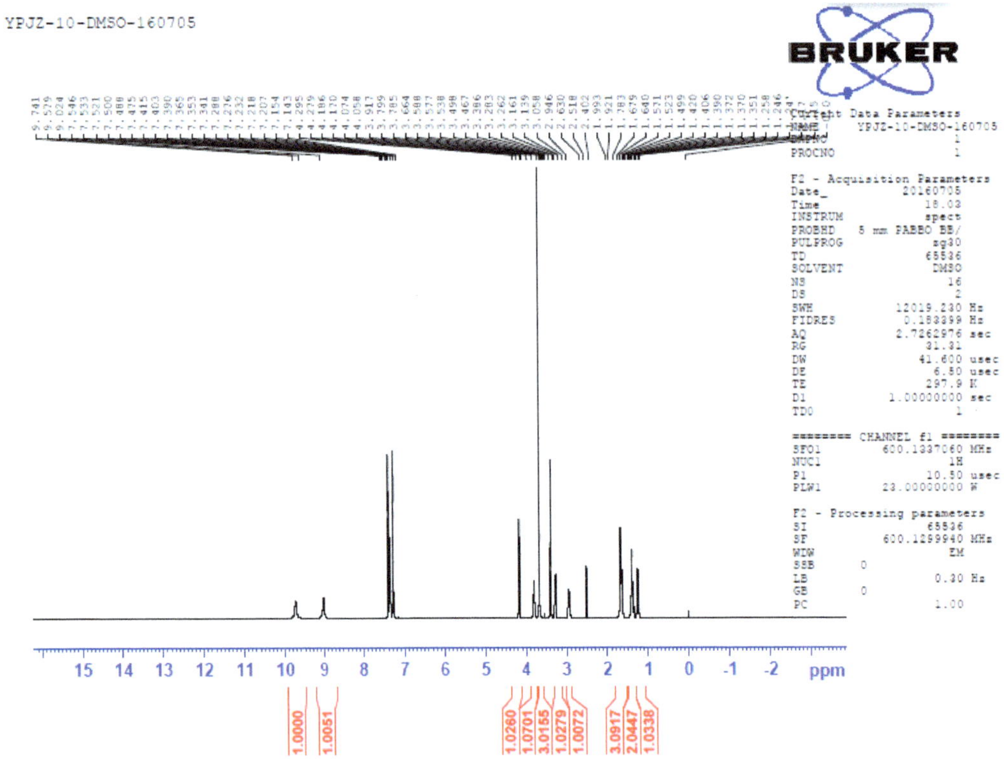

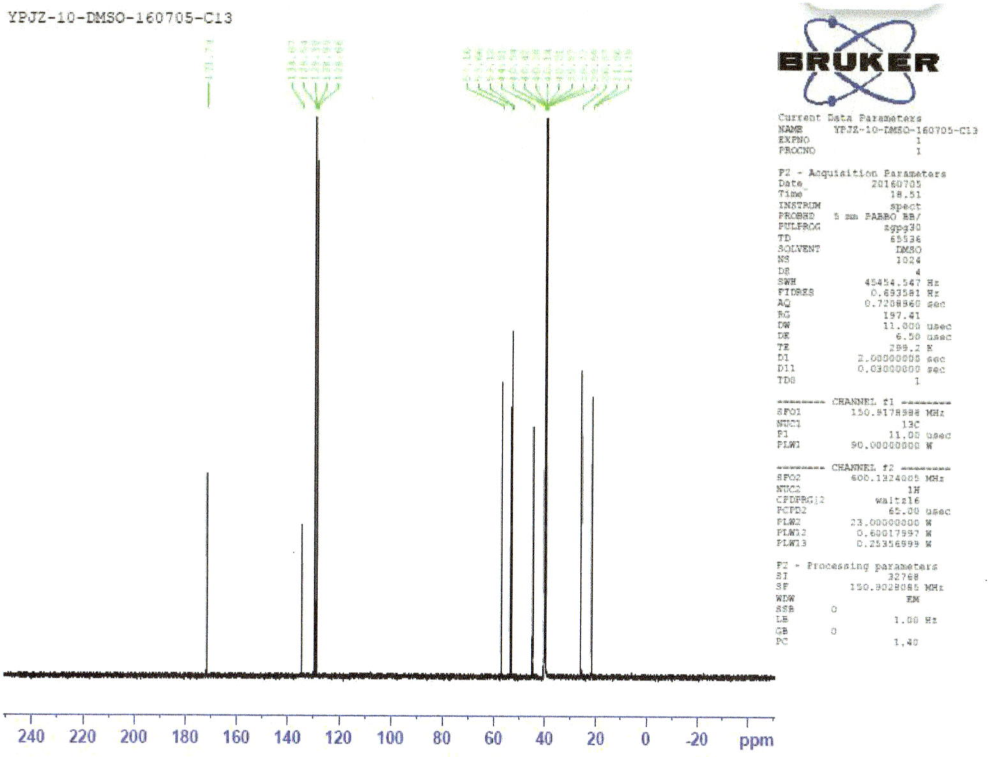

Qualitative Compound Report

Data File	20160704-010.d	**Sample Name**	YPJZ-10
Sample Type	Sample	**Position**	Vial 69
Instrument Name	Instrument 1	**User Name**	
Acq Method		**IRM Calibration Status**	Success
DA Method	MS.m	**Comment**	N20160704016

Compound Table

Compound Label	RT	Mass	Abund	Formula	Tgt Mass	Diff (ppm)
Cpd 1: C14 H19 N O2	0.32	233.1419	53275	C14 H19 N O2	233.1416	1.33

Compound Label	RT	Algorithm	Mass
Cpd 1: C14 H19 N O2	0.32	Find By Formula	233.1419

MS Zoomed Spectrum

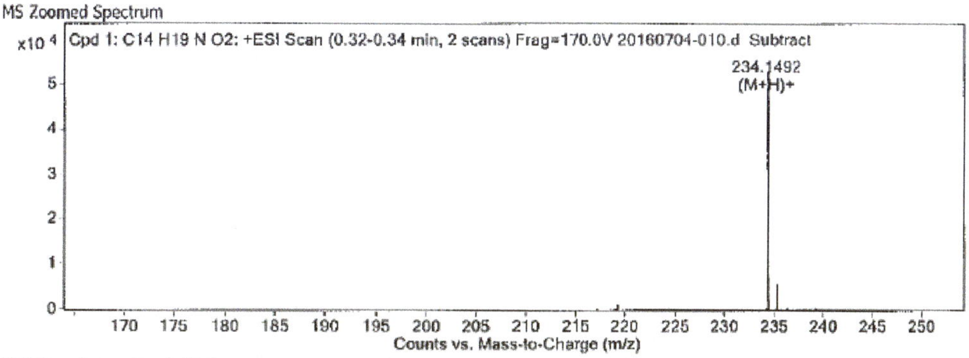

MS Spectrum Peak List

m/z	*Calc m/z*	Diff(ppm)	Abund	Formula	Ion
234.1492	234.1489	1.33	53275	C14 H20 N O2	(M+H)+

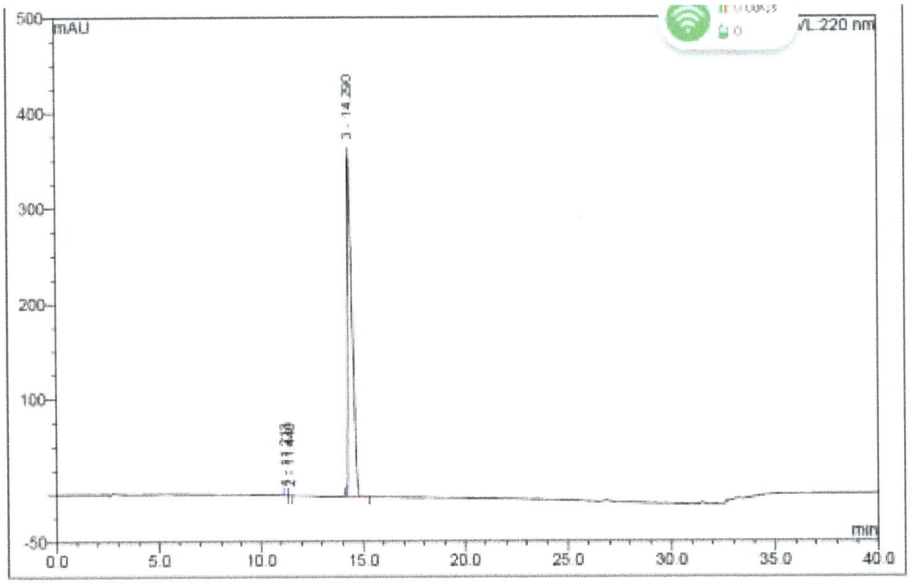

No.	Ret.Time min	Peak Name	Height mAU	Area mAU*min	Rel.Area %	Resolution	Plates
1	11.21	n.a.	0.9250	0.0774	0.08	1.79	122204
2	11.44	n.a.	1.1640	0.0919	0.10	10.30	133604
3	14.29	n.a.	366.3946	94.3027	99.82	n.a.	17697
Total:			368.484	94.472	100.00	12.093	

NMR, HRMS and HPLC spectra of dexmethylphenidate free base (1)

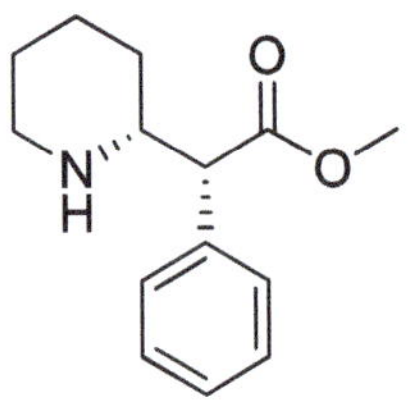

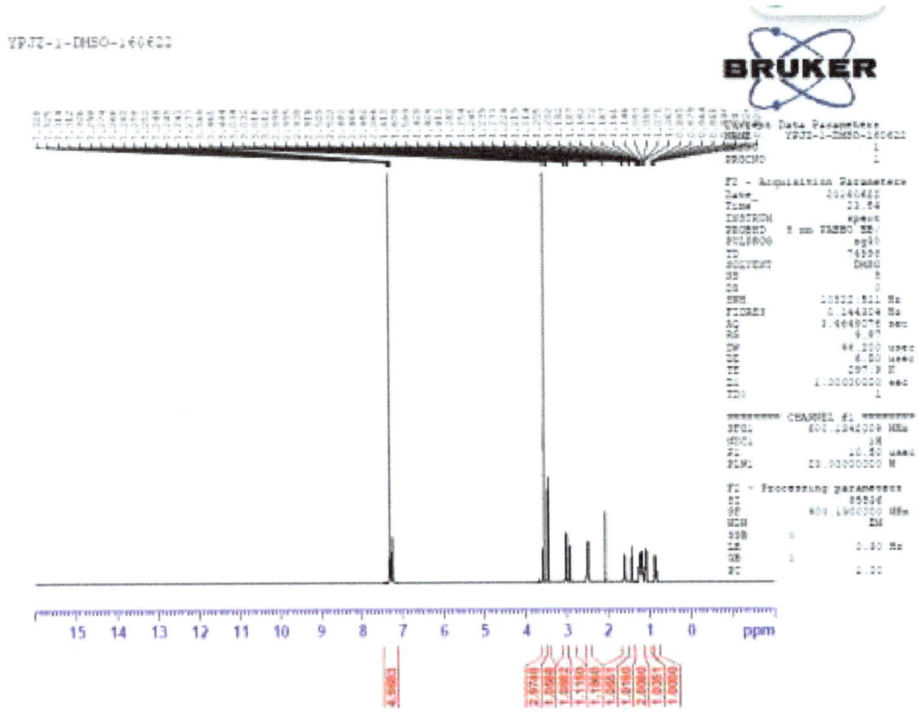

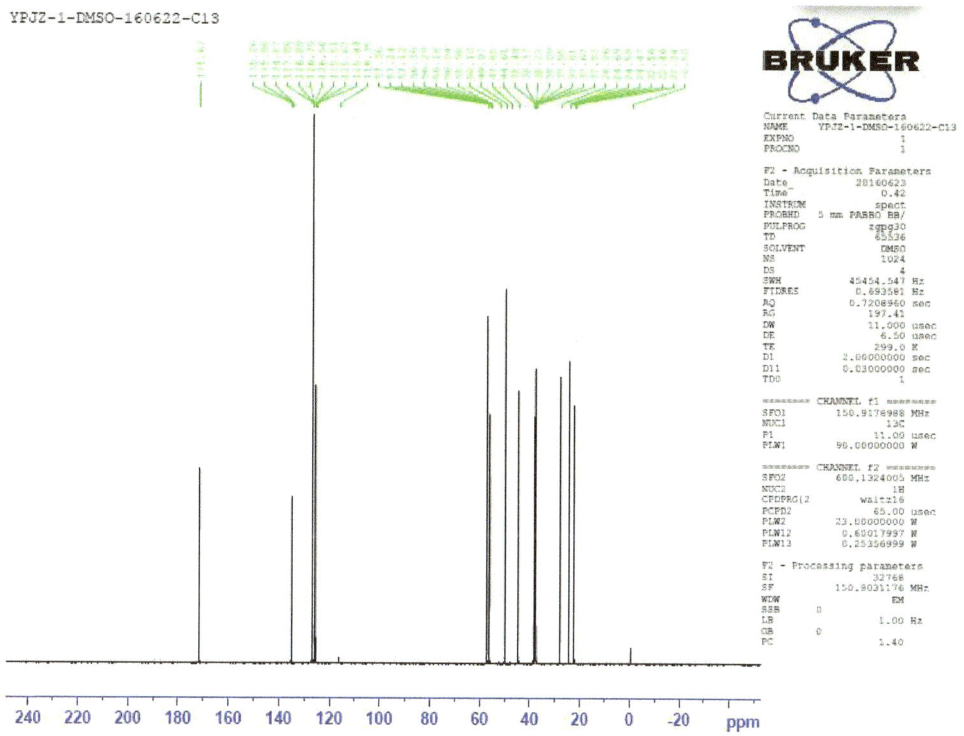

Qualitative Compound Report

Data File	20160622-005.d	**Sample Name**	YPJZ-1
Sample Type	Sample	**Position**	Vial 7
Instrument Name	Instrument 1	**User Name**	
Acq Method		**IRM Calibration Status**	Success
DA Method	MS.m	**Comment**	N20160622019

Compound Table

Compound Label	RT	Mass	Abund	Formula	Tgt Mass	Diff (ppm)
Cpd 1: C14 H19 N O2	0.03	233.1414	286354	C14 H19 N O2	233.1416	-0.78

Compound Label	RT	Algorithm	Mass
Cpd 1: C14 H19 N O2	0.03	Find By Formula	233.1414

MS Zoomed Spectrum

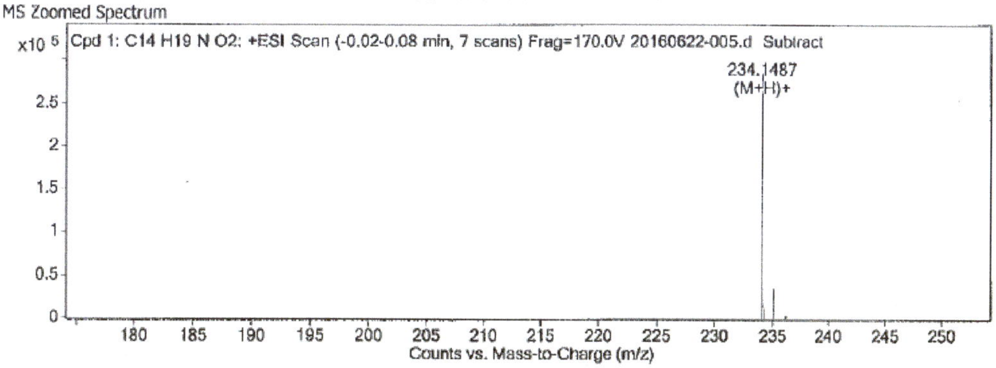

MS Spectrum Peak List

m/z	*Calc m/z*	Diff(ppm)	Abund	Formula	Ion
234.1487	234.1489	-0.78	286354	C14 H20 N O2	(M+H)+

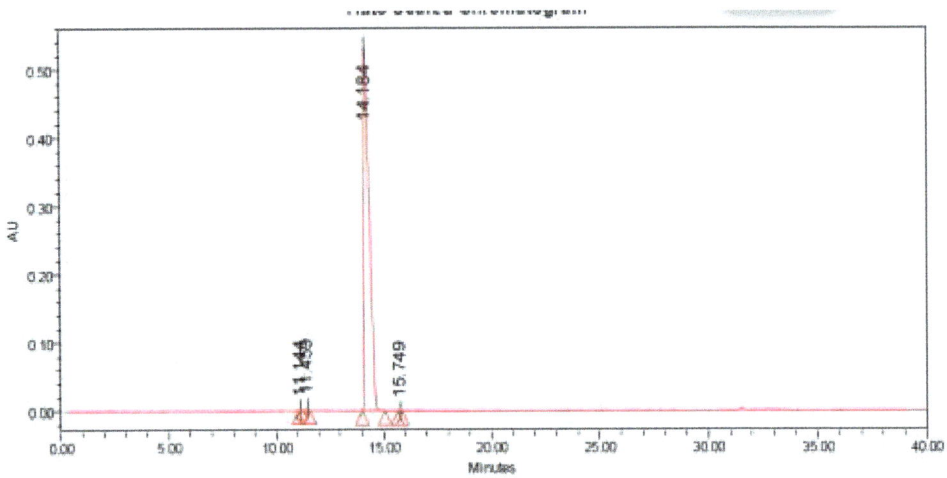

Peak Results

	Name	RT	Area	% Area	Height
1		11.144	4734	0.05	767
2		11.459	20527	0.23	3656
3		14.184	9065470	99.69	533746
4		15.749	3084	0.03	377

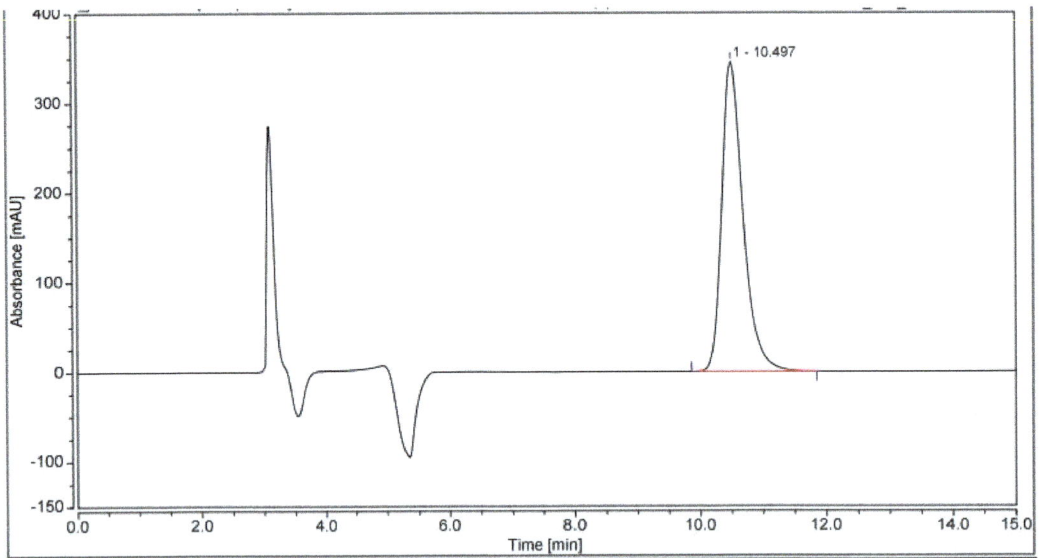

Integration Results

No.	Peak Name	Retention Time min	Area mAU*min	Height mAU	Relative Area %	Resolution (EP)	Plates (EP)
1		10.497	134.069	345.470	100.00	n.a.	5204

NMR, HRMS and HPLC spectra of (*R,R*)-2-piperdyl-2-yl-phenylacetatic acid (11)

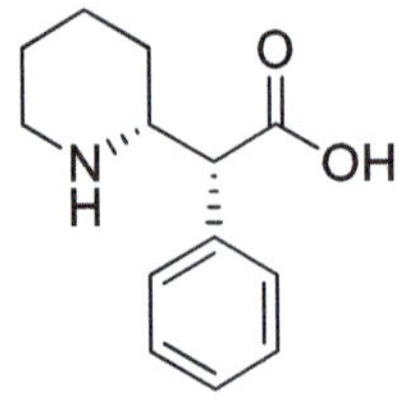

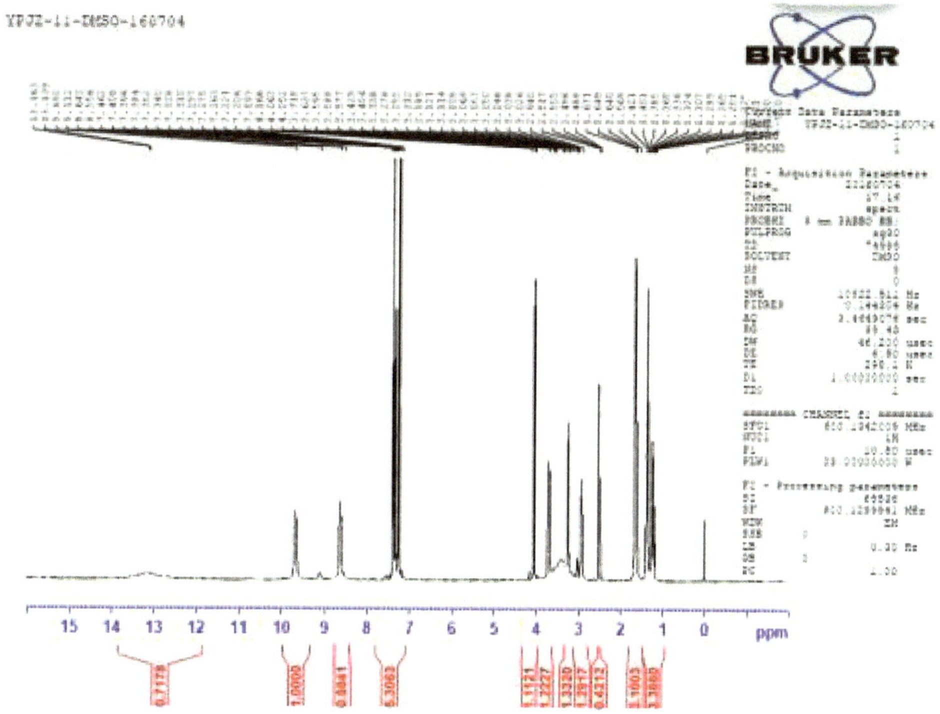
YPJZ-11-DMSO-160704
BRUKER
15 14 13 12 11 10 9 8 7 6 5 4 3 2 1 0 ppm
0.7178
1.0000
0.8841
5.3063

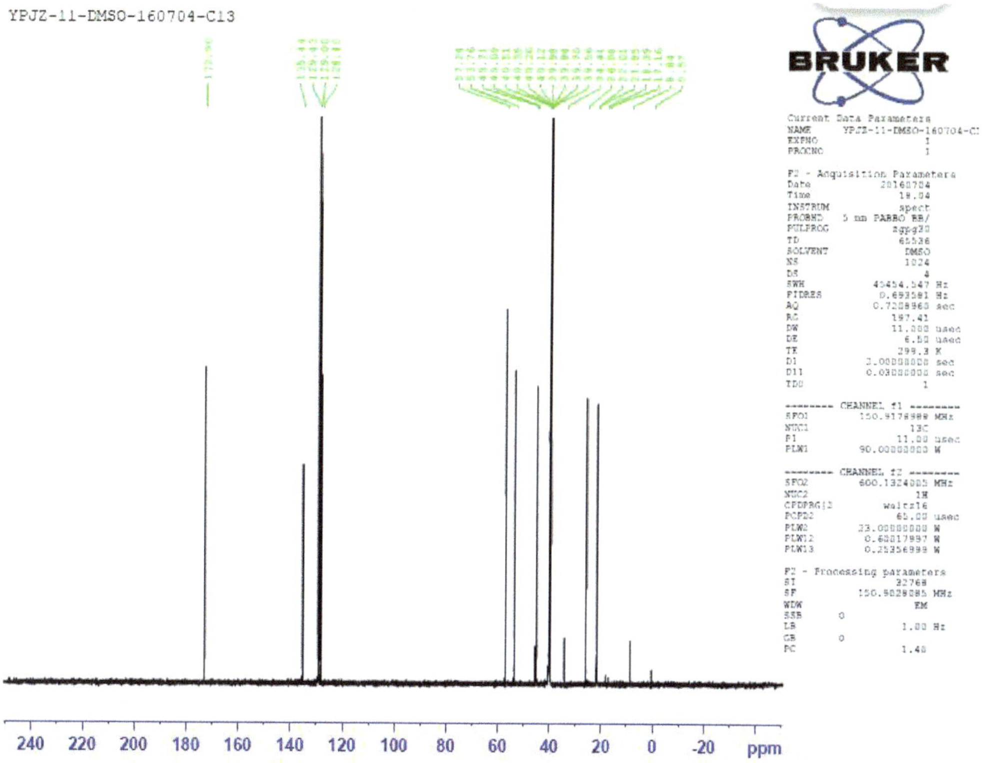
YPJZ-11-DMSO-160704-C13
BRUKER
240 220 200 180 160 140 120 100 80 60 40 20 0 -20 ppm

Qualitative Compound Report

Data File	20160622-006.d	**Sample Name**	YPJZ-11
Sample Type	Sample	**Position**	Vial 8
Instrument Name	Instrument 1	**User Name**	
Acq Method		**IRM Calibration Status**	Success
DA Method	MS.m	**Comment**	N20160622020

Compound Table

Compound Label	RT	Mass	Abund	Formula	Tgt Mass	Diff (ppm)
Cpd 1: C13 H17 N O2	0.03	219.1265	98729	C13 H17 N O2	219.1259	2.76

Compound Label	RT	Algorithm	Mass
Cpd 1: C13 H17 N O2	0.03	Find By Formula	219.1265

MS Zoomed Spectrum

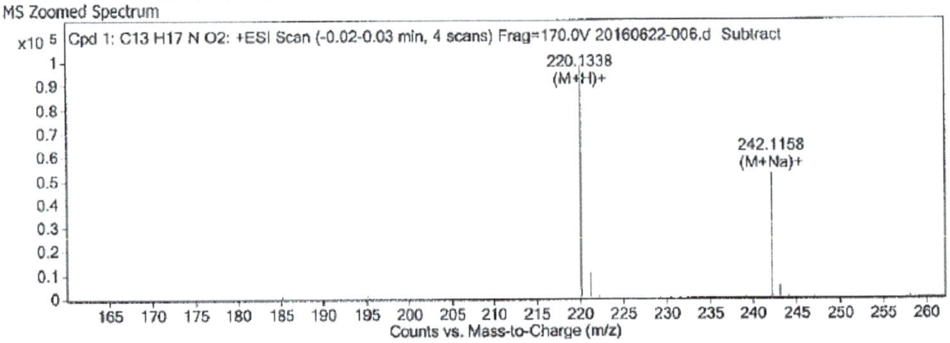

MS Spectrum Peak List

m/z	*Calc m/z*	Diff(ppm)	z	Abund	Formula	Ion
220.1338	220.1332	2.72		98729	C13 H18 N O2	(M+H)+

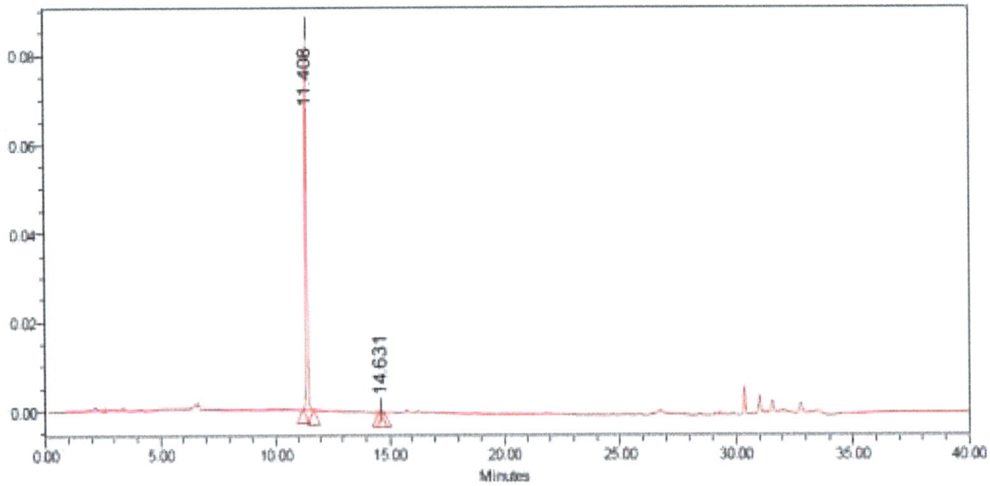

Peak Results

	Name	RT	Area	% Area	Height
1		11.408	429622	99.25	85885
2		14.631	3256	0.75	621

8

Adding a Bypass to Ensure the Chemical Plant Production

Deming Wang[1], Ying Peng[2], Zexi Wang[1]

[1]Department of Equipment Engineering, Chongqing Real Estate College, Chongqing401331, China

[2]Library, Chongqing Electric Power College, Chongqing400053, China

Correspondence: Deming Wang, Department of Equipment Engineering, Chongqing Real Estate College, Chongqing401331, China. E-mail address: dmwang4@163.com

Abstract

The chilling output of the chiller fluctuated according to the chilling load. When it dropped down below a certain value, the temperature of the supply chilled water flowing out of the chiller would drop down too much. And this would result the chiller being shut down abnormally, because of its low temperature protection. Any this shutdown of the chiller endangered the production continuity and security, and caused a certain economic losses. A bypass, which was from the outlet of the chiller's condenser to the chilled water filter inlet, was added to the chilling system. Through this bypass, some cooling water was introduced from the cooling water circulation into the chilled water circulation. Then the temperature of the supply chilled water flowing out of the chiller was promoted. With experiments, and by the aid of thermodynamics analysis, the amount of introduced cooling water, which could be introduced to promote the supply chilled water temperature for maintaining the chiller running normally, was researched. Ultimately, a certain amount of introduced cooling water, at which the chiller would no longer stop abnormally at any chilling load, was determined. And the energy lose caused by the introduced cooling water was less than 5 percent the rated output of the chiller. Compared with the chiller rated output, this energy lose was so small that it could be neglected.

Keywords: bypass, chemical plant, chilling load, chilling output, production continuity

1. Introduction

Chilling system was generally applied to many chemical plants for cooling their production processes, and their productions were often fluctuating. The chilling load (i.e., the required chilling load for production processes) varied much according to its production fluctuation. This caused the chilling output of the chiller varied much too. The chilling output could automatically adjust to fit the chilling load in an available range. But, when the chilling load was too small, the chiller could no longer adjust to fit it. The temperature of the supply chilled water flowing out of the chiller, which was monitored to prevent the chiller being frozen, dropped down and down, until to that of the chiller's low temperature protection. Then the chiller would be shut down abnormally, and restarted up in many minutes. Therefore, the security and continuity of the production were seriously affected, and the corresponding economic losses occurred at the same time. In a chemical plant, this case often occurred. In order to solve this problem, a bypass was added to the chilling system. This bypass was from the outlet of the chiller's condenser to the chilled water filter inlet. Through this bypass, some cooling water could be introduced to the chilled water circulation, to promote the supply chilled water temperature for maintaining the chiller running normally, even if the chilling load drop down below its available adjusting range.

This method of using bypass in the system was widely applied to engineering practice. In the refrigerating engineering field, Ma & Wang (2011) developed bypass check valve to enhance the overall system performance; Choi, B. Kim, Kang, & K. Kim (2011), and J. Kim, Choi, & K. Kim (2015) applied the bypass for defrosting in the heat pump; Tuo & Hrnjak (2012) showed that flash gas bypass method could improve refrigerant distribution and reduce pressure drop on refrigerant side in the mobile air conditioning system; Byun, Lee, & Jeon (2008) discovered that the hot gas bypass method was useful for retarding the formation and growth of frost at the outdoor coil and for improving the performance of the air source heat pump; Tso, Wong, Jolly, & Ng (2001) conducted simulation studies using a mathematical model to compare the performance of the hot-gas by-pass control and the suction modulation control in refrigerated shipping containers, their results showed that the suction modulation control strategy was the more energy efficient of the two; Elbel & Hrnjak (2004) found that the performance of transcritical R744 systems with direct expansion could be significantly improved by implementing a flash gas bypass; it was demonstrated by Wang, Han, Shi,

& Li (2012) that suction gas bypass was an effective regulating method of scroll compressor, and it could reduce the inner compression loss under over compression conditions and largely decrease the heating/cooling capacity of the refrigeration system; Tuo & Hrnjak (2014) used flash gas bypass for vapor-liquid separation in vapor compression systems; Huang, Li, & Yuan (2009) suggested that the hot-gas bypass defrosting could overcome the main disadvantages of the reverse-cycle defrosting method; Ding, Zhang, & Lu (2004) proved bypass two-circuit cycle freezers with one compressor were more efficient than two-evaporator in series cycle freezers in reducing the greenhouse gas emissions.

In the other engineering field, the bypass method also applied widely to the practice. For instance, W. Chen, H. Chen, Shi, Xue, Chong, & Yan (2016) installed a bypass in the low pressure district of the ejector to enhance its capacity; Zaccaria & Traverso (2016) regarded that the cold air bypass could be used for thermal management of the fuel cell. Luo, Xia, & Sun (2013) promised that bypass adjustment could improve operational efficiency and reduce heat exchanger network costs; W. Wang, X. Wang, & Liu (2011) used the bypass system for intermediate pressure start-up of a turbine unit; Yang, Pu, Wong, & Moore (2009) applied by-pass valve to reduce the overshoot and allow some exhaust compressed air to be reused in the pneumatic drive, they showed that a 12-28% saving in energy use could be acquired and meanwhile the position accuracy is maintained, compared with the traditional control of the motion of the asymmetric cylinder; research by Yoon, J. Lee, S. Lee, Tak, M. Kim, K. Kim, & Park (2013) supported that the bypass flow distribution was reduced throughout the entire core regions in block type very high temperature reactor (VHTR), and the bypass flow ratios at the inlet and the outlet were reduced by 36.19% and 14.66%, respectively; Ren & He (2009) designed a continuous circulating valve, in which a bypass was applied, to reduce oil drilling pipe accidents such as sticking and so on.

In this chemical plant, with experiments, and by the aid of thermodynamics analysis, the amount of introduced cooling water was researched. Ultimately, a certain amount of introduced cooling water, at which the chiller would no longer stop abnormally at any chilling load, was determined. And the energy lose caused by the introduced cooling water was less than 5 percent the rated output of the chiller at this introduced cooling water amount. Compared with the chiller rated output, this energy lose was so small that it could be neglected. At the same time, the security and continuity of its production were ensured, and the corresponding economic losses were avoided.

2. Case Study

The chilling system of this plant was composed of cooling water and chilled water circulations. The Cooling water circulation consisted of cooling tower, cooling water tank, filter, cooling water pump, and condenser of the chiller, etc. The chilled water circulation consisted of expansion tank, distributor, collector, filter, chilled water pump, evaporator of the chiller, as well as the production processes within which the chilling load was needed. In addition, the outflow and inflow of the chiller were called as the supply and return water, respectively. And the supply and return water pipes were shown with solid lines and dashed lines, respectively, as shown in figure 1.

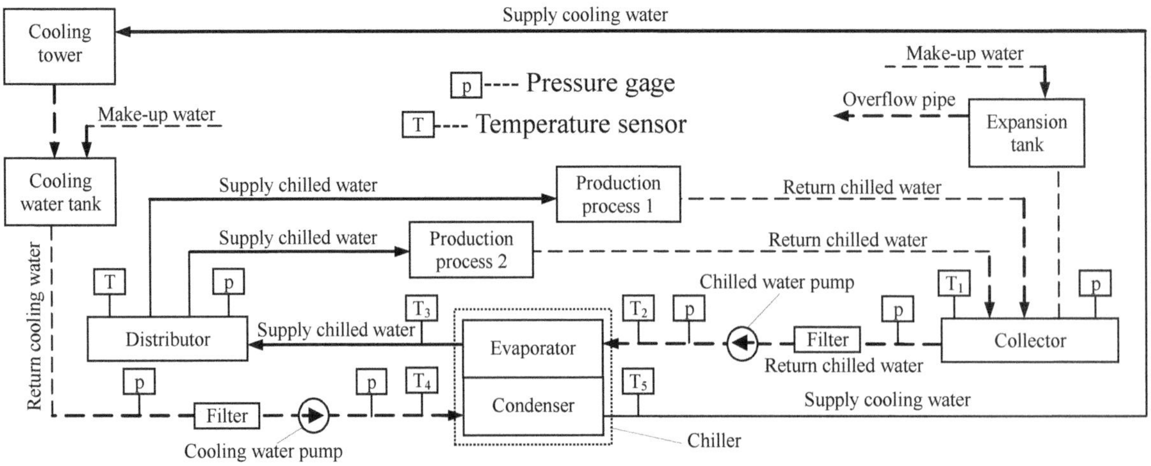

Figure1. Process of the chilling system

The design running data of the chilling system was shown in table 1. The chiller rated output was 220 kw. The temperature of the return chilled water t_1 was $10^{\circ}C$, that of the chiller condenser's inflow water t_2 was equal to t_1 in this system, and that of the supply chilled water t_3 was $5^{\circ}C$. They were measured by thermometers T_1, T_2 and T_3, respectively. The temperature of the return cooling water t_4 was $30^{\circ}C$, and that of the supply cooling water t_5 was $35^{\circ}C$. They were measured by thermometer T_4 and T_5, respectively.

Table 1. Design running data of the chilling system

Chilling output, O		Chilling load, L		Chilled water temperature		Cooling water temperature	
Rated, O_r, kw	Minimum, O_{min}, kw	Maximum, L_{max}, kw	Minimum, L_{min}, kw	Return, t_1, °C	Supply, t_3, °C	Return, t_4, °C	Supply, t_5, °C
220	66	175	30	10	5	30	35

Note: the return and supply mean the return and supply water temperatures, respectively.

The chilling output of the chiller could be automatically adjusted from 30% to 100% its rated output, i.e., from 66 kw to 220 kw; and the chilling load was fluctuating from 30 kw to 175 kw according to the production. Generally, the chilling output was nearly equal to the chilling load during production operations, because all the chilled water pipes and related equipments were thermally insulated. In fact, the chilling load often fluctuated to the value below 66 kw, even to 30 kw, while the chilling output could not drop down to that one. Then the chilling system would be running with 66 kw of the chilling output and with less than 66 kw of the chilling load, the supply chilled water temperature would be dropping down continuously, until to 3℃ and causing the chiller shutdown. This case often occurred. Any shutdown would make the chiller restart up in 20 minutes according to its procedural protection. The security and continuity of production were affected, and some economic losses occurred.

3. Research Method

In order to ensure the security and continuity of production, a bypass, which was from the outlet of the chiller's condenser to the chilled water filter inlet, was added to the chilling system, as shown with bold solid lines in figure 2. Through this bypass, some cooling water could be introduced into the chilled water circulation, to ensure the supply chilled water temperature not dropping down to 3℃, and maintain the chiller running continuously.

A modulating valve and a flow meter were installed on the bypass pipe, to modulate the introduced cooling water flow rates. With experiments, and by the aid of thermodynamics analysis, the amount of the introduced cooling water for maintaining the chiller running normally was researched.

When the supply chilled water temperature t_3 dropped down to 3℃, the chiller would shut down because of its procedural protection. During any experiment, when it dropped down to 3.5℃, it was regarded that this experiment failed.

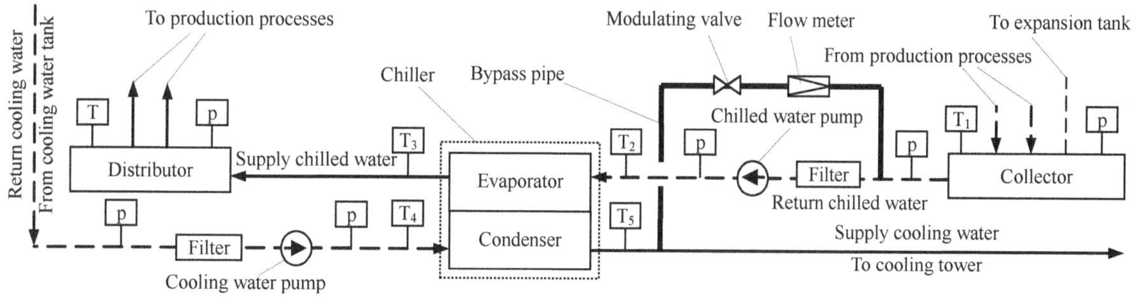

Figure 2. Process of the chilling system with bypass

On the other hand, when the supply chilled water temperature t_3 was beyond 8℃, it would dissatisfy production operation. During any experiment, when it rose up to 7.5℃, it was regarded that this experiment failed.

4. Results and Discussion

4.1 Thermodynamics Analysis

Because all the pipes and equipments, through which chilled water ran, were covered with thermal insulation layers, the heat lose could be neglected. Based on principle of energy balance according to the heat transfer edited by Yang & Tao (2006), the chilling output of the chiller with introduced cooling water can be expressed as:

$$O = \rho c \delta Q (t_5 - t_1) + \rho c (Q + \delta Q)(t_1 - t_3) \quad (1)$$

Here: O means the chilling output of the chiller, kw; ρ means the density of water, kg/L; c means the heat capacity of the water, kJ/(kg·℃); δQ means the introduced cooling water amount, L/s; t_5 means the supply cooling water temperature, ℃; t_1 means the return chilled water temperature, ℃; t_3 means the supply chilled water temperature, ℃; Q means the amount of chilled circulating water, L/s.

In the right side of equation (1), the first item denotes the load for chilling the introduced cooling water from temperature t_5 to t_1, the second item denotes the load for chilling both the introduced cooling water and chilled circulating water from temperature t_1 to t_3. At the same time, the corresponding chilling load for the production operation was required as below:

$$L = \rho c Q(t_1 - t_3) \tag{2}$$

Here: L means the chilling load for the production operation, kw. Solving equations (1) and (2), the introduced cooling water amount can be obtained as below:

$$\delta Q = \frac{O - L}{\rho c(t_5 - t_3)} \tag{3}$$

Define the chilling load ratio as:

$$\alpha = \frac{L}{O_r} \tag{4}$$

Here: α means the chilling load ratio; O_r means the chiller rated output, kw. And name the chilling output ratio as:

$$\beta = \frac{O}{O_r} \tag{5}$$

Here: β means the chilling output ratio. Then the introduced cooling water amount δQ can be expressed as:

$$\delta Q = \frac{O_r}{\rho c(t_5 - t_3)}(\beta - \alpha) \tag{6}$$

According to equation (6), the chilling output ratio β might be written as follow:

$$\beta = \frac{\rho c(t_5 - t_3)\delta Q}{O_r} + \alpha \tag{7}$$

Because ρ, c and O_r are almost constant values, the chilling output ratio β relates to t_3, t_5 and α.

4.2 Experiment Results

Experiment results were shown in figure 3 and figure 4, the chilling output ratio β were obtained according equation (7). And all those experiment results could ensure non-stop of the chiller and continuity of the production.

At the chilling load of 30 kw, i.e., $\alpha = 0.14$, it was shown in figure 3 that:

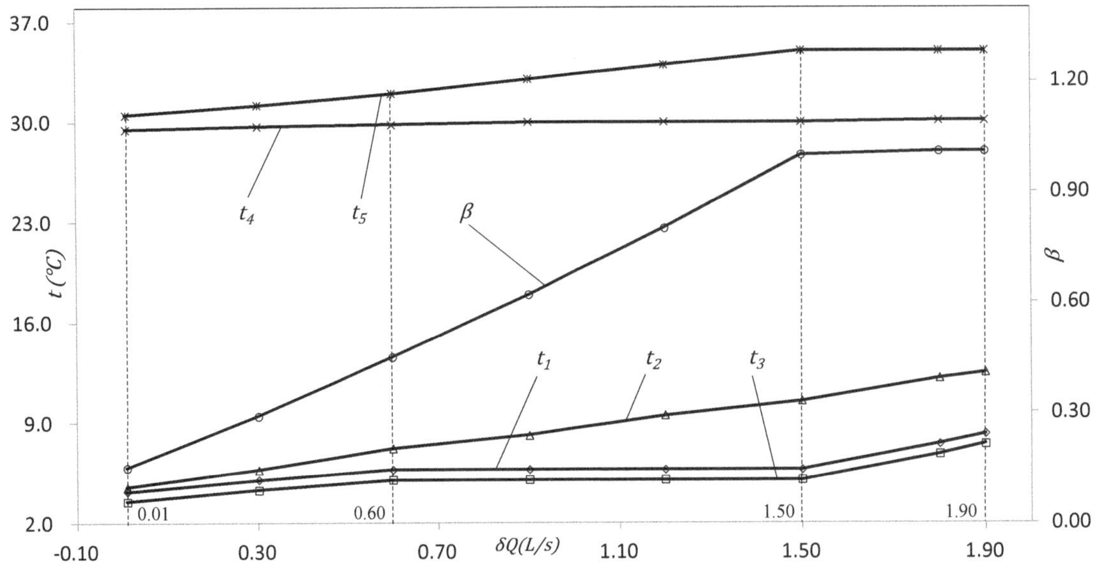

Figure 3. Experiment results while $\alpha = 0.14$

The minimum introduced cooling water amount was 0.01L/s, while the $t_3 = 3.5$℃ and the $\beta = 0.15$; and the maximum amount was 1.90L/s, while the t_3 = 7.5℃ and the β =1.00.

In the range between 0.6L/s and 1.5L/s of the δQ, the t_3 stayed the same with the δQ increasing, because the chiller could automatic adjust its output to maintain t_3 at 5℃; in the other ranges, the t_3 was rising with the δQ. The change trend of the t_1 was the same as the t_3, because the chilling load was not changed, i.e., α = 0.14. The t_2 was always rising with the δQ.

The t_4 came nearer to 30℃, because it was determined by its environment wet bulb temperature; the t_5 and β were rising before 1.50L/s of the δQ. After 1.5L/s of the δQ, the t_5 and β rose up to 35℃ and 1.00, and stayed the sames, respectively.

At the chilling load of 175 kw, i.e., α = 0.80, it was shown in figure 4 that:

The minimum introduced cooling water amount was 0.00L/s, i.e., no cooling water introduced, while the t_3 = 5.0℃ and the β = 0.80; and the maximum amount was 0.42L/s, while the t_3 = 7.5℃ and the β =1.00.

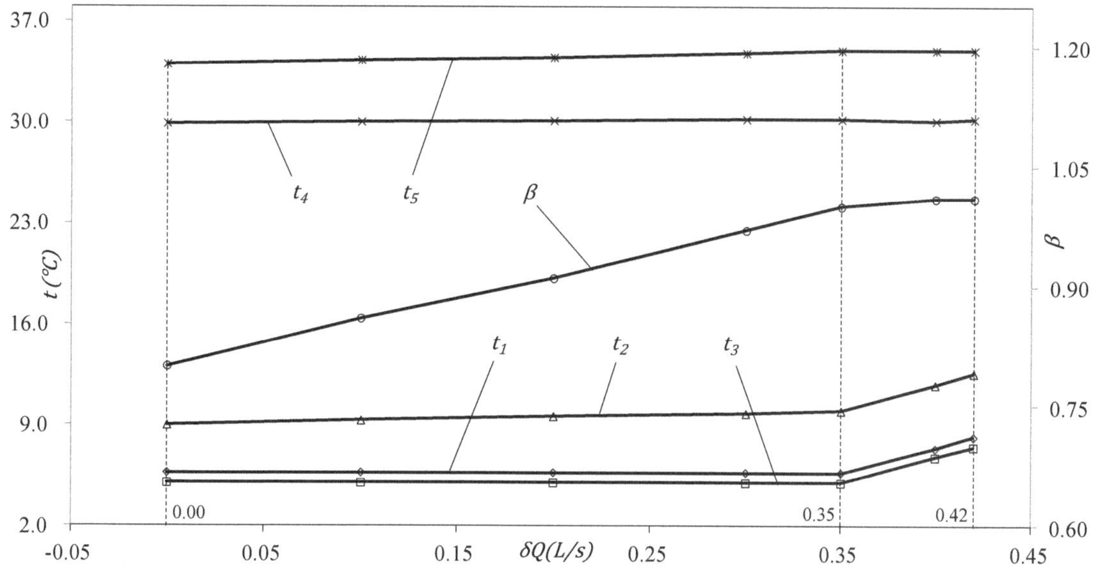

Figure 4. Experiment results while α = 0.80

The t_3 stayed the same because the chiller could automatic adjust its output to keep it at 5℃, until it began to rise after 0.35L/s of the δQ. The change trend of the t_1 was the same as the t_3, because the chilling load was not changed, i.e., α = 0.80. The t_2 was rising with the δQ, yet it was rising more sharply after 0.35L/s of the δQ.

The t_4 came nearer to 30℃, because it was determined by its environment wet bulb temperature. The t_5 and β were rising, until they rose up to 35℃ and 1.00 and stayed the sames after 0.35L/s of the δQ, respectively.

From figure 3 and figure 4, it could be obtained that the δQ between 0.01L/s and 0.42L/s could satisfy any chilling load. Considering the parameter fluctuation and energy conservation, the δQ was controlled at 0.10L/s. At this amount, the energy lose caused by the introduced cooling water was nearly 10kw, which was less than 5% the rated output of the chiller. Compared with the chiller rated output, this energy lose was so small that it could be neglected.

At the same time, because the chiller would no longer shut down abnormally, the security and continuity of the production were ensured, and the corresponding economic losses were avoided.

5. Conclusion

In some chemical plants, the production often fluctuated much, caused the chiller stopped abnormally. This seriously endangered the production. In order to keep the chiller running normally, bypass was applied to introduce some cooling water into chilled water circulation. Thermodynamics analysis and experiment study were conducted to determine the introduced amount. The chilled and cooling water temperatures and the chilling output ratio, which changed with the introduced cooling water amount, were obtained. Ultimately, a certain amount of introduced cooling water, at which the chiller would no longer stop abnormally at any chilling load, was determined. And the energy lose caused by the introduced cooling water was less than 5% the rated output of the chiller at this amount.

References

Byun, J. S., Lee, J. H., & Jeon, C. D. (2008). Frost retardation of an air-source heat pump by the hot gas bypass method. *International Journal of Refrigeration, 31*, 328-334. http://www.sciencedirect.com/science/article/pii/S01407007070008 -49

Chen, W. X., Chen, H. Q., Shi, C. Y., Xue, K. K., Chong, D. T., & Yan, J. J. (2016). A novel ejector with a bypass to enhance the performance. *Applied Thermal Engineering, 93*, 939-946. http://www.sciencedirect.com/science/article/pii /S135943111501114X

Chen, W. X., Chen, H. Q., Shi, C. Y., Xue, K. K., & Chong, D. T. (2016). Impact of operational and geometrical factors on ejector performance with a bypass. *Applied Thermal Engineering, 99,* 476–484. http://www.sciencedirect.com/science /article/pii/S1359431116300242

Choi, H. J., Kim, B. S., Kang, D., & Kim, K. C. (2011). Defrosting method adopting dual hot gas bypass for an air-to-air heat pump. *Applied Energy, 88*(12), 4544-4555. http://www.sciencedirect.com/science/article/pii/S0306261911003345

Ding, G. L., Zhang, C. L., & Lu, Z. L. (2004). Dynamic simulation of natural convection bypass two-circuit cycle refrigerator–freezer and its application: Part I: component models. *Applied Thermal Engineering, 24*, 1513-1524. http://www.sciencedirect.com/science/article/pii/S1359431103004034

Elbel, S., & Hrnjak, P. (2004). Flash gas bypass for improving the performance of transcritical R744 systems that use microchannel evaporators. *International Journal of Refrigeration, 27*, 724-735. http://www.sciencedirect.com/science/ar -ticle/pii/S0140700704001501

Huang, D., Li, Q. X., & Yuan, X. L. (2009). Comparison between hot-gas bypass defrosting and reverse-cycle defrosting methods on an air-to-water heat pump. *Applied Energy, 86*, 1697-1703. http://www.sciencedirect.com/science/article/pii /S0306261908003012

Kim, J., Choi, H. J., & Kim, K. C. (2015). A combined dual hot-gas bypass defrosting method with accumulator heater for an air-to-air heat pump in cold region. *Applied Energy, 147*, 344-352. http://www.sciencediect.com/science/article /pii/S030626191500 -2597

Luo, X. L., Xia, C. K., & Sun, L. (2013). Margin design, online optimization, and control approach of a heat exchanger network with bypasses. *Computers & Chemical Engineering, 53*, 102-121. http://www.sciencedirect.com/science/article/pii/S0098135413000446

Ma, Z. J., & Wang, S. W. (2011). Enhancing the performance of large primary-secondary chilled water systems by using bypass check valve. *Energy, 36*, 268-276. http://www.sciencedirect.com/science/article/pii/S0360544210006055

Ren, L. C., & He, D. S. (2009). Design of continuous circulation valve. *China Petroleum Machinery, 37*, 129-132. http://lib.cqvip.com/qk/94770X/200909/31438539.html

Tso, C. P., Wong, Y. W., Jolly, P. G., & Ng, S. M. (2001). A comparison of hot-gas by-pass and suction modulation method for partial load control in refrigerated shipping containers. *International Journal of Refrigeration, 24*, 544-553. http:// www.sciencedirect.com/science/article/pii/S0140700700000529

Tuo, H. F., & Hrnjak, P. (2014). Enhancement of vapor–liquid separation in vertical impact T-junctions for vapor compression systems with flash gas bypass. *International Journal of Refrigeration, 40*, 43-50. http://www.sciencedirect.com/science/article/pii/S0140700713002971

Tuo, H. F., & Hrnjak, P. (2014). Vapor-liquid separation in a vertical impact T-junction for vapor compression systems with flash gas bypass. *International Journal of Refrigeration, 40,* 189-200. http://www.sciencedirect.com/science/article/pii/S0140700713003058

Tuo, H. F., & Hrnjak, P. (2012). Flash gas bypass in mobile air conditioning system with R134a. *International Journal of Refrigeration, 35,* 1869-1877. http://www.sciencedirect.com/science/article/pii/S0140700712001272

Wang, B. L., Han, L. J., Shi, W. X., & Li, X. T. (2012). Modulation method of scroll compressor based on suction gas bypass. *Applied Thermal Engineering, 37*, 183-189. http://www.sciencedirect.com/science/article/pii/S1359431111 -006338

Wang, W. B., Wang, X. D., & Liu, D. J. (2011). IP start-up control for 300MW turbine unit with bypass system. *Procedia Engineering, 16*, 493-498. http://www.sciencedirect.com/science/article/pii/S1877705811026166

Yang, A. M., Pu, J. S., Wong, C. B., & Moore, P. (2009). By-pass valve control to improve energy efficiency of pneumatic drive system. *Control Engineering Practice, 17*, 623-628.

http://www.sciencedirect.com/science/article/pii/S096706610 -8001883

Yang, S. M., & Tao, W. Q. (2006). Heat transfer (4th ed.). Beijing: Higher education press.

Yoon, S. J., Lee, J. H., Lee, S. M., Tak, N. I., Kim, M. H., Kim, K. Y., & Park, G. C. (2013). Numerical and Experimental Investigation on Labyrinth Seal Mechanism for Bypass Flow Reduction in Prismatic VHTR Core. *Nuclear Engineering and Design, 262*, 525-534. http://www.sciencedirect.com/science/article/pii/S0029549313002938

Zaccaria, V., & Traverso, A. (2016). Transfer function development for SOFC/GT hybrid systems control using cold air bypass. *Applied Energy, 165*, 695-706. http://www.sciencedirect.com/science/article/pii/S0306261915016670

9

Cleaner and Efficient Green Chemistry Synthesis of *N,N′*-Dibenzyl or *N,N′*-(2-Hydroxybenzyl)-Ethane-1,2-Diamine,-Propane-1,3-Diamine and -1,3-Diamino-2-Propanol

Augusto Rivera[1], Ingrid Miranda-Carvajal[1], Jaime Ríos-Motta[1]

[1]Universidad Nacional de Colombia, Sede Bogotá, Facultad de Ciencias, Departamento de Química, Carrera 30 No. 45-03, Bogotá, Código Postal 111321, Colombia

Correspondence: Augusto Rivera, Universidad Nacional de Colombia, Sede Bogotá, Facultad de Ciencias, Departamento de Química, Carrera 30 No. 45-03, Bogotá, Código Postal 111321, Colombia. E-mail: ariverau@unal.edu.co

Abstract

An efficient and highly eco-friendly protocol for the preparation of *N,N′*-dibenzyl or *N,N′*-(2-hydroxybenzyl)diamines via the reduction of the corresponding di-Schiff bases that produces a good yield using water as a solvent without the need for catalysis or the azeotropic removal of water has been developed. These symmetric diimines have been reduced to their corresponding diamines with sodium borohydride using a catalyst- and solvent-free protocol with excellent yield. Mild conditions, high yields, and a simple work-up procedure are the primary benefits of this protocol.

Keywords: aqueous medium, catalyst free, green chemistry, reductive amination, di-Schiff bases, solvent-free

1. Introduction

The synthesis of cyclic amines from a carbonyl compound (aldehyde or ketone) and a diamine (or derivative) is currently the method of choice for the preparation of these compounds. In the framework of a program intended to develop 1,2,3-trisubstituted hexahydropyrimidin-5-ols (**1**) with conformational properties, we have been interested in the preparation of N,N′-bis(2-arylmethyl)-diamine compounds. In addition to interest in the stereochemical properties of these compounds and interest in their application in the asymmetric synthesis of ligands in catalysis, *N,N′*-bis(2-hydroxybenzyl) (**2**) and *N,N′*-dibenzyl (**3**) derivatives of 1,2, and 1,3-diamines are of biological importance because their derivatives with related structures act as analgesics, antiparasitics, antifungals, and antibacterials (Bisceglia, Díaz, Torres, & Orelli, 2011). Different methods have been developed to prepare 1,n diamines but the reduction of the corresponding diimines (e.g., **4** and **5**) offers the possibility to produce a wide range of such diamines (Saeed & Musad, 2009). Given our interest in the development of greener synthetic pathways, our synthesis started with the preparation of imines **10e-p** using a simple water-mediated procedure that require neither a catalyst nor any additive (Koteswara Rao et al., 2010) and the products were isolated simply by filtration. Respective di-Schiff bases **10e-p** obtained from this process were further successfully reduced with sodium borohydride under catalyst-free and solvent-free conditions (Wang, Tang, Pan, Yang, & Zhang, 2015), producing the corresponding tetrahydro-di-Schiff bases **11e-p** (Scheme 1). Two primary reduction procedures can be used to reduce C=N compounds: reduction via catalytic hydrogenation and reduction with metal hydrides. Sodium borohydride is one of the most-used reagents for the reduction of imines because of its effectiveness and its ease of handling. Most conventional imine reaction methods described to date involve treatment with reducing agents such as $NaBH_4$ in an organic solvent, such as ethanol or methanol, with the use of catalysts during reflux (Abdel-Magid, 2014). Many reactions proceed efficiently in the solid state. Indeed, in many cases, solid-state organic reactions occur more efficiently and more selectively than do their counterparts in solution; furthermore, solvent-free reactions have many advantages: reduced pollution, low cost, and simple processing and handling (Tanaka & Toda, 2000). In recent years, environmentally benign synthetic methods have received significant attention and some solvent-free protocols for imine reduction have been reported (Mohammadi & Setamdideh, 2015; Hasanloie & Setamdideh, 2014; Setamdideh & Sepehraddin, 2014; Alinezhad & Tollabian, 2010; Alinezhad, Tajbakhsh & Mahdavi, 2010; Cho & Kung, 2005). In this work we report an alternative, simple solvent-free procedure for the synthesis of tetrahydro-di-Schiff bases that uses sodium borohydride as a reducing agent.

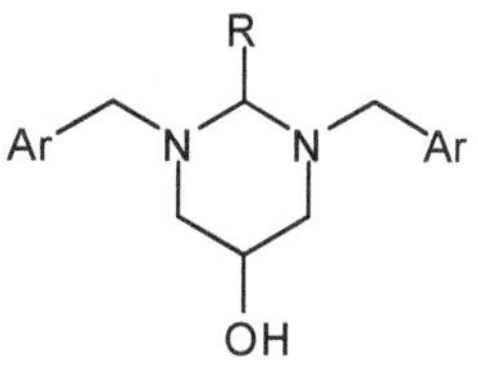

1

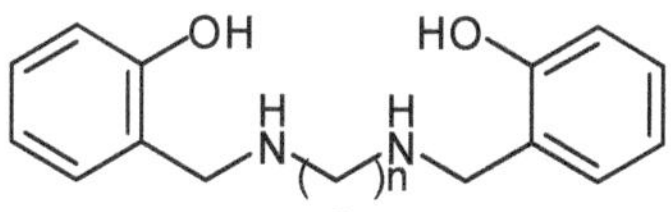

2

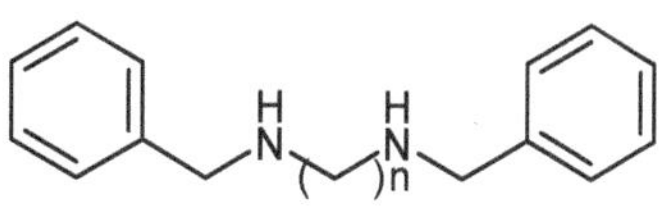

3

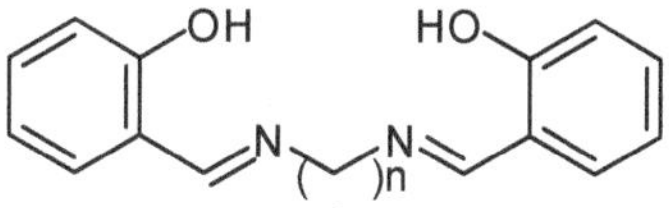

4

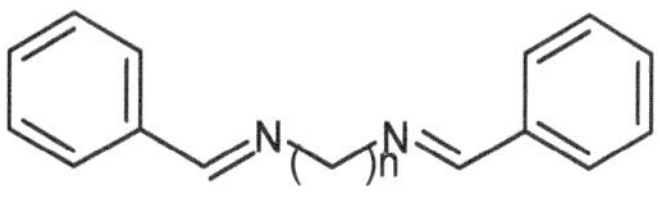

5

n = 2 or 3

2. Experimental

2.1 Materials and Equipment

All chemical reagents and starting materials were purchased from commercially available vendors and used without further purification. Reactions were monitored by thin-layer chromatography at 0.25 mm using E. Merck silica gel 60 plates (F254). Melting points were measured in open capillaries using an Electrothermal 9100 melting point apparatus and are uncorrected. FT-IR spectra were recorded in potassium bromide pellets using a Thermo Nicolet IS10 spectrophotometer. ^{1}H-NMR and ^{13}C-NMR spectra were recorded in $CDCl_3$ using a Bruker Avance AV-400 MHz spectrometer operated at 400.130 MHz for ^{1}H and at 100.634 MHz for ^{13}C. Elemental analyses (C, H, N) were performed with a Thermo Scientific Flash 2000.

*2.2 Procedure for the Synthesis of Diimines (**10e-p**)*

An appropriate aromatic aldehyde (**9a-d**) (0.68 mmol) was slowly added to a solution of the appropriate diamine (**6-8**) (0.34 mmol) in water (5 mL). After addition, the mixture was strongly stirred for 3 h at room temperature. The solid was then filtered and washed with abundant water and was dried outdoors if there was a precipitate, but if there was no precipitate, the excess water was removed under vacuum.

2.2.1 Characterization Data of the Compounds

*N,N'–bisbenzylideneethane-1,2-diamine (**10e**)*: White solid, mp 53.0–54.0 °C, (52.0–53.0 °C lit., Simion et al., 2001), yield 100%; IR (KBr): 3080–3028 cm^{-1} (Aromatic C–H stretch), 2929 (C–H stretch, CH_2), 2846 (C–H stretch, N=C–H), 1960–1675 (Overtone and combination aromatic bands), 1641 (C=N stretch) cm^{-1}. ^{1}H NMR ($CDCl_3$, 400 MHz): 3.97 (s, 4H, CH_2–CH_2), 7.38 (m, 6H, Ar–H), 7.68 (m, 4H, Ar–H), 8.28 (s ,2H, H–C=N).

*N,N'–bis(o–hydroxybenzylidene)ethane–1,2–diamine (**10f**)*: Yellow solid, mp 133.0–134.0 °C, (127.0–129.0 °C lit. Bordbar, Faal, Ahari–Mostafavi, Graragozlou, & Fazaeli, 2013), yield 88%; IR (KBr) 3451 (O–H stretch), 3051–3009 (Aromatic C–H stretch), 2930–2900 (C–H stretch, CH_2), 2868 (C–H stretch, N=C–H), 1941–1797 (Overtone and combination aromatic bands), 1635 (C=N stretch), 1283 (C–O stretch) cm^{-1}. ^{1}H NMR ($CDCl_3$, 400 MHz): 3.91 (s, 4H, CH_2–CH_2), 6.85 (td, 2H, J = 0.87 Hz; and 7.6 Hz, Ar–H7'), 7.22 (dd, 2H, J = 1.64 Hz; J = 7.6 Hz, Ar–H5'), 7.28 (td, 2H, J = 1.64 Hz; J = 7.6 Hz, Ar–H6'), 8.35 (s, 2H, H–C=N), 13.2 (s, 2H, OH).

*N,N'–bis(p–methoxybenzylidene)ethane–1,2–diamine (**10g**)*: White solid, mp 119.0–120.0 °C, (110.0–111.0 °C lit. Ünaleroğlu, Temelli, & Hökelek, 2002), yield 95%; IR (KBr) 3047–3015 (Aromatic C–H stretch), 2971 (C–H stretch, CH_2), 2920 (C–H stretch, CH_3), 2888 (C–H stretch, N=C–H), 2842 (Symmetric C–O–C stretch, Ar–OCH_3), 2025–1900 (Overtone and combination aromatic bands), 1640 (C=N stretch), 1251 (Asymmetric C–O–C stretch, Ar–OCH_3) cm^{-1}. ^{1}H NMR ($CDCl_3$, 400 MHz): 3.82 (s, 6H, OCH_3), 3.91 (s, 4H, CH_2–CH_2), 6.90 (m, 4H, Ar–H), 7.63 (m, 4H, Ar–H), 8.20 (s, 2H, H–C=N).

*N,N'–bis(p–dimethylaminobenzylidene)ethane–1,2–diamine (**10h**)*: White solid, mp 181.0–182.0 °C, yield 86%; IR (KBr) 3042 (Aromatic C–H stretch), 2910 (C–H stretch, CH_2), 2882 (C–H stretch, CH_3), 2852 (C–H stretch, N=C–H), 2000–1900 (Overtone and combination aromatic bands), 1639 (C=N stretch), 1340 (C–N stretch, $N(CH_3)_2$) cm^{-1}. ^{1}H NMR ($CDCl_3$, 400 MHz): 2.96 (s, 12H, $N(CH_3)_2$), 3.87 (s, 4H, CH_2–CH_2), 6.65 (d, 4H, J = 8.0 Hz, Ar–H5' and Ar–H7'),

7.57 (d, 4H, J = 8.3 Hz, Ar–H4' and Ar–H8'), 8.15 (s, 2H, H–C=N).

N,N'–bis(benzylidene)propane–1,3–diamine ***(10i)***: Light yellow oil, yield 82.4%; IR (KBr) 3060–3026 cm^{-1} (Aromatic C–H stretch), 2925 (C–H stretch, CH_2), 2838 (C–H stretch, N=C–H), 1958–1718 (Overtone and combination aromatic bands), 1644 (C=N stretch) cm^{-1}. ^{1}H NMR ($CDCl_3$, 400 MHz): 2.12 (m, 2H, CH_2), 3.72 (t, 4H, J = 7.1 Hz, CH_2–N), 7.37–7.42 (m, 6H, H–Ar), 7.68–7.74 (m, 4H, H–Ar), 8.28 (bs, 2H, H–C=N).

N,N'–bis(o–hydroxybenzylidene)propane–1,3–diamine ***(10j)***: Lemon yellow powder, mp 55.0–57.0 °C, (54.0–55.0 lit. Frost & Freedman, 1959) 81% yield; IR (KBr) 3424 (O–H stretch), 3049 (Aromatic C–H stretch), 2994–2945 (C–H stretch, CH_2), 2868 (C–H stretch, N=C–H), 1936–1786 (Overtone and combination aromatic bands), 1635 (C=N stretch), 1280 (C–O stretch) cm^{-1}. ^{1}H NMR ($CDCl_3$, 400 MHz): 2.11 (m, 2H, CH_2), 3.71 (t, 4H, J = 7.1 Hz, CH_2–N), 6.86 (t, 2H, J = 7.2 Hz, H–Ar), 6.95 (d, 2H, J = 8.2 Hz, H–Ar), 7.22 (d, 2H, J = 8.2 Hz, H–Ar), 7.29 (ddd, 2H, J = 8.2 Hz, H–Ar), 8.35 (s, 2H, H–C=N), 13.39 (bs, 2H, OH).

N,N'–bis(p–methoxybenzylidene)propane–1,3–diamine ***(10k)***: White solid, mp 81.0–82.0 °C, (74.0–76 °C lit. Simion et al., 2001) yield 63.4%; IR (KBr) 3006 (Aromatic C–H stretch), 2954 (C–H stretch, CH_2), 2948 (C–H stretch, CH_3), 2836 (C–H stretch, N=C–H), 2788 (Symmetric C–O–C stretch, Ar–OCH_3), 2005–1900 (Overtone and combination aromatic bands), 1640 (C=N stretch), 1259 (Asymmetric C–O–C stretch, Ar–OCH_3) cm^{-1}. ^{1}H NMR ($CDCl_3$, 400 MHz): (m, 2H, CH_2), 3.66 (t, 4H, J = 7.1 Hz, CH_2–N), 3.82 (s, 6H, O–CH_3), 6.90 (d, 4H, J = 9.0 Hz, H–Ar), 7.64 (d, 4H, J = 9.0 Hz, H–Ar), 8.20 (bs, 2H, H–C=N).

N,N'–bis(p–dimethylaminobenzylidene propane–1,3–diamine ***(10l)***: Beige solid, mp 155.0–156 °C, (144.5–145.0 °C lit. Billman & Dorman, 1962), yield 75.0%; IR (KBr) 3019 (Aromatic C–H stretch), 2918 (C–H stretch, CH_2), 2904 (C–H stretch, CH_3), 2821 (C–H stretch, N=C–H), 2000–1996 (Overtone and combination aromatic bands), 1639 (C=N stretch), 1363 (C–N stretch, $N(CH_3)_2$) cm^{-1}. ^{1}H NMR (400 MHz, $CDCl_3$) 2.05 (m, 2H, CH_2–3), 2.99 (s, 12H, $(CH_3)_2$–N), 3.61 (t, 4H, J = 6.8 Hz, CH_2–N), 6.67 (d, 4H, J = 8.4 Hz, H–Ar), 7.57 (d, 4H, J = 8.4 Hz, H–Ar), 8.12 (s, 2H, H–C=N).

N,N'–bisbenzylidene–1,3–diamino–2–propanol ***(10m)***: White solid, mp 104–105 °C, yield 74.0%; ^{1}H NMR ($CDCl_3$, 400 MHz):3.76 (ddd, 2H, J = 1.2 Hz, J = 6.8 Hz, J = 12.3 Hz, CH_2–2 or CH_2–4), 3.86 (ddd, 2H, J = 1.2 Hz, J = 5.0 Hz and J = 12.3 Hz, CH_2–2 or CH_2–4), 4.22–4.28 (m, 1H, H–C3), 7.41–743 (m, 6H, Ar–H), 7.73–776 (m, 4H, Ar–H), 8.36 (bs, 2H, H–C=N). ^{13}C NMR ($CDCl_3$, 100 MHz): 63.7, 70.1, 128.75, 129.1, 131.2, 137.0, 162.2. Elem. anal. calcd. For $C_{17}H_{18}N_2O$: C 76.69%; H 6.77%; N 10.53%, O 6.02% found C 73.00%; H 6.76%; N 10.85%.

N,N'–bis(o–hydroxybenzylidene)–1,3–diamino–2–propanol ***(10n)***: Yellow solid, mp 131–132 °C, yield 94.4%; ^{1}H NMR ($CDCl_3$, 400 MHz): 3.73 (dd, 2H, J = 6.4 Hz, J = 12.4 Hz, CH_2–2 or CH_2–4), 3.87 (dd, 2H, J = 4.4 Hz, J = 12.4 Hz, CH_2–2 or CH_2–4), 4.24–4.30 (m, 1H, H–C3), 6.89 (t, 2H, J = 7.2 Hz, J = 7.6 Hz Ar–H7'), 6.97 (d, 2H, J = 8.4 Hz, Ar–H5'), 7.28 (d, 2H, J = 1.2 Hz, Ar–H8'), 7.33 (t, 2H, J = 1.6 Hz, J = 7.8 Hz, Ar–H6') 8.41 (s, 2H, H–C=N), 13.11 (bs, 1H, OH). ^{13}C NMR ($CDCl_3$, 100 MHz): 63.47, 70.91, 117.57, 119.22, 119.33 132.11, 133.13, 161.92, 168.20. Elem. anal. calcd. For $C_{17}H_{18}N_2O_3$: C 68.46%; H 6.04%; N 9.40%, O 16.11% found C 67.53%; H 5.97%; N 9.21%.

N,N'–bis(p–methoxybenzylidene)–1,3–diamino–2–propanol ***(10o)***: White solid, mp 129–130 °C, yield 98.0%; ^{1}H NMR ($CDCl_3$, 400 MHz): 3.71 (ddd, 2H, J = 1.2 Hz, J = 6.4, 6.8 Hz, J = 12.2 Hz, CH_2–2 or CH_2–4), 3.81 (ddd, 2H, J = 1.2, Hz, J = 4.8 Hz, J = 12.4 Hz, CH_2–2 or CH_2–4), 3.84 (s, 6H, OCH_3–C6'), 4.17–4.23 (m, 1H, H–C3), 6.92 (d, J = 8.8 Hz, Ar–H), 7.69 (d, J = 8.8 Hz, Ar–H), 8.27 (bs, 2H, H–C=N). ^{13}C NMR ($CDCl_3$, 100 MHz): 55.41, 64.98, 71.52, 114.56, 130.00, 130.07, 162.68, 162.71. Elem. anal. calcd. For $C_{19}H_{22}N_2O_3$: C 69.94%; H 6.75%; N 8.59%, O 14.72% found C 69.68%; H 6.71%; N 8.52%.

N,N'–bis(p–dimethylaminobenzylidene)–1,3–diamino–2–propanol ***(10p)***: Pale yellow solid, mp 142–143 °C, yield 83.4%; ^{1}H NMR ($CDCl_3$, 400 MHz): 3.00 (s, 12H, $N(CH_3)_2$), 3.67 (dd, 2H, J = 6.4 Hz, J = 11.6 Hz, CH_2–2 or CH_2–4), 3.77 (dd, 2H, J = 5.2 Hz, J = 11.6 Hz, CH_2–2 or CH_2–4), 4.15–4.21 (m, 1H, H–C3), 6.70 (d, 4H, J = 4.0 Hz, Ar–H5' or Ar–H7'), 7.61 (d, 4H, J = 8.4 Hz, Ar–H4' or Ar–H8'), 8.20 (s, 2H, H–C=N), 9.74 (s, 1H, OH). ^{13}C NMR ($CDCl_3$, 100 MHz): 40.11, 64.97, 71.61, 112.09, 130.10, 152.90 163.27, 190.96. Elem. anal. calcd. For $C_{21}H_{28}N_4O$: C 71.59%; H 7.96%; N 15.91%, O 4.55% found C 71.15%; H 7.92%; N 15.74%

2.3 General Procedure of the Reduction Reactions ***(11e-p)***

The diimine (0.10 mmol) and sodium borohydride (0.20 mmol) were ground together with an agate mortar, and then heated to between 307 and 361 K for 15 min. After cooling to room temperature, the residue was dissolved in ethanol 95% until it stopped bubbling, and it was then extract in chloroform. The organic layer was dried with (Na_2SO_4) and the solvent was removed under vacuum.

2.3.1 Characterization Data of the Compounds

N,N'–bis(benzyl)ethane–1,2–diamine ***(11e)***: Yellow oil, yield 82.0%; ^{1}H NMR ($CDCl_3$, 400 MHz): 1.75 (bs, 2H, NH)

2.74 (s, 4H, N–CH_2–), 3.87 (s, 4H, Ph–CH_2–N), 7.23–7.27 (m, 10H, Ar–H).

N,N'–bis(o–hydroxybenzyl)ethane–1,2–diamine ***(11f)***: Orange oil, yield 88.0%; ^{1}H NMR ($CDCl_3$, 400 MHz): 1.25 (bs, 1H), 2.85 (s, 4H, CH_2–CH_2), 3.99 (s, 4H, CH_2–Ar), 6.78 (td, 2H, J = 1.2 Hz; J = 7.2 Hz, Ar–H7'), 6.83 (dd, 2H, J = 1.2 Hz; J = 8.4 Hz, Ar–H5'), 6.98 (dd, 2H, J = 1.2 Hz; J = 7.4 Hz, Ar–H8'), 7.18 (td, 2H, J = 1.6 Hz; J = 7.6 Hz, Ar–H6').

N,N'–bis(p–methoxybenzyl)ethane–1,2–diamine ***(11g)***: Yellow solid, mp 119–120 °C, yield 83.1%; ^{1}H NMR ($CDCl_3$, 400 MHz): 2.65 (s, 2H, NH), 2.83 (s, 4H, CH_2–CH_2), 3.81 (s, 4H, CH_2–Ar), 3.76 (s, 6H, OCH_3), 6.87–6.92 (m, 4H, Ar–H), 7.21–7.25 (m, 4H, Ar–H).

N,N'–bis(p–dimethylaminobenzyl)ethane–1,2–diamine ***(11h)***: Orange oil, yield 59.0%; ^{1}H NMR ($CDCl_3$, 400 MHz): 2.26 (bs, 2H, NH), 2.77 (s, 4H, CH_2–CH_2), 2.93 (s, 12H, $(CH_3)_2$–N), 3.68 (s, 4H, CH_2–Ar), 6.70 (d, 4H, J = 8.4 Hz, Ar–H5'and Ar–H7'), 7.18 (d, 4H, J = 8.8 Hz, Ar–H4' and Ar–H8').

N,N'–bis(benzyl)propane–1,3–diamine ***(11i)***: Yellow oil, yield 67.0%; ^{1}H NMR ($CDCl_3$, 400 MHz): 1.66 (bs, 2H, NH), 1.73 (q, 2H, J = 6.8 Hz, CH_2–3), 2.71 (t, 4H, J = 6.8 Hz, CH_2–N), 3.78 (s, 4H, CH_2–Ar), 7.30–7.37 (m, 10H, H–Ar).

N,N'–bis(o–hydroxybenzyl)propane–1,3–diamine ***(11j)***: Yellow solid, mp 110–111°C, yield 48.0%; ^{1}H NMR ($CDCl_3$, 400 MHz): 1.25 (bs, 2H, NH), 1.78 (q, 2H, J = 7.2 Hz, CH_2–3), 2.75 (t, 4H, J = 7.2 Hz, CH_2–N), 3.99 (s, 4H, CH_2–Ar), 6.77 (td, 2H, J = 1.2 Hz, J = 7.4 Hz, Ar– H7'), 6.82 (dd, 2H, J = 0.8 Hz, J = 8.4 Hz, Ar– H5'), 6.98 (dd, 2H, J = 1.2 Hz, J = 7.6 Hz, Ar– H8'), 7.16 (td, 2H, J = 0.8, J = 7.6 Hz, Ar– H6').

N,N'–bis(p–methoxybenzyl)propane–1,3–diamine ***(11k)***: Yellow oil, yield 74.0%; ^{1}H NMR ($CDCl_3$, 400 MHz): 1.69 (bs, 2H, NH), 1.72–1.75 (m, 2H, CH_2–3), 2.68 (t, 4H, J = 13.6 Hz, CH_2–N), 3.71 (s, 4H, CH_2–Ar), 3.79 (s, 6H, CH_3–O), 6.85 (d, 4H, J = 8.4 Hz, Ar–H5' and Ar–H7'), 7.21 (d, 4H, J = 8.4 Hz, Ar–H4' and Ar–H8').

N,N'–bis(p–dimethylaminobenzyl)propane–1,3–diamine ***(11l)***: Yellow oil, yield 64.0%; ^{1}H NMR ($CDCl_3$, 400 MHz): 1.73 (m, 2H, CH_2–3), 2.04 (bs, 2H, NH), 2.69 (t, 4H, J = 6.8 Hz, CH_2–N), 2.93 (s, 12H, $(CH_3)_2$–N), 3.68 (s, 4H, CH_2–Ar), 6.70 (d, 4H, J = 8.8 Hz, Ar– H5' and Ar– H7'), 7.16 (d, 4H, J = 8.8 Hz, Ar– H4' and Ar– H8').

N,N'–bisbenzyl–1,3–diamino–2–propanol ***(11m)***: Yellow oil, yield 95.0%; ^{1}H NMR ($CDCl_3$, 400 MHz): 1.91 (bs, 2H, NH), 2.61 (dd, 2H, J = 7.6 Hz, J = 12.0 Hz, CH_2–2 or CH_2–4), 2.76 (dd, 2H, J = 3.6 Hz, J = 12.0 Hz, CH_2–2 or CH_2–4), 3.75–3.82 (m, 4H, CH_2–Ar), 3.83–3.84 (m 1H, H–C(OH)), 7.29–7.34 (m, 10H, H–Ar). ^{13}C NMR ($CDCl_3$, 100 MHz): 48.81, 49.58, 62.07, 128.31, 128.64, 129.99, 131.40. Elem. anal. calcd. For $C_{17}H_{22}N_2O$: C 75.55%; H 6.66%; N 10.37%, O 5.92%, found C 78.74%; H 8.66%; N 9.80%.

N,N'–bis(o–hydroxybenzyl)–1,3–diamino–2–propanol ***(11n)***: Lemon yellow oil, yield 86.0%; ^{1}H NMR ($CDCl_3$, 400 MHz): 2.63–2.74 (m, 4H, CH_2–2 and CH_2–4), 3.94–3.99 (m, 4H, CH_2–Ar), 4.03 (bs, 1H, H–C(OH)), 4.43 (bs, 2H, NH), 6.78 (t, 2H, J = 7.6 Hz, Ar–H7'), 6.81 (d, 2H, J = 8.0 Hz, Ar–H5'), 6.97 (dd, 2H, J = 1.0 Hz, J = 7.4 Hz, Ar–H8'), 7.16 (td, 2H, J = 1.6 Hz, J = 8.0 Hz, Ar–H6'). ^{13}C NMR ($CDCl_3$, 100 MHz): 51.19, 51.45, 68.28, 115.64, 118.56, 121.54, 127.86, 128.23, 157.39. Elem. anal. calcd. For $C_{17}H_{22}N_2O_3$: C 67.55%; H 7.28%; N 7.28%, O 15.89% , found C 56.94%; H 6.04%; N 7.18%.

N,N'–bis(p–methoxybenzyl)–1,3–diamino–2–propanol ***(11o)***: Yellow oil, yield 93.0%; ^{1}H NMR ($CDCl_3$, 400 MHz): 1.25 (bs, 2H, NH), 2.58 (dd, 2H, J = 7.6 Hz, J = 12.0 Hz, CH_2–2 or CH_2–4), 2.73 (dd, 2H, J = 3.6 Hz, J = 12.0 Hz, CH_2–2 or CH_2–4), 3.68–3.77 (m, 4H, CH_2–Ar), 3.79 (s, 6H, OCH_3), 3.80–3.84 (m, 1H, H–C(OH)), 6.85 (d, 4H, J = 8.8 Hz, Ar–H5' and Ar–H7'), 7.22 (d, 4H, J = 8.4 Hz, Ar–H4' and Ar–H8'). ^{13}C NMR ($CDCl_3$, 100 MHz): 51.65, 52.03, 54.17, 67.24, 113.11, 128.71, 131.24, 158.29. Elem. anal. calcd. For $C_{19}H_{26}N_2O_3$: C 69.09%; H 7.88%; N 8.48%, O 14.54% , found C 75.25%; H 8.61%; N 9.12%.

N,N'–bis(p–dimethylaminobenzyl)–1,3–diamino–2–propanol ***(11p)***: Brown oil, yield 85.0%; ^{1}H NMR ($CDCl_3$, 400 MHz): 2.23 (bs, 2H, NH), 2.60 (dd, 2H, J = 6.4 Hz, J = 12.0 Hz, CH_2–2 or CH_2–4), 2.81 (dd, 2H, J = 2.4 Hz, J = 12.0 Hz, CH_2–2 or CH_2–4), 2.93 (s, 12H, $N(CH_3)_2$), 3.67 (d, 2H, J = 12.8 Hz, CH_2–Ar), 3.75 (d, 2H, J = 12.8 Hz, CH_2–Ar), 3.86 (bs, 1H, H–C(OH)), 6.69 (d, 4H, J = 8.8 Hz, Ar–H5' and Ar–H7'), 7.17 (d, 4H, J = 8.8 Hz, Ar–H4' and Ar–H8'). ^{13}C NMR ($CDCl_3$, 100 MHz): 39.46, 39.52, 64.33, 111.75, 111.86, 128.00, 129.38, 149.44. Elem. anal. calcd. For $C_{21}H_{32}N_4O$: C 70.68%; H 8.97%; N 15.70%, O 4.49%, found C 67.04%; H 8.30%; N 12.93%.

3. Results and Discussions

As shown in Scheme 1, we quickly and efficiently synthesized a series of diimine derivatives (**10e-p**) by condensing some aromatic aldehydes (**9a-d**) and diamines –namely ethane-1,2-diamine (**6**), propane-1,3-diamine (**7**) and 1,3-diamino-2-propanol (**8**)– using water as the reaction medium at room temperature. Compared to current procedures (Patil & Adimurthy, 2013), most of which employ various catalysts or use harsh conditions or vigorous drying of the solvents, this procedure is less hazardous, non-toxic, cheap, and benign to the environment. We found that the reactions occur smoothly, producing di-Schiff bases with yields of 63–100% (Table 1). The reaction times were short (12 h). All di-Schiff bases (except **10i**, which was viscous) were solids isolated by simple filtration without further purification. In methods used to prepare di-Schiff bases such as **4** and **5**, the amine is generally reacted with a carbonyl source catalyzed by protic or Lewis acids and a dehydrating agent; for example, the mixture is refluxed in heptane in the presence of acetic acid, or an azeotrope of the mixture with benzene is processed in a Dean-Stark apparatus in the presence of sulfuric acid (H_2SO_4) (Love, Boston, Nguyen, & Rorer, 1999; Ryabukhin et al., 2012). Our method is advantageous because the reactions occur very efficiently in water, and neither an acid catalyst nor an aromatic solvent are needed. The structures of di-Schiff bases **10e-p** were established with spectral data. IR spectra of the corresponding condensed product (**10e-p**) displayed the characteristic absorption bands within 1613-1636 cm^{-1} due to the stretching of the C=N bond. In the ^{1}H NMR spectra of **10e-p**, the aromatic hydrogens of the phenylene ring were observed as multiplets at δ 6.39-7.90. The CH=N hydrogen resonated as a singlet at δ 8.40-8.72. ^{13}C NMR chemical shifts for compounds **10e-p** were observed in their expected regions.

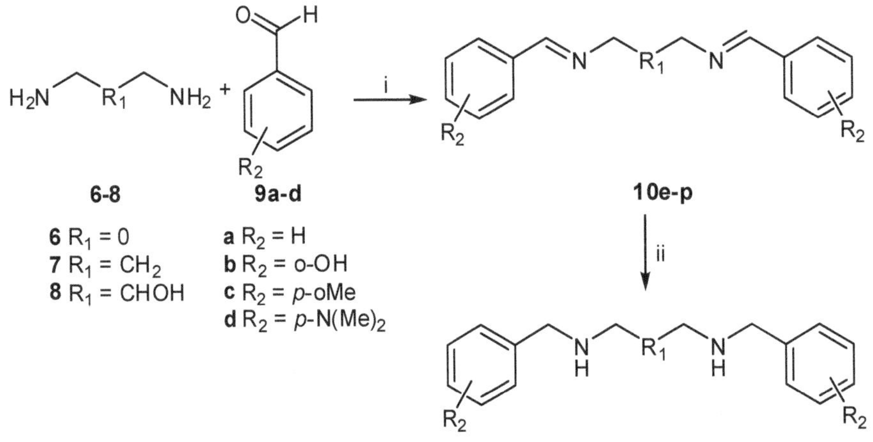

11e-p

Scheme 1. *Reagents and conditions*: (i) water at room temperature for 3 h; (ii); $NaBH_4$ ground gently at room temperature for 10 min

Table 1. Synthesis of **10e-p** and **11e-p** using green chemistry procedures

Entry	diamine	aldehyde	diimine	Yield (%)	product	Yield (%)
1	6	9a	10e	100.0	11e	82.0
2	6	9b	10f	88.0	11f	88.0
3	6	9c	10g	95.0	11g	83.1
4	6	9d	10h	86.0	11h	59.0
5	7	9a	10i	82.4	11i	67.0
6	7	9b	10j	81.0	11j	48.0
7	7	9c	10k	63.4	11k	74.0
8	7	9d	10l	75.0	11l	64.0
9	8	9a	10m	74.0	11m	95.0
10	8	9b	10n	94.4	11n	86.0
11	8	9c	10o	98.0	11o	93.0
12	8	9d	10p	83.4	11p	85.0

Reduction of the di-Schiff bases **10e-p** by sodium borohydride under catalyst-free and solvent-free conditions produced tetrahydro-di-Schiff bases **11e-p**. The direct solid-solid reductive aminations were carried out by grinding the corresponding di-Schiff base and sodium borohydride with an agate mortar and pestle at room temperature. Then, the reactions were conducted by heating at 60-90°C on a hot plate and were complete within 15 minutes, producing compounds **11e-p** in 48–95% yields. The reactions produced an oily residue, which was extracted with cold chloroform. From the perspective of green chemistry, it is preferable not to use solvents containing halogens. Therefore, we tried to use solvents such as ethyl ether that do not contain halogen atoms, but our attempts were unsuccessful. Chloroform facilitates isolation of the product. Removal of the solvent produced tetrahydro-di-Schiff bases **11e-p** as viscous liquids needing no further purification in most cases. This technique has an easy procedure and a short reaction time, and it does not require significant effort to isolate the products with a high-percentage yield. The structures of products **11e-p** were established by IR and NMR experiments. The preparation of 1,3-bis((E)-benzylideneamino)propan-2-ol **11m** is representative of the general procedure employed. Various temperatures, reaction times and reaction equivalents were explored until we obtained good isolated yields of **11m.** These experiments led to the use of a 1:2 ratio of di-Schiff bases (**10e-p**) with $NaBH_4$, heated at 60–90 °C for 15 min.

In conclusion, we have developed a facile, efficient and green method for the synthesis of 1,2 and 1,3-diimines by the condensation of aldehydes with 1,2- and 1,3-diamines in water. Then, in a second step, the resulting Schiff bases were treated with $NaBH_4$ to produce *N,N′*-dibenzyl-diamines (**1**) or *N,N′*-(2-hydroxybenzyl)-diamines (**2**) under solvent-free conditions. This reaction proceeds in the absence of a solvent and a catalyst. Compared to the previously reported methodologies, our protocol offers considerable benefits, including that it has a simple procedure, is environmentally benign, produces high yields, does not require the use of a catalyst, and allows for the product to be synthesized on the gram scale. Further studies regarding this protocol are ongoing in our laboratory.

Acknowledgments

We acknowledge the financial support provided to us by the Dirección de Investigaciones, Sede Bogotá (DIB) at the Universidad Nacional de Colombia through the research project No. 28427 (QUIPU Code 201010023412). I.M.-C. is also grateful to COLCIENCIAS for her doctoral scholarship.

References

Abdel-Magid, A. F. (2014). Reduction of C=N to CH-NH by metal hydrides. In P. Knochel & G. A. Molander (Eds): *Comprehensive Organic Synthesis* (2nd ed.) (Vol.8; pp 85–149). Amsterdam, Netherlands: Elsevier Ltd.

Alinezhad, H., & Tollabian, Z. (2010). One-pot reductive amination of carbonyl compounds using sodium borohydride-cellulose sulfuric acid. *Bulletin of the Korean Chemical Society*, *31*(7), 1927-1930. http://dx.doi.org/10.5012/bkcs.2010.31.7.1927

Alinezhad, H., Tajbakhsh, M., & Mahdavi, N. (2010). One-pot reductive amination of carbonyl compounds using sodium borohydride-Amberlyst 15. *Synthetic Communications*, *40*(7), 951–956. http://dx.doi.org/10.1080/00397910903026731

Billman, J. H., & Dorman, L. C. (1962). Hexahydropyrimidines I. Preparation of 2-substituted-1,3-bis(*p*-dimethylaminobenzyl)-hexahydropyrimidines. *Journal of Pharmaceutical Sciences*, *51*(11), 1071–1073. http://dx.doi.org/10.1002/jps.2600511113

Bisceglia, J. A., Díaz, J. E., Torres, R. A., & Orelli, L. R. (2011). 1,n-Diamines. Part 3: Microwave-assisted synthesis of N-acyl-N′-arylhexahydropyrimidines and hexahydro-1,3-diazepines. *Tetrahedron Letters*, *52*(41), 5238–5240. http://dx.doi.org/10.1016/j.tetlet.2011.07.131

Bordbar, M., Faal, A. Y., Ahari–Mostafavi, M., Graragozlou, M., & Fazaeli, R. K. (2013). Multi-wavelength spectrophotometric determination of acidity constants of some salicylaldimine derivatives. *Journal of Molecular Liquids*, *178*, 70–77. http://dx.doi.org/10.1016/j.molliq.2012.10.039

Cho, B. T., & Kung, S. K. (2005). Direct and indirect reductive amination of aldehydes and ketones with solid acid-activated sodium borohydride under solvent-free conditions. *Tetrahedron*, *61*(24), 5725–5734. http://dx.doi.org/10.1016/j.tet.2005.04.039

Frost, A. E., & Freedman, H. H. (1959). Addition of hydrogen cyanide to aromatic Schiff bases. *Journal of Organic Chemistry*, *24*(12), 1905-1907. http://dx.doi.org/10.1021/jo01094a021

Hasanloie S. T., & Setamdideh, D. (2014). $NaBH_4$/C: A convenient system for reductive amination of aldehydes. *Oriental Journal of Chemistry*, *30*(1), 341-344. http://dx.doi.org/10.13005/ojc/300145

Koteswara, R. V., Subba, R. S., Satheesh, K. B., Reddi, M., Naidu, K., Naga, R. C., & Ghosh, S. K. (2010). Synthesis of Schiff's bases in aqueous medium: a green alternative approach with effective mass yield and high reaction rates. *Green Chemistry Letters and Reviews*, *3*(3), 217-223. http://dx.doi.org/10.1080/17518251003716550

Love, B. E., Boston, T. S., Nguyen, B. T., & Rorer, J. R. (1999). A comparison of imine forming methodologies. *Organic Preparation and Procedures International*, *31*(4), 399–406. http://dx.doi.org/10.1080/00304949909355728

Mohammadi, N., & Setamdideh, D. (2015). $NaBH_4/PhCO_2H$: An efficient system for reductive amination of aldehydes. *Oriental Journal of Chemistry*, *31*(4), 2435-2437. http://dx.doi.org/10.13005/ojc/310476

Patil, R. D., & Adimurthy, S. (2013). Catalytic methods for imine synthesis. *Asian Journal of Organic Chemistry*, *2*(9), 726–744. http://dx.doi.org/10.1002/ajoc.201300012

Ryabukhin, S. V., Panov, D. M., Plaskon, A. S., Chuprina, A., Pipko, S. E., Tolmachev, A. A., & Shivanyuk, A. N. (2012). Combinatorial synthesis of chemical building blocks 1. Azomethines. *Molecular Diversity*, *16*(4), 625-637. http://dx.doi.org/10.1007/s11030-012-9407-9

Saeed, B. A., & Musad, I. A. (2009). Synthesis of symmetrical and non-symmetrical diimines from dimedone.

Molecules, *14*(6), 2278-2285. http://dx.doi.org/10.3390/molecules14062278

Setamdideh, D., & Sepehraddin, F. (2014). Convenient reductive amination of aldehydes by $NaBH_4$/Cation exchange resin. *Journal of the Mexican Chemical Society*, *58*(1), 22-26. Retrieved from http://www.scielo.org.mx/scielo.php?script=sci_arttext&pid=S1870-249X2014000100005&lng=es

Simion, A., Simion, C., Kanda, T., Nagashima, S., Mitoma, Y., Yamada, T.,...Tashiro, M. (2001). Synthesis of imines, diimines and macrocyclic diimines as possible ligands, in aqueous solution. *Journal of the Chemical Society, Perkin Transactions, 1*, 2071-2078. http://dx.doi.org/10.1039/B102749M

Tanaka, K., & Toda, F. (2000). Solvent-free organic synthesis. *Chemical Reviews*, *100*(3), 1025-1074. http://dx.doi.org/10.1021/cr940089p

Ünaleroğlu, C., Temelli, B., & Hökelek, T. (2002). A facile synthesis and characterization of the novel 2,2-dichloro-1-[3-(2,2-dichloro-acetyl)-2-(4-methoxy-phenyl)-imidazolidin-1-yl]-ethanone. *Synthetic Communications*, *32*(21), 3255–3261. http://dx.doi.org/10.1081/SCC-120014029

Wang, C. C., Tang, Y., Pan, Y. H., Yang, J., & Zhang, Y. M. (2015). A novel one-pot synthesis of vicinal diamines with benzaldehyde and aniline derivatives mediated by $BiCl_3$/Zn. *Tetrahedron Letters*, *56*(22), 2863–2866. http://dx.doi.org/10.1016/j.tetlet.2015.04.067

10

The Main Group Elements, Fragments, Compounds and Clusters Obey the 4n Rule and Form 4n Series: They are Close relatives to Transition Metal Counterparts via the 14n Linkage

Enos Masheija Rwantale Kiremire

Correspondence: Enos Masheija Rwantale Kiremire, Department of Chemistry and Biochemistry, University of Namibia, Private Bag 13301, Windhoek, Namibia. Email: kiremire15@yahoo.com

Abstract

The paper presents numbers which were derived from 4n-based series in a matrix table. The numbers agree precisely with the total number of valence electrons surrounding the respective skeletal elements. The series approach focuses mainly on the number of skeletal elements and their respective number of valence electron content regardless of the origin of the electrons and the type of skeletal elements. For instance, any 6 skeletal elements of transition metal carbonyls surrounded by 86 valence electrons coded as (6,86), series S = 14n+2 normally adopt an octahedral geometry whereas (6,26) series S = 4n+2 for main group elements also tend to adopt an octahedral shape. The transition metal carbonyl cluster series were extensively covered in our previous articles. This paper demonstrates that the main group fragments, clusters and molecules which we normally explain by terms such as valency, valence electrons and octet rule also obey the 4n-based series. The fragments, molecules and clusters of the main group elements correspond well to those of respective transition metal clusters especially the carbonyls if the masking electrons are removed from them. Hence, the series approach is a qualitative method that acts as a unifier of some transition metal clusters with some main group elements clusters.

Keywords: unifying cluster series, fragments, Isolobal fragments and series, valence electron content, valence matrix table, skeletal elements, octet and eighteen electron rules, k-values, multiple bonds, coding system, counting bonds using series

1. Introduction

The discovery of Isolobal principle was a milestone in the history of chemical sciences (Hoffmann, 1982). The principle ably links the concepts of organic, inorganic and organometallic chemistry (Hoffmann, 1982) by using chemical fragments. The fragments are extremely useful in explaining numerous carbonyl and borane shapes of clusters which have attracted the attention of scientists for many years(Lipscomb, 1963, Wade, 1971,1976, Rudolph, 1976, Pauling, 1977, Jensen, 1978, Cotton, et al, 1980, Mingos,1972, 1984, Douglas, et al, 1994, King, 2002, Jemmis, 2001a, 2001b,2002, 2003, 2005,2006, 2008, Bowen, et, al, 2007, King, et al, 2009, Welch, 2013). It was recently discovered that transition metal carbonyl clusters could readily be categorized using a matrix table (Kiremire, 2016a, 2016b) derived from 14n-based series. What is interesting is that the valence electron content of a given carbonyl cluster, correspond precisely with a number that is found in the matrix table or its extension. For instance, the set of numbers (6, 86) represents six skeletal atoms of a cluster surrounded by a total of 86 valence electrons from the skeletal elements, cluster ligands and the charge if present. As we know, there is a wide range of clusters that that qualify for this code including $M_6(CO)_{16}$, M = Co, Rh and Ir; $Os_6(CO)_{18}^{2-}$, $Ru_6(C)(CO)_{17}$, and $Co_6(CO)_{15}^{2-}$. Using the code notation (6, 86), it becomes easy to categorize the clusters by first removing the 'backbone electrons' of the series, 6x14 = 84 from the total 86. The remaining 2 electrons are added for balancing the electron valence content. This can be formulated as S = 14n+2 = 86 where n = 6. This set of numbers represents CLOSO series. Furthermore, a carbonyl cluster of 6 skeletal atoms surrounded by 86 valence electrons normally adopt an octahedral shape (O_h)(Cotton and Wilkinson, 1980). It has been established that the carbonyl clusters follow the 14n±q series while those of the main group elements follow the parallel series 4n±q where q is an even integer (Kiremire, 2015). In the previous work, emphasis was placed mainly on carbonyl clusters. In this paper, the focus will be placed mainly on clusters of the main group elements covering the range of small to large clusters. An attempt will be made to correlate them with the transition metal carbonyls where necessary.

2. Results and Discussion

Derivation of Numbers in the Matrix Table 1.

Since use of series to categorize clusters and fragments is not well known, it is important to explain briefly their origins and applications. It was recently found that categorization of carbonyl clusters is readily done using 14n-based series. A sample of the series is represented in the matrix form shown in Table 1. Let us consider the series S = 14n+0, when n = 1, S value is 14, when n= 1, S =14+2 = 14; n =1, and S = 14n+4 =18, and n =1, S =14n+6 = 20 and so on. These numbers are given in first row of Table 1. The numbers of second row (n =2) of the Table and other rows are generated similarly. The hypothetical numbers in Table 1 have been derived from cluster series covering the range S = 14n-2 to 14n+12. For transition metal carbonyl series, most non-capped clusters lie within the series S= 14n+2 (CLOSO), 14n+4 (NIDO), 14n+6 (ARACHNO), 14n+8(HYPHO), and 14n+10 (KLAPO). Although the capping range stops at S= 14n-12, a large number of palladium carbonyls such as $Pd_{23}(CO)_{20}L_{10}$(L = PMe_2Ph), S = 14n-32, $C^{17}C$[M-6] and $Pd_{38}(CO)_{28}L_{12}$, S = 14n-52, $C^{27}C$[M-11] fall in the range of cluster series far below S = 14n-12 (Kiremire, 2015, 2016). The corresponding numerical numbers based upon 4n series for main group elements is shown in Table 2 which is the counterpart of Table 1 for transition metal carbonyl clusters. Table 3 gives a few selected examples of metal carbonyl complexes which belong to different family members (family branches) of the 'clan series' S = 14n+2 (CLOSO), 14n+4(NIDO) and 14n+6 (ARACHNO). Table 4 is the counterpart of Table 3 and gives a wide range of selected fragments of different families of the clan series S = 4n+0 (MONO-CAPPED), 4n+2(CLOSO), 4n+4(NIDO), 4n+6 (ARACHNO) up to S = 4n+10. Table 4 covers simple two skeletal element clusters such as C_2 CN^+ to more complex seven skeletal element clusters such as As_7^{3-} (Zintyl ion) and Sb_7^{3-}. A comparison between 14n+4 and 4n+4 for n values 1 to 10 giving selected example of each n-value to underpin the isolobal corresponding relationship between family members for transition metal fragments and main group elements. This is given in Table 5. The table also indicates that fragments which correspond to similar families have the same k values. For instance, $Rh_2(Cp)_2(CO)_2$ (S = 14n+4, k = 2) and C_2H_4 (S = 4n+4, k =2). This means we have a Rh=Rh double bond in the former and another double bond in the later C=C in the latter. This study clearly indicates that the compounds and fragments of the main group elements tend to obey the law of series based upon 4n skeletal backbone while the transition metal clusters especially the carbonyl series obey the law of series based upon 14n skeletal backbone.

The 18-Electron Rule and the Cluster Series Trees.

The 18 electron rule which is extremely useful and has been widely used (Tolman, 1972; Jensen 2005) as can be seen from Table 1 belongs to the series S = 14n +4(NIDO). The 18 valence electron content occurs when n =1 is inserted into the formula S = 14n+4. This means that in this case, each cluster fragment has one skeletal element surrounded by a total of 18 valence electrons. This figure of 18 is just one slot or box in the first row (n = 1) in Table 1. A large number of transition metal complexes especially carbonyls fall into this box numbered 18. These include among others, $M(CO)_5$, M = Fe, Ru, Os; $Ni(CO)_4$,$Cr(C_6H_6)_2$, $M(C_5H_5)_2$, Fe,Ru, Os; $M(CO)_6$,M =Cr, Mo, W, $(C_4H_4)Fe(CO)_3$, $V(CO)_6^-$, $Mn(CO)_6^+$,$Fe(CO)_4^{2-}$,$(Cp)_2Ti(CO)_2$,$Mn(CO)_6^+$,$Fe(CO)_4^{2-}$,$(Cp)_2Ti(CO)_2$,$(C_4H_6)Fe(CO)_3$,$(C_6H_6)Cr(CO)_3$,$(Cp)Co(CO)_2$,(Cp)$V(CO)_4$,and $M(NH3)_6^{3+}$, M=Co, Rh, Ir. If we regard n = 1 for S = 14n+4 as a branch of a large tree whose base is S = 14n+4, and n =2 as a second branch and n =3 as a third branch, and so on, then we can have a good understanding what the series represent. We can regard the tree as representing a CLAN TREE of cluster series with branches representing families with individual clusters as family members of that branch. Hence, the above complexes may be regarded as members of the cluster family of the first branch (n =1) for the 'clan series' S = 14n+ 4 (NIDO). Sketches of the cluster trees are shown below T-1 for S= 14n+4, T-2 for S = 14n+2 and T-3 for S = 14n+6. More clan trees of other series can be designed in the similar manner to include series for main group elements in T-4 and T-5 as well as the selected families for bi-capped series T-6, mono-capped series T-7, closo series T-8, nido series T-9 and arachno series T-10. As an illustration of the origin of the clan tree of series, let us consider the NIDO series S = 14n+4 (T-1). The numbers 18(n=1), 32(n=2), 46(n=3), 60(n=4) and 74(n=5) are simply extracted from the Matrix Table 1 under the column labeled S = 14n+4.The numerical values are simply obtained by substituting the value of n into the series formula = 14n+4.The sequence of the numbers is such that from one number to the next, there is a difference of 14. This is clearly a result of the change in n value by on since the series formula S = 14n+4 remains constant. The significance of these numbers can be illustrated by S = 14n+4 = 18 when n =1. The value S = 18 for n =1 represents a large number of transition metal complexes which have one central metal atom (skeletal element) surrounded by a total of 18 valence electrons. THIS IS THE BASIS OF THE 18 ELECTRON RULE which was put forward by Langmuir (Langmuir, 1921). Examples of complexes which obey the 18 electron rule are many including $Fe(CO)_5$, $Ni(CO)_4$, $Fe(\eta^5\text{-}C_5H_5)_2$, and $Co(NH_3)_6^{3+}$. The selected numbers shown in other clan trees (T-2 to T-10) of series were also extracted from Table 1 in a similar manner.

The Applications of Series to Predict Possible Shapes and Linkages of Molecules and Clusters

The carbonyl clusters of low nuclearity, there is a correlation between the k-value and the linkages or bonds. For example, $Rh_2(Cp)_2(CO)_2$, S = 14n+4 ≡4n+4, k = 2n-2 = 2(2)-2 = 2. A sketch of the complex is shown in S-1 indicating the presence of a double bond. Similarly $Mo_2(Cp)_2(CO)_4$, S = 14n+2 ≡4n+2, k = 2n-1 = 2(2)-1 = 3. This means, the complex is expected to have an Mo≡Mo triple bond (Meissler, et al, 2014). A sketch of the triple bond is shown in S-2 as well as other multiple bonds shown in S-3 and S-4 for $Mo(Cp)(CO)_2GeR$ and $Ga_2R_2^{2-}$(Robinson, et al,1997, 1998). We can also predict the skeletal shape of a cluster based on the value of n and k value as presented in the cluster trees. For instance, k = 2 for n = 2 in T-1. This means two skeletal atoms are linked with a double bond as shown in S-1. For n = 2 and k = 3, this gives us a triple bond as in S-2 while in T-3, k = 1 for n = 2 giving us a single bond as in $Mn_2(CO)_{10}$ complex. For the same series T-1 (14n+4), when n = 4, k = 6. The characteristic shape for this is a tetrahedral (T_d) as sketched in S-5. For n = 5 and k = 8 for S = 4n+4, the characteristic shape is a square pyramid sketched in S-6 and k =9 for M-5 gives a trigonal bipyramid as a characteristic shape (S-7) for S = 14n+2 closo series and finally for k = 11 and n = 6 for S = 14n+2, the characteristic shape is O_h(S-8).

Derivation of Cluster Formulas from Cluster Series Formula

The cluster series formula carries some important information. The formula gives us the valence electron content of a cluster formula in question. It also gives the cluster number k which is related to cluster linkages and for ordinary molecules the k value is the same as the number of bonds linking up the skeletal atoms. For instance k = 3 for N_2 means that the two nitrogen atoms are linked with a triple bond. Furthermore, a knowledge of the series can assist us derive the appropriate cluster formula of the fragment for a given number of skeletal atoms. Let us consider a specific example as an illustration. Let the series be S = 14n+6 (ARACHNO) and n = 2. The figure [14] in the series formula represents the backbone fragment electrons of the series, n = the number of skeletal atoms in the formula and 6 = the excess 'determinant' electrons of the series after 14n. In this particular case, 6 tells us we are dealing with ARACHNO series. If the determinant number was 2, we will be dealing with CLOSO series and so on. The following scheme (Scheme 1) summarizes how the appropriate carbonyl clusters of different transition metal atoms can be generated from the series formula S = 14n+6 for n = 2 for the generation of the $Mn_2(CO)_{10}$, $Fe_2(CO)_9$ and $Co_2(CO)_8$ carbonyls. In Scheme 2, it is demonstrated how the carbonyl clusters, $Fe_2(Cp)_2(CO)_4$, $Rh_4(CO)_{12}$,$Rh_6(CO)_{16}$ and the main group element clusters $B_6H_6^{2-}$, and the hypothetical Zintyl ion type of clusters M_6^{2-}, M = C, Si, Ge,Sn, and Pb can derived from a series formula.

Keeping the Number of Skeletal Elements Constant and Varying the Number of Valence Electron Content

Let us consider keeping the number of skeletal elements constant and changing the number of valence electron around them. This is simply illustrated in Table 6. As indicated in Table 6, two skeletal elements have been selected. If the skeletal atoms are transition metals in carbonyl clusters are surrounded by 28 electrons (S = 14n+0, n = 2), then k = 2n-0 = 2(2)-0 = 4. The increase in valence electrons by 2 from 28 to 30 changes the series to S = 14n+2 with k = 2n-1 = 2(2)-1 = 3. A good example for this is $Mo_2(Cp)_2(CO)_4$ which has an Mo≡ Mo triple bond(Meissler, et al, 2014).When the valence electrons are increased again by 2 to S = 32 the series formula becomes S = 14n+4 with k= 2n-2 = 2(2)-2 = 2. This means that the two skeletal atoms get linked up by a double bond. This is what is indeed observed in $Rh_2(Cp)_2(CO)_2$ carbonyl cluster complex where a Rh=Rh bond is present(Meissler, et al, 2014). Further increase of the valence electrons by 2 the series formula becomes S = 14n+6 and k = 2n-3 = 2(2)-3 = 1. Thus, the bond order has decreased further to 1. The linkage between the two skeletal atoms is one. This is what is observed in $Mn_2(CO)_{10}$ where Mn—Mn single bond is present(Meissler, et al, 2014). A similar variation in bond order is experienced in moving from C_2H_6(S= 4n+6 = 14, k =2n-3= 2(2)-3 = 1) where C—C single bond is known to C_2H_4(S= 4n+4 = 12, k =2n-2= 2(2)-2= 2) where we get a C=C double bond. Further decrease in valence electrons by 2 to C_2H_2(S= 4n+2 = 10, k =2n-1= 2(2)-1= 3) where we get a C≡C triple bond. Finally, when 2 more valence electrons are removed to produce C_2(S= 4n+0 = 12, k =2n-0= 2(2)-0= 0) where we get a controversial C≣C quadruple bond (Shaik, et al, 2012). Therefore, according to the series approach, if the number of skeletal atoms are kept constant and the number of valence electrons is increased, the k-values or skeletal linkages are decreased and vice versa. In principle, according to the series analysis, if we were to add 2 more electrons to $Mn_2(CO)_{10}$ (S = 14n+6, k = 1), we would get $[Mn_2(CO)_{10}]^{2-}$, S = 14n+8, k = 2n-4 = 2(2)-4 = 0. This implies that $[Mn_2(CO)_{10}]^{2-}$ is unstable and decomposes into $2[Mn(CO)_5]^-$ ions which belong to the series S = 14n+4 that obeys the 18 electron rule. In the same way, adding 2 electrons to C_2H_6(S = 4n+6, k =2n-3 = 2(2)-3 = 1) we would get $[C_2H_6]^{2-}$(S = 4n+8, k = 2n-4 = 2(2)-4 = 0).What this implies by having k = 0, is that the system$[C_2H_6]^{2-}$(k = 0) is unstable and it decomposes to form $2[CH_3]^-$ ions. These are equivalent to having 2 CH_4 molecules which obey the series S = 4n+4 and this is consistent with the octet rule.

Table 1. Matrix Numbers for Valence Electron Content for Transition Metals

14n-12	14n-10	14n-8	14n-6	14n-4	14n-2	14n+0	14n+2	14n+4	14n+6	14n+8	14n+10	14n+12	n
2	4	6	8	10	12	14	16	18	20	22	24	26	1
16	18	20	22	24	26	28	30	32	34	36	38	40	2
30	32	34	36	38	40	42	44	46	48	50	52	54	3
44	46	48	50	52	54	56	58	60	62	64	66	68	4
58	60	62	64	66	68	70	72	74	76	78	80	82	5
72	74	76	78	80	82	84	86	88	90	92	94	96	6
86	88	90	92	94	96	98	100	102	104	106	108	110	7
100	102	104	106	108	110	112	114	116	118	120	122	124	8
114	116	118	120	122	124	126	128	130	132	134	136	138	9
128	130	132	134	136	138	140	142	144	146	148	150	152	10
142	144	146	148	150	152	154	156	158	160	162	164	166	11
156	158	160	162	164	166	168	170	172	174	176	178	180	12
170	172	174	176	178	180	182	184	186	188	190	192	194	13
184	186	188	190	192	194	196	198	200	202	204	206	208	14
198	200	202	204	206	208	210	212	214	216	218	220	222	15
212	214	216	218	220	222	224	226	228	230	232	234	236	16
226	228	230	232	234	236	238	240	242	244	246	248	250	17
240	242	244	246	248	250	252	254	256	258	260	262	264	18
254	256	258	260	262	264	266	268	270	272	274	276	278	19
268	270	272	274	276	278	280	282	284	286	288	290	292	20
282	284	286	288	290	292	294	296	298	300	302	304	306	21
296	298	300	302	304	306	308	310	312	314	316	318	320	22
310	312	314	316	318	320	322	324	326	328	330	332	334	23
324	326	328	330	332	334	336	338	340	342	344	346	348	24
338	340	342	344	346	348	350	352	354	356	358	360	362	25
352	354	356	358	360	362	364	366	368	370	372	374	376	26
366	368	370	372	374	376	378	380	382	384	386	388	390	27
380	382	384	386	388	390	392	394	396	398	400	402	404	28
394	396	398	400	402	404	406	408	410	412	414	416	418	29
408	410	412	414	416	418	420	422	424	426	428	430	432	30

Table 2. Matrix Numbers for Valence Electron Content for Main Group Elements

4n-12	4n-10	4n-8	4n-6	4n-4	4n-2	4n+0	4n+2	4n+4	4n+6	4n+8	4n+10	4n+12	n
-8	-6	-4	-2	0	2	4	6	8	10	12	14	16	1
-4	-2	0	2	4	6	8	10	12	14	16	18	20	2
0	2	4	6	8	10	12	14	16	18	20	22	24	3
4	6	8	10	12	14	16	18	20	22	24	26	28	4
8	10	12	14	16	18	20	22	24	26	28	30	32	5
12	14	16	18	20	22	24	26	28	30	32	34	36	6
16	18	20	22	24	26	28	30	32	34	36	38	40	7
20	22	24	26	28	30	32	34	36	38	40	42	44	8
24	26	28	30	32	34	36	38	40	42	44	46	48	9
28	30	32	34	36	38	40	42	44	46	48	50	52	10
32	34	36	38	40	42	44	46	48	50	52	54	56	11
36	38	40	42	44	46	48	50	52	54	56	58	60	12
40	42	44	46	48	50	52	54	56	58	60	62	64	13
44	46	48	50	52	54	56	58	60	62	64	66	68	14
48	50	52	54	56	58	60	62	64	66	68	70	72	15
52	54	56	58	60	62	64	66	68	70	72	74	76	16
56	58	60	62	64	66	68	70	72	74	76	78	80	17
60	62	64	66	68	70	72	74	76	78	80	82	84	18
64	66	68	70	72	74	76	78	80	82	84	86	88	19
68	70	72	74	76	78	80	82	84	86	88	90	92	20
72	74	76	78	80	82	84	86	88	90	92	94	96	21
76	78	80	82	84	86	88	90	92	94	96	98	100	22
80	82	84	86	88	90	92	94	96	98	100	102	104	23
84	86	88	90	92	94	96	98	100	102	104	106	108	24
88	90	92	94	96	98	100	102	104	106	108	110	112	25
92	94	96	98	100	102	104	106	108	110	112	114	116	26
96	98	100	102	104	106	108	110	112	114	116	118	120	27
100	102	104	106	108	110	112	114	116	118	120	122	124	28
104	106	108	110	112	114	116	118	120	122	124	126	128	29
108	110	112	114	116	118	120	122	124	126	128	130	132	30

Table 3. Selected Examples conforming to Series of Transition Metals

SERIES(S) S = 14n+q	n VALUE	S-VALUE = V	CODE	EXAMPLES	k VALUE k = 2n-q/2
14n+2	1	16	(1,16)	$Os(CO)_4$, $RhCl(L)_3$, L= PPh_3, $Pt(H)L_2$, $PtCl(C_2H_4)^-$, $(Cp)_2TiCl_2$, $Ir(Cl)L_2(CO)$	1
14n+2	2	30		$Mo_2(Cp)_2(CO)_4$	3
14n+2	3	44	(3,44)	$Os_3(CO)_{10}$	5
14n+2	4	58	(4,58)	$Os_4(CO)_{13}$	7
14n+2	5	72	(5,72)	$Os_5(CO)_{16}$	9
14n+2	6	86	(6,86)	$Os_6(CO)_{18}^{2-}$, $Co_6(CO)_{16}$, $Rh_6(CO)_{16}$, $Ir_6(CO)_{16}$, $Rh_6(C)(CO)_{15}^{2-}$, $Re_6(C)(CO)_{19}^{2-}$	11
14n+2	7	100	(7,100)	$Os_7(H)_2(CO)_{21}$, $Rh_7(N)(CO)_{15}^{2-}$ $Co_7(N)(CO)_{15}^{2-}$	13
14n+4	1	18	(1,18)	$M(CO)_5$, M = Fe, Ru, Os; $Ni(CO)_4$,$Cr(C_6H_6)_2$, $M(C_5H_5)_2$, Fe,Ru, Os; $M(CO)_6$,M =Cr, Mo, W, $(C_4H_4)Fe(CO)_3$, $V(CO)_6^-$, $Mn(CO)_6^+$,$Fe(CO)_4^{2-}$,$(Cp)_2Ti(CO)_2$, $(C_4H_6)Fe(CO)_3$,$(C_6H_6)Cr(CO)_3$, $(Cp)Co(CO)_2$,$(Cp)V(CO)_4$,$M(NH3)_6^{3+}$, M=Co, Rh,Ir	0
14n+4	2	32	(2,32)	$Rh_2(Cp)_2(CO)_2$,$Re_2(Cp)_2(CO)_4$, $Re_2H_2(CO)_8$	2
14n+4	3	46	(3,46)	$Os_3H_2(CO)_{10}$,$Re_3H_3(CO)_{10}^{2-}$	4
14n+4	4	60	(4,60)	$Os_4(CO)_{14}$, $Os_4H_4(CO)_{12}$, $Co_4(CO)_{12}$, $Rh_4(CO)_{12}$, $Ir_4(CO)_{12}$, $Re_4H_6(CO)_{12}^{2-}$	6
14n+4	5	74	(5,74)	$Os_5H_2(CO)_{16}$, $Os_5(C)CO)_{15}$, $Os_5(C)CO)_{14}^{2-}$, $Os_5(C)CO)_{15}$, $Ru_5(C)CO)_{15}$, $Ru_5(N)CO)_{14}^-$,	8
14n+6	1	20	(1,20)	$Os(CO)_6$	-1
14n+6	2	34	(2,34)	$M_2(CO)_{10}$,M = Mn, Tc, Re $Fe_2(CO)_9$,$M_2(CO)_8$, M =Co, Rh, Ir; $Os_2(CO)_8^{2-}$,$Mo_2(Cp)_2(CO)_6$, $M_2(Cp)_2(CO)_4$,M =Fe,Os $Cr_2(Cp)_2(CO)_6$, $Cr_2(Cp)_2(NO)_4$, $FeCo(Cp)(CO)_6$	1
14n+6	3	48	(3,48)	$M_3(CO)_{12}$,M = Fe, Ru, Os $Re_3H_3(CO)_{12}$, $Os_3H_2(CO)_{11}$	3
14n+6	4	62	(4,62)	$Os_4(CO)_{15}$, $Re_4(CO)_{16}^{2-}$,	5
14n+6	5	76	(5,76)	$Os_5(CO)_{18}$, $Rh_5(CO)_{15}^-$, $Ni_5(CO)_{12}^-$, $Ni_3Cr_2(CO)_{16}^{2-}$, $Ni_3Mo_2(CO)_{16}^{2-}$, $Ni_3W_2(CO)_{16}^{2-}$, $RuIr_4(CO)_{15}^{2-}$, $PtRh_4(CO)_{14}^{2-}$	7
14n+6	6	90	(6,90)	$Co_6(C)(CO)_{15}^{2-}$, $Os_6(P)(CO)_{18}^-$, $Rh_6(C)(CO)_{15}^{2-}$, $Rh_6(N)(CO)_{15}^-$,	9

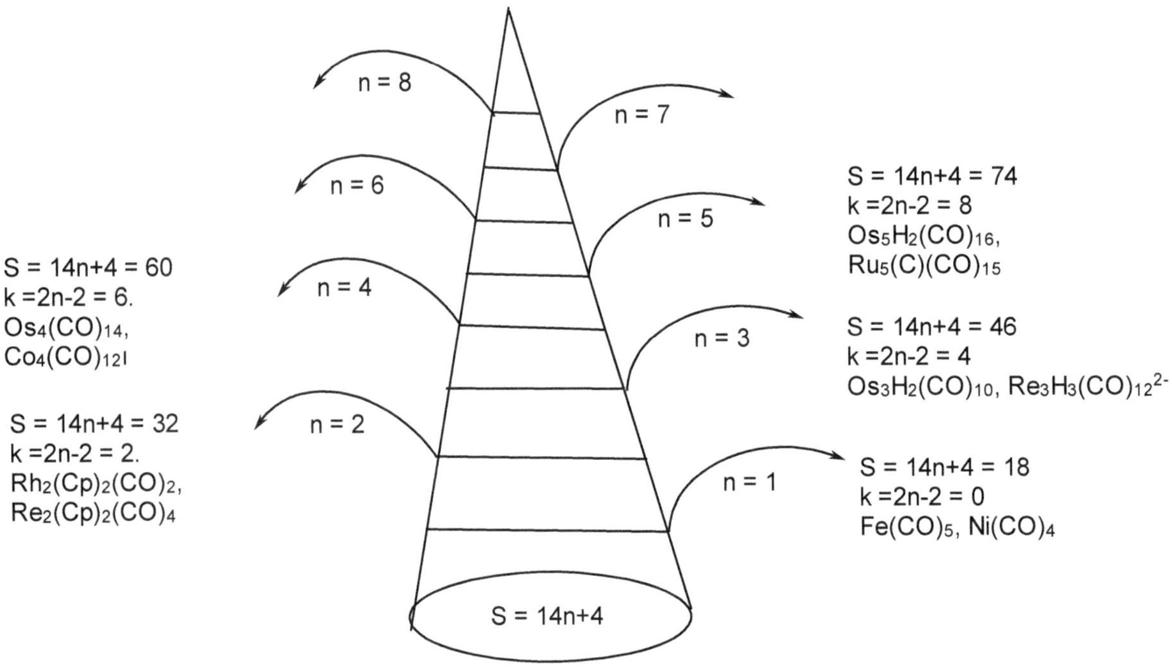

T-1. Sketch of Clan Tree of S = 14n+4 Cluster Series with some of its Family Cluster Branches

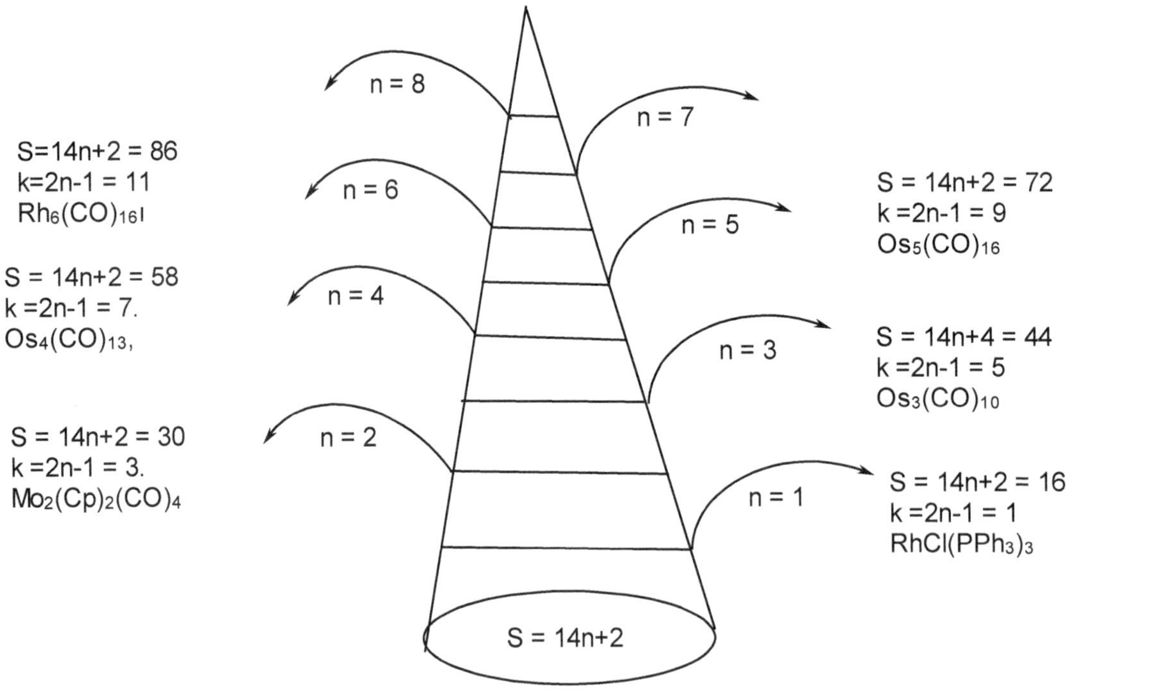

T-2. Sketch of Clan Tree of S = 14n+2 Cluster Series with some of its Family Cluster Branches

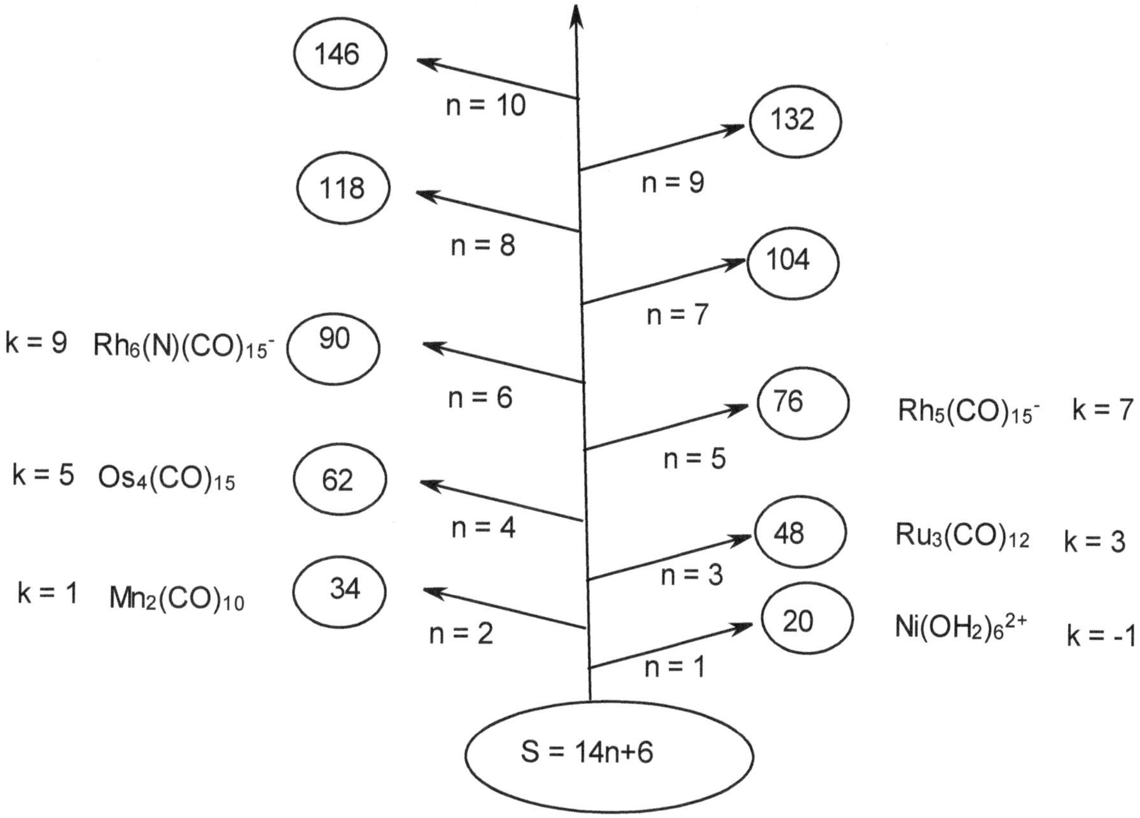

T-3 ARACHNO SERIES

1. $Rh_2(Cp)_2(CO)_2$

14+0
2[Rh(Cp)] ⟶ 4n+0

9+5
2CO ⟶ 0+4

n = 2 S = 4n+4

k = 2n-2 = 2(2)-2 = 2

SKELETAL SHAPE

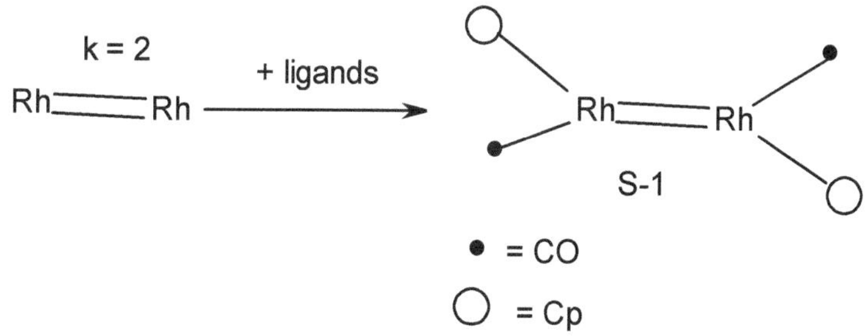

2. $Mo_2(Cp)_2(CO)_4$

SKELETAL SHAPE

14+1
$2[Mo(Cp)(CO)_2] \longrightarrow 4n+2$
6+5+4

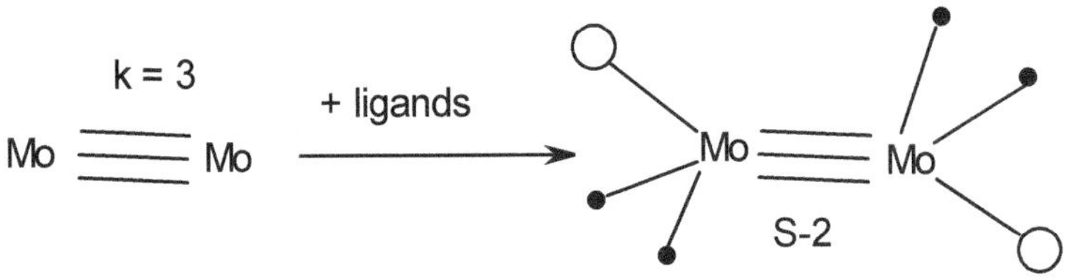

S-2

n = 2 S = 4n+2

k = 2n-1 = 2(2)-1 = 3

• = CO
○ = Cp

3. $Mo(Cp)(CO)_2GeR$

SKELETAL SHAPE

14+1
$1[Mo(Cp)(CO)_2] \longrightarrow 4n+1$
6+5+4

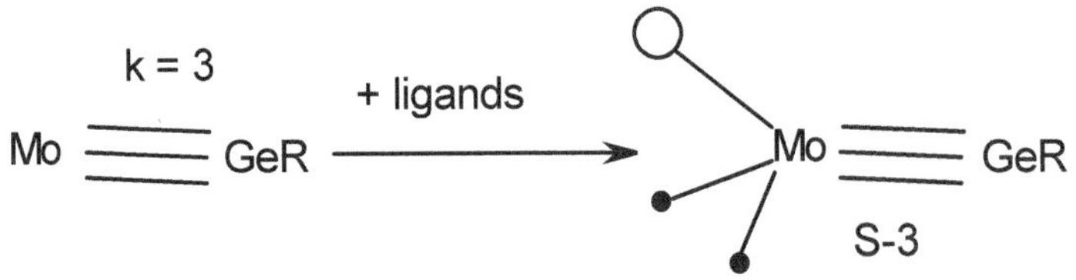

S-3

$1[GeR] \longrightarrow 4n+1$
4+1

• = CO
○ = Cp

n = 2 S = 4n+2

k = 2n-1 = 2(2)-1 = 3

4. $[Ga_2R_2]^{2-}$

4+0
$2[GaR] \longrightarrow 4n+0$
3+1
$q \longrightarrow 0+2$

q = CHARGE

S = 4n+2 k = 2n-1 = 2(2)-1 = 3

SKELETAL SHAPE

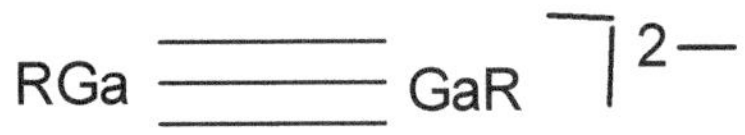

S-4

Tetrahedral, T_d

k =6, M-4

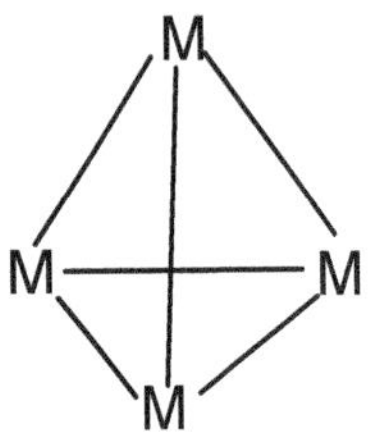

S-5

Square pyramid, C_{4v}

k = 8, M - 5

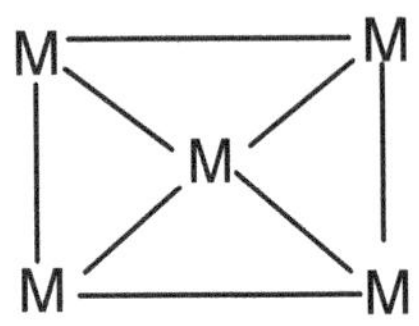

S-6

Trigonal bipyramid, D_{3h}

k = 9, M-5

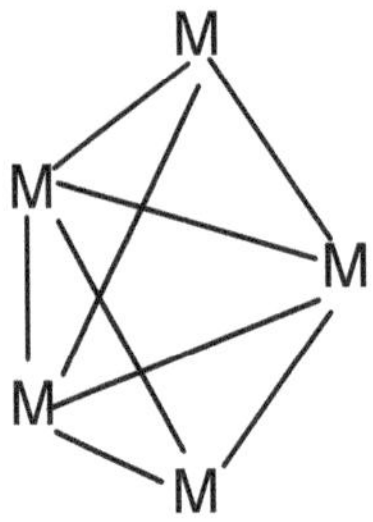

S-7

Octahedral, O_h

k = 11, M - 6

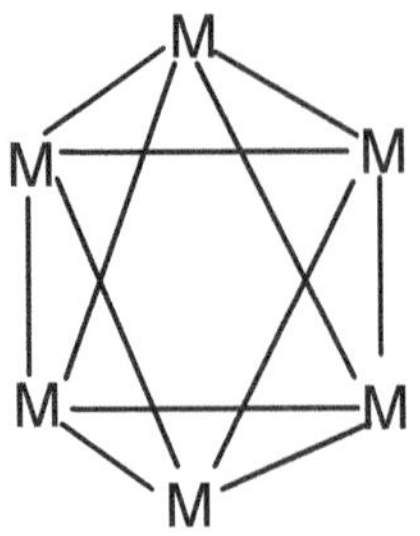

S-8

Table 4. Selected Examples of fragments conforming to Series of Main Group Elements

SERIES(S)	n VALUE	S-VALUE = V	CODE	EXAMPLES	k VALUE
S = 4n+q					k = 2n-q/2
4n+0	1	4	(1,4)	C, BH, BeX_2, X=H,F, Cl, Br, I	2
4n+0	2	8	(2,8)	C_2, B_2H_2, CN^+, CB^-	4
4n+2	1	6	(1,6)	BX_3, X = H,F, Cl, Br, I; O, NH, $SnCl_2$	1
4n+2	2	10	(2,10)	C_2H_2, CO, CN^-,N_2, NO^+, BNR_2	3
4n+2	5	22	(5,22)	$B_5H_5^{2-}$, M_5^{2-}, M = Sn, Pb Bi_5^{3+}, $C_2B_3H_5$	9
4n+2	6	26	(6,26)	$B_6H_6^{2-}$	11
4n+4	1	8	(1,8)	MX_4, M= C, Si, Ge, Sn, Pb; X =H,F, Cl, Br, I;H_2CO, HX, X = F, Cl, Br, I;H_2M, M = O,S, Se MX_4, M = C, Si, Ge, Sn, Pb; X = H, F, Cl, Br, I	0
4n+4	2	12	(2, 12)	C_2H_4, NO^-, O_2, S_2, B_2H_6, C_3	2
4n+4	3	16	(3, 16)	C_3H_4, BN_2^{3-}	4
4n+4	4	20	(4, 20)	M_4, M =N, P,As, Sb, Bi, $InBi_3^{2-}$, $GaBi_3^{2-}$	6
4n+4	5	24	(5,24)	B_5H_9, N_5^+	8
4n+6	1	10	(1,10)	PF_5, ClF_3, SO_2, POF_3, SF_4,SOF_2, SF_4, ICl_4^+, XeF_2	-1
4n+6	2	14	(2,14)	X_2, X = F,Cl, Br, I; C2H6, M_2X_4, M = N, P, As, Sb; X = H, F, Cl, Br, I; O_2^{2-}, S_2X_2, X =F,Cl,Br	1
4n+6	3	18	(3,18)	S_3, C_3H_6	3
4n+6	4	22	(4,22)	B_4H_{10}, C_4H_6, Te_4^{2+}	5
4n+6	6	30	(6,30)	C_6H_6, $B_3N_3H_6$, C_5H_5N	9

SERIES(S)	n VALUE	S-VALUE = V	CODE	EXAMPLES	k VALUE
S = 4n+q					k = 2n-q/2
4n+8	1	12	(1,12)	SF_6, ClF_5, XeF_4, BrF_6^+, ICl_4^-, ClO_3^-, XeF_3^+, XeO_2, XeF_5^+	-2
4n+8	3	20	(3,20)	C_3H_8, BrF_2^+	2
4n+8	4	24	(3,24)	C_4H_8	4
4n+10	1	14	(1,14)	IF_7, AsF_6^-, $XeOF_4$, XeF_5^-, XeF_6, XeO_2F_2	-3
4n+10	4	26	(2,26)	Se_4^{2-}, C_4H_{10}	3
4n+10	5	30	(5,30)	C_5H_{10}	5
4n+10	6	34	(6,34)	S_6^{2+}, $B_2N_4^{8-}$,C_6H_{10}	7
4n+10	7	38	(7,38)	M_7^{3-}, M = As, Sb; P_4S_3	9

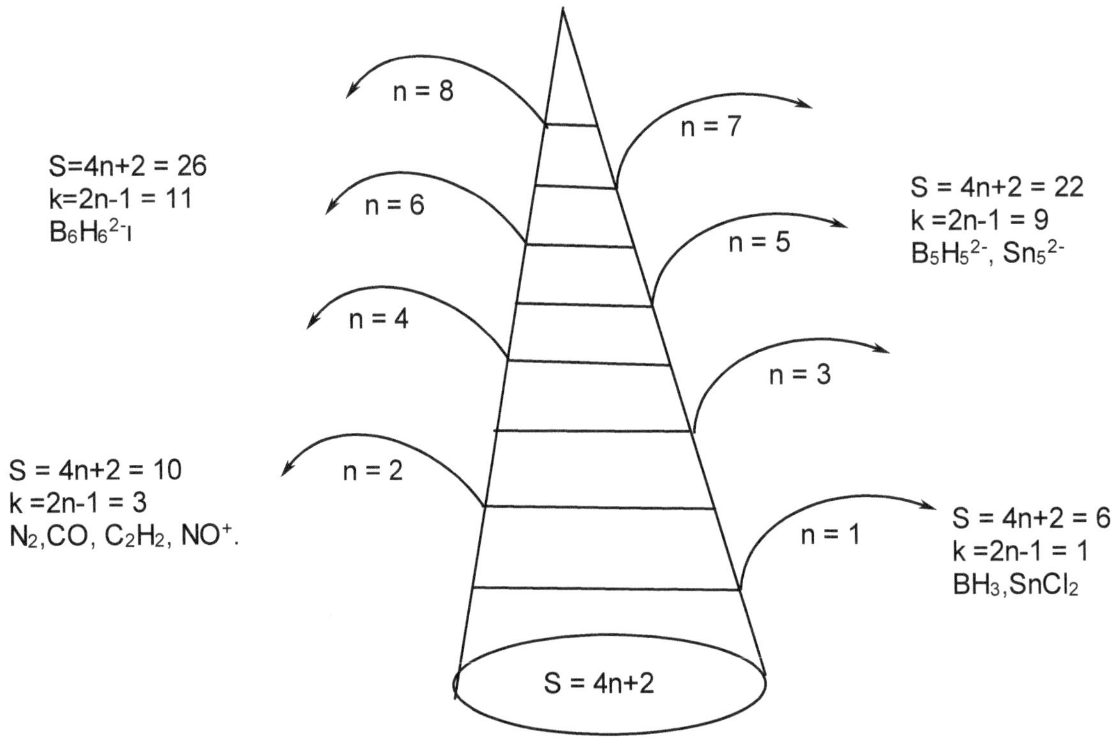

T-4. Sketch of Clan Tree of S = 4n+2 Cluster Series with some of its Family Cluster Branches

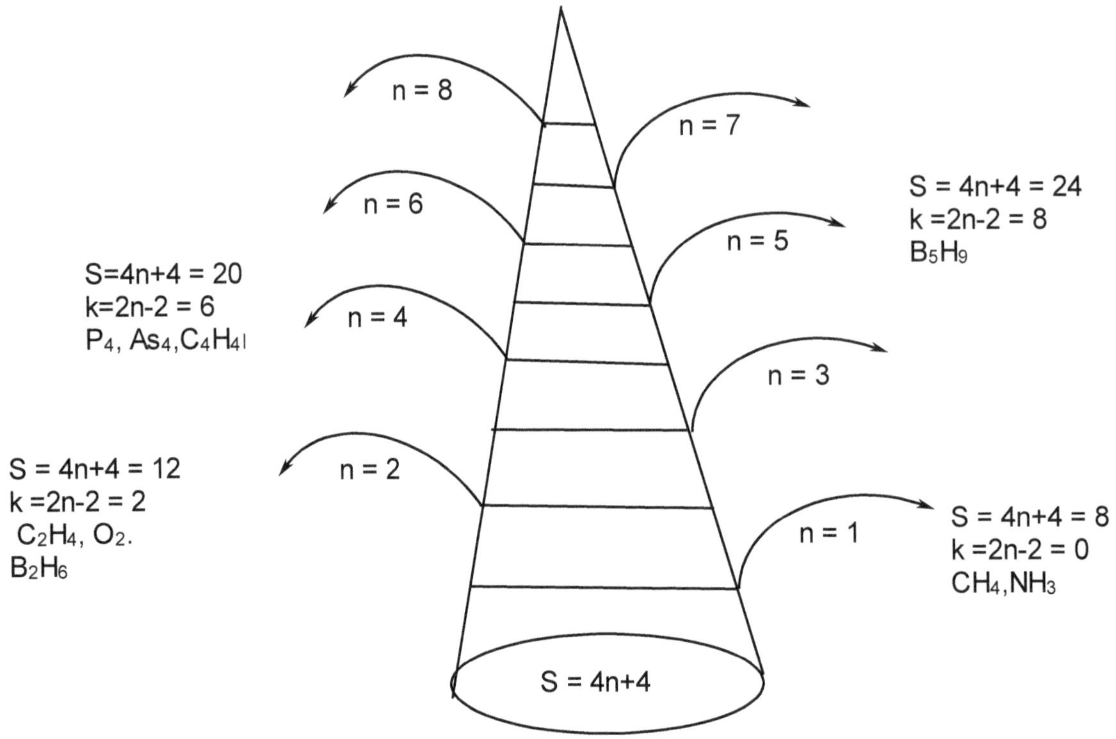

T-5. Sketch of Clan Tree of S = 4n+4 Cluster Series with some of its Family Cluster Branches

Table 5. The Variation of k value with for selected values of n for S = 14n+4 and S = 4n+4 series

n	14n+4	k value	4n+4	TM*	MG*
		k =2n-2		14n+4	4n+4
1	18	0	8	$Fe(CO)_5$	CH_4
2	32	2	12	$Rh_2(Cp)_2(CO)_2$	C_2H_4
3	46	4	16	$Re_3H_3(CO)_{12}^{2-}$	C_3H_4
4	60	6	20	$Ir_4(CO)_{12}$	C_4H_4, P_4
5	74	8	24	$Ru_5(C)(CO)_{15}$	B_5H_9
6	88	10	28	$Os_6H_2(CO)_{19}$	B_6H_{10}
7	102	12	32	$Os_6Pt(CO)_{18}H_8$	B_7H_{11}
8	116	14	36	$Os_8(CO)_{26}$	B_8H_{12}
9	130	16	40	$Co_9(P)(CO)_{21}^{2-}$ $Ni_9(C\ )(CO)_{17}^{2-}$	Ge_9^{4-}
10	144	18	44	$Os_{10}(CO)_{32}$	$B_{10}H_{14}$

Table 6. Sample to show the variation of k value with Valence Electrons(S-value) when n is constant

	S-value							S-value	
	V_T	Example	SERIES		k	k value		V_{MG}	Example
n =2	34	$Mn_2(CO)_{10}$	14+6	ARACHNO	2n-3	1	4n+6	14	C_2H_6
	32	$Rh_2(Cp)_2(CO)_2$	14n+4	NIDO	2n-2	2	4n+4	12	C_2H_4
	30	$Mo_2(Cp)_2(CO)_4$	14n+2	CLOSO	2n-1	3	4n+2	10	C_2H_2
	28	$Os_2(CO)_6$	14n+0	MONOCAP	2n-0	4	4n+0	8	C_2

TM = Transition metal cluster, MG = Main group element cluster.

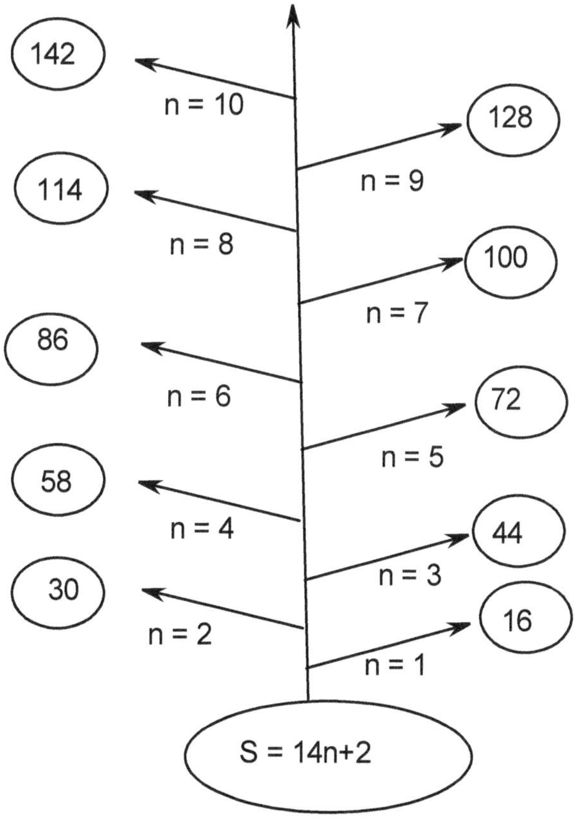

T-8 CLOSO SERIES

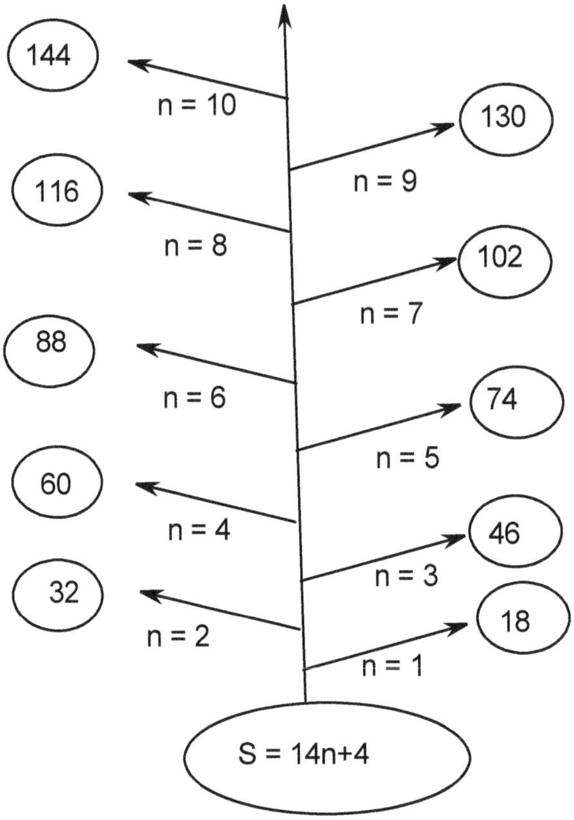

T-9 NIDO SERIES

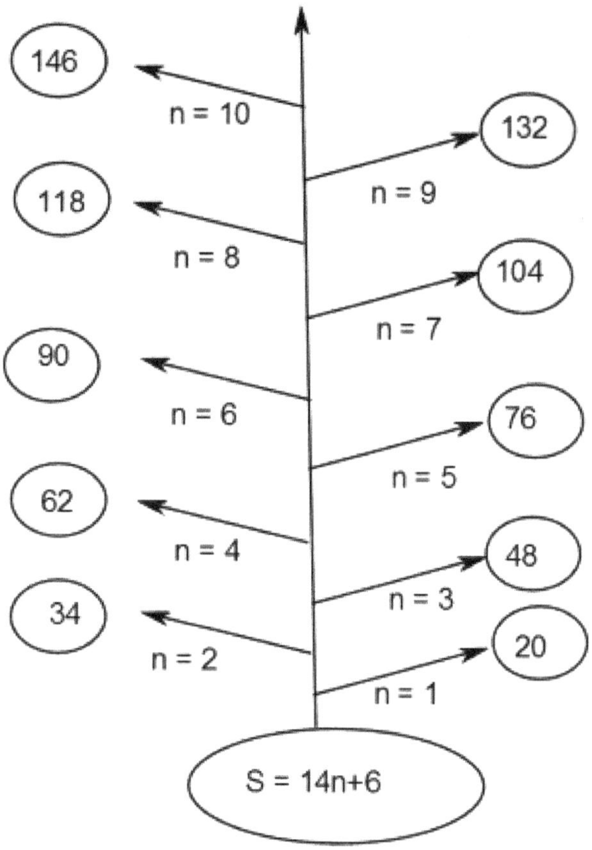

T-10 ARACHNO SERIES

Mn ⟶ [14] = [$Mn(H)(CO)_3$], n = 2, S = 14n+6

F_{MN} = [$Mn(H)(CO)_3$](2)+3CO = $Mn_2(H)_2(CO)_6$+3CO = $Mn_2(CO)_7$+3CO = $Mn_2(CO)_{10}$

Fe ⟶ [14] =[$Fe(CO)_3$], n=2, S=14n+6

F_{Fe} = [$Fe(CO)_3$](2)+3CO = $Fe_2(CO)_6$+3CO = $Fe_2(CO)_9$

Co ⟶ [14] = [$Co(H)(CO)_2$](2)+3CO =$Co_2(H)_2(CO)_4$+3CO = $Co_2(CO)_5$+3CO = $Co_2(CO)_8$

ISOLOBAL HYDROCARBON

F_C= 4n+6, n =2

C ⟶ [4]; F_c = [C](2)+6H = C_2H_6

HENCE, $Mn_2(CO)_{10}$, $Fe_2(CO)_9$, $Co_2(CO)_8$ ARE ISOLOBAL TO C_2H_6.

SCHEME 1. CONVERSION OF SERIES FORMULA INTO CLUSTER FORMULAS

1. S =14n+6, n = 2, Fe CARBONYL CONTAING, Cp LIGAND

Fe ⟶ [14] ⟶ [Fe(Cp)(H)]

F_{Fe} = [Fe(H)(Cp)](2)+3CO = $Fe_2(Cp)_2(H)_2$+ 3CO = $Fe_2(Cp)_2(H)_2$+3CO
$Fe_2(Cp_2(CO)$ +3CO = $Fe_2(Cp)_2(CO)_4$

2. S =14n+4, n = 4, Rh CARBONYL WITH Rh AND CO LIGANDS ONLY

Rh ⟶ [14] = $[Rh(H)(CO)_2]$

F_{Rh} = [Rh(H)(CO)2](4)+2CO = $Rh_4(H)_4(CO)_8$+2CO = $Rh4(CO)_2(CO)_8$+2CO = $Rh_4(CO)_{10}$+2CO
= $Rh_4(CO)_{12}$

3. S =14n+2, n = 6, Rh CO CLUSTER WITH Rh AND CO LIGANDS ONLY

Rh ⟶ [14] = $[Rh(H)(CO)_2]$

F_{Rh} = [Rh(H)(CO)2](6)+2CO = $Rh_6(H)_6(CO)_{12}$+CO = $Rh_6(CO)_3(CO)_{12}$+2CO = $Rh_6(CO)_{15}$+CO
= $Rh_6(CO)_{16}$

4. F_B= 4n+2 = [BH](6) +2 = $B_6H_6^{2-}$

5. F_C = [C](6)+2 = C_6^{2-} = C_6H_2 F_C = C_6^{2-} = Si_6^{2-} = Ge_6^{2-} = Sn_6^{2-} = Pb_6^{2-}

SCHEME 2. MORE CONVERSION EXAMPLES

3. Conclusion

Just as in transition metal carbonyl clusters which form according to 14n-based series,the elements of the main group periodic table form compounds, ions, fragments and clusters which follow the 4n-based series. Then (n) in the series formula represents the number of skeletal elements in a fragment. For the main group elements a carbon atom(C) or (BH) fragment are good building blocks and each fragment comprises of 4 valence electrons. In the case of transition metal carbonyl clusters the building block or skeletal backbone fragment comprises of 14 valence electrons. A series formula S = 14n+q or S = 4n+q may be regarded as representing many clusters, molecules and fragments. The numerical values corresponding to n =1, 2, 3, 4, and so forth may be considered to be family cluster branches for S= 14n+q or 4n+q. Thus, we may regard, S = 14n+q or 4n+q as a 'cluster clan tree' out of which various cluster families emanate. We may regard a carbonyl cluster complex of specific number of skeletal elements and valence electrons to join its family members on any suitable branch corresponding to a particular family clan tree. Thus, the chemical fragments precisely belong to specific cluster families of specific clan tree series. The 18 electron rule(S = 14n+4, n =1, NIDO CLAN TREE) or 16 electron rule(S = 14n+2, n = 1, CLOSO CLAN TREE) or octet rule(S= 4n+4, n = 1, NIDO CLAN TREE, MAIN GROUP ELEMENTS) are special cases of cluster families of specific clan trees. They may be considered to be the first families we encounter at the beginning of the clan trees of series. In summary, the transition metal complexes especially carbonyls have a tendency of forming clusters and fragments in accordance with 14n series whereas the main group element clusters or compounds and fragments are formed according to 4n series.

Acknowledgements

The author wishes to express gratitude to the University of Namibia for the provision of facilities and NAMSOV, Namibia for the financial support, Mrs Merab Kambamu Kiremire for her continued encouragement to write the articles and Mr. Philipus Hishimone for assisting in drawing the sketches of chemical structures.

References

Cotton, F. A., Wilkinson, G. (1980). *Advanced Inorganic Chemistry, 4th Ed.*, John Wiley and Sons, New York, 1980.

Douglas, B., McDaniel, D., & Alexander, J. (1994). *Advanced Concepts and Models of Inorganic Chemistry, 3rd Ed.*, John-Wiley& Sons, New York, 1994.

Grubisic, A., Li, X., Stokey, S. T., Cordes, J., Ganteför, B. K., Kiran, H., Jensen, B., Burgert, P., & Schnöckel, R. (2007). Closo-alanes: A new chapter in Aluminium Hydride Chemistry. *J. Am. Chem. Soc., 129,* 5969-5975. http://dx.doi.org/10.1021/ja0700404

Hoffmann, R. (1982). Building Bridges between Inorganic and Organic Chemistry. *Angew. Chem. Int. Ed. Engl., 21,* 711-724. http://dx.doi.org/10.1002/anie.198207113

Jemmis, E. D., Balakrishnarajan, M. M., & Pancharatna, P. D. (2001a). Unifying electron counting rule for Macropolyhedral Boranes, Metallaboranes, and Metallocenes. *J. Am. Chem. Soc., 123*(18), 4313-4323. http://dx.doi.org/10.1021/ja003233z

Jemmis, E. D., & Balakrishnarajan, M. M. (2001b). Polyhedral boranes and elemental boron. Direct structural relations and diverse electronic requirements. *J. Am. Chem. Soc., 123*, 4324-4330. http://dx.doi.org/10.1021/ja0026962

Jemmis, E. D., Balakrishnarajan, M. M., & Pancharatna, P. D. (2002). Electronic Requirements for Macropolyhedral Boranes. *Chem. Rev., 102*(1), 93-144. http://dx.doi.org/10.1021/cr990356x

Jemmis, E. D., & Jayasree, E. G. (2003). Analogies between boron and carbon. *Acc. Chem. Res., 36*, 816-824. http://dx.doi.org/10.1021/ar0300266

Jemmis, E. D. (2005). Building relationships between polyhedral boranes and elemental boron. *Inorg. Chem., 18*, 620-628.

Jemmis, E. D., Jayasree, E. G., & Parameswaran, P. (2006). Hypercarbons in polyhedral structures. *Chem. Soc. Rev., 35*, 157-168. http://dx.doi.org/10.1039/B310618G

Jemmis, E. D., & Prasad, D. L. V. K. (2008). Unknowns in the chemistry of Boron. *Current Science, 95*(10), 1277-1283.

Jensen, W. B. (1978). Electron Orbital Counting Rules and Bonding Topologies. *De Education Quimica, 3,* 210-222.

Jensen, W. B. (2005). The Origin of the 18-Electron Rule. *J. Chem. Ed., 82*, 28. http://dx.doi.org/10.1021/ed082p28

King, R. B. (2002). Spherical Chemical Bonding Models for three-diamensional aromaticity in deltahedral boranes. *PINSA, 68, A,* 473-486.

Kiremire, E. M. R. (2016a). The categorization and Structural Prediction of Transition Metal Carbonyl Clusters Using the 14n Series Numerical Matrix. *Int. J. Chem., 8*(1), 109-125. http://dx.doi.org/10.5539/ijc.v8n1p109

Kiremire, E. M. R. (2016b). Generating Formulas of Transition Metal Carbonyl Clusters of Osmium, Rhodium and Rhenium. *Int. J. Chem., 8*(1), 126-144. http://dx.doi.org/10.5539/ijc.v8n1p126

Kiremire, E. M. R. (2015a). Capping and Decapping Series of Boranes. *Int. J. Chem., 7*(2), 186-197. http://dx.doi.org/10.5539/ijc.v7n2p186

Kiremire, E. M. (2015b). Classification of Transition Metal Carbonyl Clusters Using the 14n Rule Derived from Number Theory. *Orient. J. Chem., 31*(2), 605-618. http://dx.doi.org/10.13005/ojc/310201

Kiremire, E. M. (2015c). A Unique Bypass to the Carbonyl Cluster Nucleus Using the 14n Rule. *Orient. J. Chem., 31*(3), 1469-1476. http://dx.doi.org/10.13005/ojc/310326

Kiremire, E. M. (2015d). Unification and Expansion of Wade-Mingos Rules with Elementary Number Theory. *Orient. J. Chem., 31*(1), 387-392. http://dx.doi.org/10.13005/ojc/310146

Kiremire, E. M. (2015e). Boranes Hiding Inside Carbonyl Clusters. *Orient. J. Chem., 31*(Spl,Edn), 121-127.

Langmuir, I. (1921). Types of Valence. *Science, 54*, 59-67. http://dx.doi.org/10.1126/science.54.1386.59

Lipscomb, W. N. (1963). *Boron Hydrides*. W. A. Bejamin, Inc., New York.

Miessler, G., Fischer, P., & Tarr, D. (2014). *Inorganic Chemistry, 5th Edition*, Pearson Education, Inc., Upper Saddle River.

Mingos, D. M. P. (1972). A General Theory for Cluster and Ring Compounds of the Main Group and Transition Elements. *Nature(London), Phys. Sci., 236*, 99-102. http://dx.doi.org/10.1038/physci236099a0

Mingos, D. M. P. (1984). *Polyhedral Skeletal Electron Pair Approach. Acc. Chem. Res., 17*(9), 311-319. http://dx.doi.org/10.1021/ar00105a003

Pauling, L. (1977). Structure of Transition Metal Cluster Compounds: Use of an additional orbital resulting from f, g character of spd bond orbitals. *Proc. Natl. Acad. Sci., USA, 74*(12), 5235-5238. http://dx.doi.org/10.1073/pnas.74.12.5235

Shaik, S., Danovich, D., Wu, W., Su, P., Rzepa, H. S., & Hiberty, P. C. (2012). Quadruple bonding in C_2 and analogous eight valence electron species. *Nature Chemistry, 4*, 195-200. http://dx.doi.org/10.1038/nchem.1263

Su, J., Li, X. W., Crittendon, R. C., & Robinson, G. H. (1997). How short is -Ga:Ga- Triple Bond? Synthesis and Molecular Structre of $Na_2[Mes^*{}_2C_6H_3\text{-Ga:Ga-}C_6H_3Mes^*{}_2](Mes^*{}_2 = 2,4,6\text{-}iPr_3C_6H_2)$: The First Gallyne. *J. Am. Chem. Soc., 119*(23), 5471-5472. http://dx.doi.org/10.1021/ja9700562

Tolman, C. A. (1972). The 16 and 18 Electron Rule in Organometallic Chemistry and Homogeneous catalysis. *Chem. Soc. Rev*., 337-353. http://dx.doi.org/10.1039/cs9720100337

Wade, K. (1971). The structural significance of the number of skeletal bonding electron-pairs in carboranes, the higher boranes and borane ions and various transition metal carbonyl cluster compounds. *Chem. Commun.,* 792-793. http://dx.doi.org/10.1039/c29710000792

Wade, A. (1976). Structural and Bonding Patterns in Cluster Chemistry. *Adv. Inorg. Chem. Radiochem.*, *18*, 1-66. http://dx.doi.org/10.1016/s0065-2792(08)60027-8

Xie, Y., Grev, R. S., Gu, J., Schaefer, H. F., Schleyer, P.v.R., Su, J., Li, X. W., & Robinson, G. H. (1998). The Nature of the Gallium-Gallium Triple Bond. *J. Am. Chem. Soc., 120*, 3773-3780. http://dx.doi.org/10.1021/ja9739300

Xu, B., Li, Q. S., King, R. B., & Schaefer III, H. F. (2009). Heptacarbonyldiosmium and Hexacarbonyldiosmium: Two Highly Unsaturated Binuclear Osmium. *Croat. Chem. Acta., 82*(1), 207-218.

11

Unusual underground Capping Carbonyl Clusters of Palladium

Enos Masheija Rwantale Kiremire

Correspondence: Enos Masheija Rwantale Kiremire, Department of Chemistry and Biochemistry, University of Namibia, Private Bag 13301, Windhoek, Namibia. E-mail:kiremire15@yahoo.com

Abstract

Transition metal carbonyls form unlike boranes a wide range of clusters. High nuclearity carbonyl clusters have a tendency to form capped clusters. Using the method of series explained in this paper, many capped carbonyl clusters have been identified for group 7, 8, 9 and 10 transition metals such as rhenium, osmium, rhodium and palladium. The series have discovered that palladium form exclusively capped carbonyl clusters. Furthermore, it has been discovered that some of the capped clusters have negative nuclear closo function. Such carbonyls have been regarded as capping underground. This paper presents the unique characteristic of high nuclearity capping carbonyl clusters of palladium.

Keywords: polyhedral skeletal electron pair theory, palladium capping series, underground capping clusters, clusters of high nuclearity, carbonyls of rhenium, osmium, rhodium, and palladium, hoffmann and isolobal series.

1. Introduction

The polyhedral skeletal electron pair theory (PSEPT) has been exceedingly useful in categorizing clusters especially boranes and carbonys for many years(Mingos, 1972; Wade, 1976; Cox, 1981; Mingos, 1984;Lee, et al, 1989;Jemmis, et al, 1984;Jemmis et al, 2001; 2002; 2003; Balakrishnarajan and Hoffmann, 2004; Jemmis, 2005; 2005; Wales, 2005; Jemmis, et al., 2008; Welch, 2013). In our previous extensive work on clusters we found that the 4n and 14n rules could also be utilized to categorize the clusters and predict the shapes carbonyl cluster series (Kiremire, 2014; Kiremire, 2015a; Kiremire, 2015b) especially those of small to medium size level. Furthermore, we found that using 4n series cluster molecular formulas could be decomposed into fragments from which cluster series can be derived and their geometrical shapes predicted (Kiremire, E. M. R., 2015c). Unlike the PSEPT which was derived from molecular orbital theory, the 14n±q and 4n±q (where q is an integer) rules were empirically derived by analysis of carbonyl, hydrocarbons and borane cluster formulas and known shapes.

1.1 Authenticity of the 4n And 14n Series

Since the use of series categorize clusters is not well known, it is extremely important to explain this approach to cluster analysis. Recently, it has been found that Hoffmann's isolobal fragments (Hoffmann, 1982) could assigned series(Kiremire,2015d). For a given cluster formula, skeletal elements of the cluster are identified. The formula is then decomposed into suitable mono-skeletal fragments. If a selected skeletal element is a main group element of the periodic table, then the 4n rule is applied, if it is a transition metal, the 14n rule is applied. For example, if carbon (C) is taken as a fragment, it has [4] valence electrons. According to the series approach used in this paper, it will be classified as belonging to 4n(n = 1) series . If the fragment is [CH], then the electron content will be [4+1] and this will be categorized as being a member of 4n+1 series. A phosphorus atom P with a valence electron content of five will be categorized in the same way. That is, [5] = [4+1] = 4n+1. Thus, the four electrons must be separated first to form the backbone of the series and the rest will be used in the categorization of the series and the cluster. The fragment [BH] = [3+1] = [4] will belong to 4n +0 series. In the case of transition metal carbonyl clusters, the classification is based upon 14n rule. Take the case of the fragment F = [$Os(CO)_3$], the valence electron content is [8+3x2] = [14]. This means we can classify the fragment as belonging to 14n+0 series. It has been found that the 14n is isolobal to 4n. Therefore it is easier to use 4n instead of 14n for classification. If we get a fragment F = [$Os(CO)_4$], then the valence content is [16], then we can split this into [14+2] = 14n+2 series. This is the same as 4n+2 due to isolobal relationship. In case of the fragment F = [$Rh(CO)_3$], the valence content is [9+3x2] = [15] = [14+1]. This means we can classify the fragment as belonging to 14n+1 series or to 4n+1 due to isolobal relationship. If a fragment has less than 14, then its valence content can be transformed into a form consistent with 14n series. For instance, the fragment, such as F= [Pd(CO)] has [10+2] valence electron content. This is then transformed into [10+2] = [14-2] = 14n -2 which equivalent to 4n-2 due to isolobal relationship. In fact, the fragment F= [Pd(CO)] is equivalent to F = $Os(CO)_2$ fragment. The isolobal relationship may simply be represented as given here 14n±q is isolobal to or simply 14n±q = 4n±q.

2. Results and Discussion

Using the series approach, a large number of known carbonyl clusters have been categorized and the results came out precisely as the ones obtained from literature (Hughes and Wade,2000). These findings are given In Table 1. Using the same method, a good collection of different 19 examples have been analyzed using fragments and series and the results are given in the following section so as to expose the reader to a variety of applying series for categorization of clusters.

2.1 Capping Clusters

In our earlier work (Kiremire, 2015a, Kiremire 2015b), a capping symbol C^nC (n = 1, 2, 3,.) was proposed. The capping symbol of clusters has been covered in previous work. But will briefly be repeated here for the purpose of getting readers be more familiar with it. For mono-capped closo, series S = 14n+0, the capping symbol C^1C, n has a value of 1 signifying one cap. In the case of bi-capped cluster, S = 14n-2, the capping symbol is Cp = C^2C(n = 2). This means, if S = 14n-2 = 14n+1(-2), then the capping is Cp = $C^1+C^1 = C^2C$. The first C^1 comes from the 14n component of the series as it implies a mono-cap. The second C^1 comes from 1(-2) component of the series. According to the series, for every negative integer of two (-2), there is an additional capping. Thus, S = 14n-4 = 14n+2(-2), the capping symbol is Cp = $C^1+C^2= C^3C$ (tri-capped). Similarly, for S = 14n-6 = 14n+3(-2), the capping symbol is Cp = $C^1+C^3= C^4C$(tetra-capped). Some of the capped carbonyl clusters are given in Table 1. They include among others, $Re_4H_4(CO)_{12}$,S = 14n, C^1C[M-3];$Os_6(CO)_{18}$, S = 14n, C^1C[M-5]; $Os_7(CO)_{21}$, S = 14n, C^1C[M-6]; and $Os_8(CO)_{22}{}^{2-}$, S = 14n-2, C^2C[M-6]. If we analyze $Re_4H_4(CO)_{12}$ S = 14n, C^1C[M-3] cluster system closely, C^1 means that one skeletal atom is capping on three others which constitute an inner core closo cluster that obeys the 14n+2 or simply 4n+2 rule. The building block of the rhenium cluster must have a content of [14] electrons according to the series while the capping cluster must have a valence electron content of [12]. The suitable fragments of rhenium will be [Re(H)$(CO)_3$] and [Re(H)$(CO)_2$] respectively. The [M-3] symbol means that the inner core cluster is a member of S = 4n+2 series and n = 3. We can construct the formula of the inner core as follows: F = 4n+2 = [Re(H)$(CO)_3$] (3)+CO = $Re_3(H)_3(CO)_9$ + CO = $Re_3(H)_3(CO)_{10}$. This is the inner closo cluster. In order to derive the entire cluster formula, we must add the mono-cap, that is, $Re_3(H)_3(CO)_{10}$ + [Re(H)$(CO)_2$] = $Re_4(H)_4(CO)_{12}$. Let us discuss a few more examples to illustrate how fragments are utilized to generate series. Consider $Os_4(CO)_{16}$ cluster. There are 4 skeletal osmium atoms. Hence, we can decompose this into 4[$Os(CO)_3$] fragments and 4CO ligands. The valence electron content of [$Os(CO)_3$] fragment = 8+3x2 = 14. In using the series method, it makes the work easier to create fragments whole valence electron content are [14] for transition metals or [5] for main group elements. If not the values must be close to these values less or above. The fragment F = [$Os(CO)_3$] belongs to 14n series. Therefore, 4 [14n] = 14n just as carbon fragment [C] fragment S = 4n+0 series (n = 1) and C_4 = 4[C] = 4[4n] = S = 4n+0 series (n = 4). Each of the CO ligand donates 2 valence electrons and therefore 4CO ligands will contribute 8 valence electrons. Therefore the net cluster series S = (14n+0)+8 = 14n+8. This cluster belongs to the HYPHO series. Since 14n is isolobal to 4n, then the series can be regarded to be the same as S = 4n+8. The simplicity of working with 4n rather than 14n, makes 4n more user friendly. Take another example, $Os_9(CO)_{24}{}^{2-}$ cluster. This cluster has 9 skeletal osmium atoms. It is easier to aim to decompose the cluster into nine [14] valence electron fragments. This is done as follows, $Os_9(CO)_{24}{}^{2-}$ = 9[$Os(CO)_3$]-3CO+2. This gives us 27CO ligands. Since we have only 24CO ligands in the cluster, we need to subtract 3CO ligands to balance the ligands. This gives rise to subtracting 3x2 = 6 valence electrons. The cluster has a charge, q of -2. This means that q also supplies 2 more electrons to the valence electron content. To sum up, 9[$Os(CO)_3$] = 9[14] = 9[14n] = 14n+0 which is the same as 4n+0. We need to remove the valence electrons of 3CO and add 2 electrons from the charge. This gives us S= (4n+0)-(6)+(2) = 4n-6+2 = 4n-4. That is, S = 4n-4 = 14n+2(-2), Cp = $C^1+C^2 = C^3C$[M-6]. This is a tri-capped cluster. This means that three of the nine osmium atoms are capped and the 6 comprise of nuclear octahedral closo cluster. The complex is a tri-capped octahedral cluster. The symbol {M-6] means this belongs to 14n+2 or 4n+2 series. We can generate the formula of this inner cluster using the electron building block [14] of the carbonyl clusters and the n =6 value of the closo cluster as follows F = [$Os(CO)_3$] (6)+2 = $Os_6(CO)_{18}{}^2$. We could also have used CO ligand instead of converting +2 into a charge of (2—).ˉ The complex $Os_6(CO)_{18}{}^{2-}$ as is found is indeed an octahedral closo complex(Cotton and Wilkinson,1980). Let us consider a few members of S = 14n+0 series starting with $Os(CO)_3$ fragment. Since F = $Os(CO)_3$ then S = 4n+0 and Cp = C^1C[M-0]. According to series, the building block of the series is based upon the [14] valence electron content. The capping fragment is based upon the [12] valence electron content.The capping is centered upon the closo inner nucleus. This means [M-0] = 4n+2 and F = 4n+2 = [$Os(CO)_3$(0)+CO since n = 0. This gives F = CO for the first member of 14n+2 or 4n+2 series. In order to get the formula of the 14n or 4n series we have to add the value of the inner core [M-0] to the value of the capping fragment. That is, F = [M-0] +C^1. The value of C^1 = [$Os(CO)_2$]. Hence, F = CO + [$Os(CO)_2$] = $Os(CO)_3$. This means that CO is the closo nucleus of $Os(CO)_3$ while $Os(CO)_2$ is the capping component of $Os(CO)_3$. The combination of the two generates the first member of the fragment of the 4n series. The fragment $Os_2(CO)_6$ is the second member of S = 4n+0 series. This means that it is also a mono-capped cluster. That is, Cp = C^1C[M-1] meaning that there is one skeletal capping atom while the other forms the inner closo core. We need to define what [M-1] means in terms of the series. The closo symbol [M-1] = 4n+2 = [$Os(CO)_3$(1)+CO = $Os(CO)_3$ + CO = $Os(CO)_4$. Let us recall that for osmium, C^1 = $Os(CO)_2$. Hence, F = C^1 + [M-1] = $Os(CO)_2$ + $Os(CO)_4$ = $Os_2(CO)_6$. It is quite clear that the series easily explain why $Os(CO)_3$ and $Os_2(CO)_6$ are members of the first mono-capping series. The third member of the 4n series is $Os_3(CO)_9$. The corresponding capping symbol for this is Cp = C^1C[M-2]. The symbol, [M-2] represents 4n+2 content = [$Os(CO)_3$](2)+CO = $Os_2(CO)_6$+CO = [$Os_2(CO)_7$].Therefore the cluster formula is generated by adding the capping fragment to the closo nuclear component C^1 + [M-2] = [$Os(CO)_2$]+ [$Os_2(CO)_7$] = $Os_3(CO)_9$. In the same manner, it can easily be shown that $Os_4(CO)_{16}$ is a member of S = 4n series and Cp = C^1C[M-3] where C^1 = $Os(CO)_2$ and [M-3] = $Os_3(CO)_{10}$. The clusters, $Os_6(CO)_{18}$ and $Os_7(CO)_{21}$ which are also members of the C^1C series have already been discussed.

2.2 Palladium Underground Capping Carbonyl Clusters

From the discussion presented above, it is quite clear that series are have the power to analyze clusters of all types based on fragments and series derived from them using the valence electron content. An analysis on palladium carbonyl clusters was made on a large collection of carbonyl clusters using series. The results are given in Table 2. A large number of these palladium carbonyls have been produced by Dahl research team (Mednikov, Dahl, 2010). It is quite amazing that palladium forms a wide range of capped carbonyl clusters. They range from closo [M-11] to closo [M^{-76}]. What was found interesting according to series is that for palladium there are clusters which cap using negative closo nuclear clusters. It is proposed that such capping clusters with negative closo values such as [M^{-76}] be regarded as undergoing clapping underground. This is similar to highways which are constructed for underground traffic travel. Consider capping at [M-0]. This means capping at base level high way. The symbol [M-0] for palladium carbonyl means [M-0] = 4n+2 = [$Pd(CO)_2$](0)+CO = CO. That is, the cluster has a closo nucleus with no skeletal atom (n = 0) and the skeletal building block is based on [14] valence electrons. In the case of palladium a suitable fragment is F = [$Pd(CO)_2$]and a capping fragment of twelve valence electrons [12], [is F =Pd(CO)]. The integer (+2) means two valence electrons added to 14n or 4n series. Likewise, capping at [M^{-1}] means = 4n+2 = [Pd(CO)2(-1)+CO = - $Pd(CO)_2$+CO = -Pd(CO). That is, capping at underground highway level n = -1. The negative sign means that closo value function will have a negative value since n = -1. This negative value must be added to the value of the capping function to generate the cluster molecular formula. Since [M^{-1}] has a negative power, n = -1, as in the case of $Pd_{54}(CO)_{40}L_{14}$, C^{55}C[M^{-1}] cluster. The capping fragment F= [Pd(CO)] and hence C^{55} = 55[Pd(CO)] = $Pd_{55}(CO)_{55}$ and M^{-1}] = -[Pd(CO)]. Therefore the molecular formula is obtained by adding the two functions as F = $Pd_{55}(CO)_{55}$ -[Pd(CO)] = $Pd_{54}(CO)_{54}$. This cluster is equivalent to $Pd_{54}(CO)_{40}L_{14}$ in terms of valence electron content as CO and L each donate two electrons.

2.3 Capping carbonyl Clusters above the Ground and Underground

An X-Y type of axis has been designed to explain the impression derived from the capping cluster carbonyls of palladium. This is shown in Figure 1. As can be seen from Table 2, there are clusters whose closo nucleus starts from [M-0] to [M-11]. The closo [M-0] may be regarded as the beginning of being above the ground and those below the [M-0] similar to the X-axis may be taken as capping below the ground (underground). Those capping above the ground include [M-0], osmium cluster $Os_{20}(CO)_{40}^{2-}$, C^{20}C[M-0] is used as a reference, [M-1] closo carbonyl, $Pd_{34}(CO)_{24}L_{12}$,C^{33}[M-1]. This closo is based on the first highway above the ground. On the other hand the [M-2] series, C^1, C^{30}, C^{33},and C^{35} are based on the highway level 2 above the ground. Other examples include the [M-5] series C^3, C^9, C^{11}, and C^{25}. The [M-6] series are C^{17}, C^{23}, C^{24}, and C^{36}. Members of [M-7] series are C^3, C^{25} and C^{29}. The last series to include is the [M-9], C^3 and C^8. The capping image here is similar to that of athletics where runners in lanes compete in jumping series of huddles. In the example of [M-6] lane or highway, one cluster is at C^{17} huddle or series, another at C^{23}, another at C^{24}, and the last at C^{36}. From the capping symbol of a cluster, its molecular formula may be derived. The palladium cluster with a capping code C^{36}C[M-6] is $Pd_{33}Ni_9$ $(CO)_{41}L_6^{4-}$ and this can be readily be converted into its equivalent cluster. The capping fragment has an electron content of [12] and in the case of palladium carbonyl clusters this corresponds to [Pd(CO)] and in for skeletal building block contains [14] skeletal electrons. Therefore,C^{36} = 36[Pd(CO)] = $Pd_{36}(CO)_{36}$. The closo inner core [M-6] = 4n+2 = [$Pd(CO)_2$](6)+CO = $Pd_6(CO)_{12}$+CO = $Pd_6(CO)_{13}$. The formula of the cluster will be the sum of the capping value and that of the inner core cluster $Pd_{36}(CO)_{36}$ +$Pd_6(CO)_{13}$ = $Pd_{42}(CO)_{49}$. In the cluster $Pd_{33}Ni_9$ $(CO)_{41}L_6^{4-}$,Pd = Ni in terms of valence electron content. Also the ligand CO = L in terms of electron content. The charge of (-4) has electron content of 2CO ligands. Hence, the cluster $Pd_{33}Ni_9$ $(CO)_{41}L_6^{4-}$ = 33Pd+9Pd+41CO +6CO +2CO = $Pd_{42}(CO)_{49}$. The underground capping cluster is also readily identified by the series. Take the case of $Pd_{59}(CO)_{36}L_{14}$,S = 4n-136, Cp = C^{69}C[M^{-10}]. According to the series the symbol, C^{69} means that there 69 capping fragments each of 12 electrons. That is, C^{69} = 69[Pd(CO)] = $Pd_{69}(CO)_{69}$. The closo inner core [M^{-10}] cluster means that it belongs to the closo series S = 4n+2. Therefore [M^{-10}] = 4n+2 = [$Pd(CO)_2$](-10)+CO = -$Pd_{10}(CO)_{20}$ +CO = -$Pd_{10}(CO)_{19}$. In terms of the series, this means that in order to generate the observed cluster formula of the carbonyl complex, we must the value of the inner core from the value of the capping function. This leads us to the following formula of the complex F = $Pd_{69}(CO)_{69}$ - $Pd_{10}(CO)_{19}$ = $Pd_{59}(CO)_{50}$. In terms of valence electron content $Pd_{59}(CO)_{36}L_{14}$, L = CO and hence the cluster is the same as $Pd_{59}(CO)_{36}(CO)_{14} = Pd_{59}(CO)_{50}$. Just as in developed nations complex road system, this cluster could be viewed as having been formed (capping) along the high way 10 $[M^{-10}]$ underground. Let us consider the next underground capping cluster $Pd_{69}(CO)_{36}L_{18}$, 4n-168, and $Cp = C^{85}C[M^{-16}]$. The capping formula can be utilized to recreate the cluster formula of the carbonyl complex. Following the same procedure as in the above example, $C^{85} = 85[Pd(CO)] = Pd_{85}(CO)_{85}$. The closo inner core $[M^{-16}] = 4n+2 = [Pd(CO)_2](-16)+ CO = -Pd_{16}(CO)_{32}+CO = -Pd_{16}(CO)_{31}$. In order to get the formula of the cluster complex we have to subtract the inner core from the capping function. This gives us $F = Pd_{85}(CO)_{85} - Pd_{16}(CO)_{31} = Pd_{69}(CO)_{54}$. The cluster $Pd_{69}(CO)_{36}L_{18}$ has the same valence electron content as $Pd_{69}(CO)_{54}$. Consider the cluster $Pd_{165}(CO)_{60}L_{30}$, S = 4n-480, $Cp = C^{241}C[M^{-76}]$. The capping function, $C^{241} = 241[Pd(CO)] = Pd_{241}(CO)_{241}$. The closo function, $[M^{-76}] = 4n+2 = [Pd(CO)2](-76)+CO = -Pd_{76}(CO)_{151}$. The molecular formula is $Pd_{241}(CO)_{241} - Pd_{76}(CO)_{151} = Pd_{165}(CO)_{90}$. This is the same as $Pd_{165}(CO)_{60}L_{30}$. The cluster series formula such as 4n-480 can also be used to calculate the molar mass. If we put n = 241 into the series, we get S = 4(165)-480 =180. Since 14n is isolobal to 4n, then the difference is 10n. This means for transition metal we must add 10n electrons we will get the value of valence electron cluster of the palladium cluster. That is, to the 165x10 = 1650 we must add 180 =1830. The valence electron content of the cluster $Pd_{165}(CO)_{60}L_{30}$ is 165x10+90x2 = 1860. What is fascinating is the discovery by use of series that the cluster, $Ni_{38}Pt_6(CO)_{48}(H)^{5-}$, S= 4n-74; $Cp = C^{38}C[M-6]$ has an octahedral closo nuclear cluster comprising of six platinum atoms (Rossi, 2011) at the centre of a large cluster. The rhodium cluster(Hughes and Wade, 2000), $Rh_{22}(CO)_{37}^{4-}$, S = 4n-38, $Cp = C^{20}C[M-2]$, has two atoms in the centre of the cluster of 22 skeletal atoms. For the cluster(Hughes and Wade, 2000), $Os_{17}(CO)_{36}^{2-}$, S = 4n-28, $Cp = C^{15}C[M-2]$. According to the series, this cluster has 2 atoms at the centre of a cluster of 17 skeletal elements. Thus, both the rhodium and osmium clusters are placed on [M-2] highway above the ground. The osmium is at 'city' number 15 along the [M-2] highway while the rhodium is at city number 20(C^{20}) along the [M-2] highway above the ground. Clearly, the palladium element mainly form unique carbonyl clusters, those which do capping above the ground and those which cap underground.

2.4 The Generating Function of the Palladium Carbonyl Clusters

It has been found that one of the best way of generating the molecular formulas of known and unknown carbonyl clusters is to use the [4] or [14] function and multiples. That is, the multiples of 4n or 14n series generate the backbone of the series. The addition of CO ligands to the backbone members gives the clusters series with positive integers such as 4n+2 closo, 4n+4 nido, and 4n+6 arachno. The capping clusters begin with digit zero (0) after 4n, that is, S = 4n+0(mono-capped), the next capping has the series S = 4n-2(bi-capped), 4n-6(tri-capped) and so on. These have negative integers after 4n. Using 4n instead of 14n is much easier and faster. This is the approach that has been used to categorize the clusters from main group and transition metals. Since palladium forms exclusively capping clusters, generating functions that give capping clusters have been designed for palladium. A selected number of these clusters and fragments have been created and are given in Table 3. A number of known and hypothetical clusters were identified in the table. Among others, these include $Pd_{12}(CO)_{15}L_7$,$Pd_{10}(CO)_{14}L_4$, $Pd_3(CO)_3L$, $Pd_8(CO)_{14}$, $Pd_{10}(CO)_{16}$, $Pd_{10}(CO)_{12}L_6$, and $Pd_{16}(CO)_{13}L_9$.

Table 1. Comparison of Categorization of Clusters Using 4n Series and the known Literature results

	SERIES	SERIES	LITERATURE*	CAP SYMBOL
$Os_2(CO)_{10}$	4n+8	HYPHO	HYPHO	
$Re_3(H)(CO)_{14}$	4n+8	HYPHO	HYPHO	
$Os_4(CO)_{16}$	4n+8	HYPHO	HYPHO	
$Re_4(H)_4(CO)_{16}$	4n+8	HYPHO	HYPHO	
$Os_5(CO)_{19}$	4n+8	HYPHO	HYPHO	
$Re_2(CO)_{10}$	4n+6	ARACHNO	ARACHNO	
$Os_2(CO)_8^{2-}$	4n+6	ARACHNO	ARACHNO	
$Re_3(H)_3(CO)_{12}$	4n+6	ARACHNO	ARACHNO	
$Os_4(CO)_{12}$	4n+6	ARACHNO	ARACHNO	
$Re_4(CO)_{16}^{2-}$	4n+6	ARACHNO	ARACHNO	
$Os_5(CO)_{18}$	4n+6	ARACHNO	ARACHNO	
$Os_6(CO)_{21}$	4n+6	ARACHNO	ARACHNO	
$Re_2(H)_2(CO)_{10}$	4n+4	NIDO	NIDO	
$Re_3(H)_3(CO)_{10}^{2-}$	4n+4	NIDO	NIDO	
$Os_3(H)_2(CO)_{12}$	4n+4	NIDO	NIDO	
$Os_4(CO)_{14}$	4n+4	NIDO	NIDO	
$Re_4(H)_6(CO)_{12}^{2-}$	4n+4	NIDO	NIDO	
$Os_5(H)_2(CO)_{16}$	4n+4	NIDO	NIDO	
$Re_5(C)(H)(CO)_{16}^{2-}$	4n+4	NIDO	NIDO	
$Re_6(H)_8(CO)_{18}^{2-}$	4n+4	NIDO	NIDO	
$Os_6(H)_2(CO)_{19}$	4n+4	NIDO	NIDO	
$Os_7(H)_2(CO)_{22}$	4n+4	NIDO	NIDO	
$Re_4(H)_5(CO)_{12}^{-}$	4n+2	CLOSO	CLOSO	
$Os_5(CO)_{16}$	4n+2	CLOSO	CLOSO	
$Re_6(C)(CO)_{19}^{-2}$	4n+2	CLOSO	CLOSO	
$Os_6(CO)_{18}^{2-}$	4n+2	CLOSO	CLOSO	
$Re_4(H)_4(CO)_{12}$	4n+0	MONOCP	MONOCP	$C^1C[M-3]$
$Os_6(CO)_{18}$	4n+0	MONOCP	MONOCP	$C^1C[M-5]$
$Os_7(CO)_{21}$	4n+0	MONOCP	MONOCP	$C^1C[M-6]$
$Re_7(C)(CO)_{21}^{-3}$	4n+0	MONOCP	MONOCP	$C^1C[M-6]$
$Re_8(C)(CO)_{24}^{-2}$	4n-2	BICP	BICP	$C^2C[M-6]$
$Os_8(CO)_{22}^{2}$	4n-2	BICP	BICP	$C^2C[M-6]$
$Os_9(CO)_{24}^{2}$	4n-4	TRICP	TRICP	$C^3C[M-6]$
$Pd_6Ru_6(CO)_{24}^{2-}$	4n-10	HEXACP	HEXACP	$C^6C[M-6]$

$B_{12}H_{12}^{2-}$ ↓   q = charge

12[BH] ⟶ 12[4] ⟶ 4n+0

q ⟶ 0+2

S = 4n+2 ⟶ CLOSO

This closo cluster will have the well known standard shape of $B_{12}H_{12}^{2-}$

$NB_{11}H_{12}$ ↓

1[NH] ⟶ 1[5+1] =1[4+2] ⟶ 4n+2

11[BH] ⟶ 11[BH] = 11[3+1] = 11[4] ⟶ 4n+0

S = 4n+2 ⟶ CLOSO

n = 1+11 = 12   F = 4n+2 = [BH](12)+2 = $B_{12}H_{12}^{2-}$

This means that the cluster will adopt a shape similar to that of $B_{12}H_{12}^{2-}$.

6. $C_2B_8H_{10}$

↓

2[C] → 2[4] → 4n+0

8[BH] → 8[3+1] = 8[4] → 4n+0

2H → 0+2

S = 4n+2 → CLOSO

$F = 4n+2 = [BH](10)+2 = B_{10}H_{10}^{2-}$ n = 2+8 = 10

This means that the shape of the cluster will be similar to that of $B_{10}H_{10}^{2-}$

7. $C_2B_8H_{10}$

↓

2[C] → 2[4] → 4n+0

8[BH] → 8[3+1] = 8[4] → 4n+0

2H → 0+2

S = 4n+2 → CLOSO

8. $F = 4n+2 = [BH](10)+2 = B_{10}H_{10}^{2-}$ n = 2+8 = 10

This means that the shape of the cluster will be similar to that of $B_{10}H_{10}^{2-}$

B_6H_{10}

↓

6[BH] → 6[3+1] = 6[4] → 4n+0

4H → 0+4

S = 4n+4 → NIDO

n = 6, this means that it has a shape derived from that of Closo $B_7H_7^{2-}$

The shape will be a pentagonal bipyramid.

$C_2B_8H_{10}$

↓

2[C] → 2[4] → 4n+0

8[BH] → 8[3+1] = 8[4] → 4n+0

2H → 0+2

S = 4n+2 → CLOSO

$F = 4n+2 = [BH](10)+2 = B_{10}H_{10}^{2-}$ n = 2+8 = 10

This means that the shape of the cluster will be similar to that of $B_{10}H_{10}^{2-}$

B_6H_{10}

↓

6[BH] → 6[3+1] = 6[4] → 4n+0

4H → 0+4

S = 4n+4 → NIDO

n = 6, this means that it has a shape derived from that of Closo $B_7H_7^{2-}$

The shape will be a pentagonal bipyramid.

$(Cp^*{}_2Rh_2B_3H_7$

↓

$2[Cp^*Rh] \longrightarrow 2[5+9] \longrightarrow 2[14] = 2[4] \longrightarrow 4n+0$

$3[BH] \longrightarrow 3[3+1] = 3[4] \longrightarrow 4n+0$

$4H \longrightarrow 0+4$

$S = 4n+4 \longrightarrow$ NIDO

$n = 2+3 = 5$ $F = 4n+4 = [BH](5)+4H = B_5H_9$

The ideal shape of the cluster is that B_5H_9 which is a delivative of CLOSO $B_6H_6^{2-}$.

The cluster will have a square pyramid shape.

9.. $Rh_4(CO)_{12}$

↓

$4[Rh(CO)_3] = 4[9+6] = 4[14+1] = 4[4+1] \longrightarrow 4n+4$

$n = 4$

$S = 4n+4 \longrightarrow$ NIDO

$k = 2n-2 = 2(4)-2 = 6$

$V = 4n+4 = 4(4)+4 = 20$

$TM = 20+40 = 60$

$F = 4n+4 = [BH](4)+4H = B_4H_4+4H = B_4H_8$

10. $Rh_5(CO)_{15}^{-}$

↓

$5[Rh(CO)_3] \longrightarrow 5[4+1] = 4n+5$

$q \longrightarrow 0+1$

$S = 4n+6$ $V = 4(5)+6 = 26$, $TM = 26+50 = 76$

$n=5$ $k = 2n-3 = 2(5)-3 = 7$

$F = 4n+6 = [BH](5)+6H = B_5H_5+6H = B_5H_{11}$

11. q = charge

$Fe_5(C)(CO)_{14}^{2-}$

↓

$5[Fe(CO)_3] \longrightarrow 4n+0$

$-CO \longrightarrow 0-2$

$C \longrightarrow 0+4$

$q \longrightarrow 0+2$

$n = 5$ $S = 4n+4 \longrightarrow$ NIDO

$S = V = 4(5)+4 = 24$, $TM = 24+50 = 74$

$k = 2n-2 = 2(5)-2 = 8$

$F = 4n+4 = [BH](5)+4H = B_5H_5+4H = B_5H_9$

Derived from octahedral $B_6H_6^{2-}$

Square pyramid geometry.

12.

$[Os(CO)_3] \longrightarrow 4n+0 = C^1C[M\text{-}0]$

$n = 1$

$[M\text{-}0] = 4n+2 = [Os(CO)_3](0)+CO = CO$

$C^1 = Os(CO)_2$ $F = Os(CO)_2 + CO = Os(CO)_3$

$Os_6(CO)_{18} \longrightarrow 4n = C^1C[M\text{-}5]$

$[M\text{-}5] = 4n+2 = [Os(CO)_3(5)+CO = Os_5(CO)_{15}+CO = Os_5(CO)_{16}$

$F = Os(CO)_2 + Os_5(CO)_{16} = Os_6(CO)_{18}$

$n = 6$

$S = V = 4n = 4(6) = 24$, $TM = 24+60 = 84$

13. $Rh_6(CO)_{16}$

↓

$6[Rh(CO)_3] \longrightarrow 6[9+6] = 6[14+1] = 6[4+1] = 4n+6$

$-2CO \longrightarrow 0-4$

$S = 4n+2 \longrightarrow$ CLOSO

$n = 6$ $k = 2n-1 = 2(6)-1 = 11$

14. $Os_6(CO)_{18}^{2-}$ q = charge

↓

$6[Os(CO)_3] \longrightarrow 6[8+6] = 6[14] = 6[4] = 4n+0$

$q \longrightarrow 0+2$

$n = 6,$ $S = 4n+2 \longrightarrow$ CLOSO

$F = 4n+2 = [BH](6)+2 = B_6H_6^{2-}$

Shape will be similar to that of $B_6H_6^{2-}$. This is a regular octahedral shape.

15.. $Os_7(CO)_{21}$ $Os_6(CO)_{19}$

↓

$7[Os(CO)_3] = 7[14] = 7[4] = 4n+0 = C^1C[M\text{-}6]$

16.. $Os_8(CO)_{22}^{2-}$

↓

$8[Os(CO)_3] \longrightarrow 4n+0$

$-2CO \longrightarrow 0-4$

$q \longrightarrow 0+2$

$S = 4n-2 \longrightarrow Cp = C^1 + C^1 = C^2C[M\text{-}6]$

$[M\text{-}6] = 4n+2 = [Os(CO)_3](6)+CO = Os_6(CO)_{19}$

$C^2 = 2[Os(CO)_2] = Os_2(CO)_4$

$F = Os_2(CO)_4 + Os_6(CO)_{19} = Os_8(CO)_{23}$

$$Os_8(CO)_{23} = Os_8(CO)_{22}^{2-}.$$

17.. $Pd_{23}(CO)_{20}L_{10}$

$$\downarrow$$

$$23[PdCO]^{14-2}_{10+2} \longrightarrow 4n-46$$

$$-3CO \longrightarrow 0-6$$

$$10L \longrightarrow 0+20$$

$$S = 4n-32$$

$$Cp = C^{1}+C^{16} = C^{17}C[M\text{-}6]$$

$$[M\text{-}6] = 4n+2 = [Pd(CO)_2](6)+CO = Pd_6(CO)_{12}+CO = Pd_6(CO)_{13}$$

$$C17 = 17[PdCO] = Pd_{17}(CO)_{17}$$

$$F = C^{17}+ [M\text{-}6] = Pd_{17}(CO)_{17} + Pd_6(CO)_{13} = Pd_{23}(CO)_{30} = Pd_{23}(CO)_{20}L_{10}$$

The k value of the cluster is given by k = 2n+16 since S = 4n-32. Since n = 23, k = 2(23)+16 = 62.The capping fragment PdCO has 12 valence electrons and hence S = [12] = [14-2] = [4-2] = 4n-2. For S = 4n-2, k =2n+1 = 2(1)+1 = 3. Thus, for every capping, k increases by 3. For F = C^{17}+[M-6], k = 17x3+ 11 = 51+11 = 62.

18.. $Pd_{59}(CO)_{32}L_{21}$

$$\downarrow$$

$$59[PdCO]^{14-2}_{10+2} \longrightarrow 4n-118$$

$$-27CO \longrightarrow 0-54$$

$$21L \longrightarrow 0+42$$

$$S = 4n-130$$

$$Cp = C^{1}+ C^{65} = C^{66}C[M^{-7}]$$

$$[M^{-7}] = 4n+2 = [Pd(CO)_2](-7)+CO = -Pd_7(CO)_{14}+CO = -Pd_7(CO)_{13}$$

$$C^{66} = 66[PdCO] = Pd_{66}(CO)_{66}$$

$$F = -Pd_7(CO)_{13} + Pd_{66}(CO)_{66} \longrightarrow Pd_{59}(CO)_{53} = Pd_{59}(CO)_{32}L_{21}$$

19. $Pd_{165}(CO)_{60}L_{30}$

$$\downarrow$$

$$165[PdCO]^{14-2}_{10+2} \longrightarrow 4n-330$$

$$-105CO \longrightarrow 0-210$$

$$30L \longrightarrow 0+60$$

$$S = 4n-480$$

$$Cp = C^{1}+ C^{240} = C^{241}C[M^{-76}]$$

$$[M^{-76}] = 4n+2 = [Pd(CO)_2](-76)+CO = -Pd_{76}(CO)_{152}+CO = -Pd_{76}(CO)_{151}$$

$$C^{241} = 241[PdCO] = Pd_{241}(CO)_{241}$$

$$F = -Pd_{76}(CO)_{151} + Pd_{241}(CO)_{241} \longrightarrow Pd_{165}(CO)_{90} = Pd_{165}(CO)_{90} = Pd_{165}(CO)_{60}L_{30}$$

Table 2. Derived Series of some Palladium carbonyl Clusters

CLUSTER	SERIES	CAPPING SYMBOL
$Pd_{38}(CO)_{28}L_{12}$	4n-52	$C^{27}C[M-11]$
$Pd_{20}Ni_{26}(CO)_{54}^{6-}$	4n-70	$C^{36}C[M-10]$
$Pd_{16}Ni(CO)_{22}L_4^{2-}$	4n-14	$C^{8}C[M-9]$
$Pd_{12}(CO)_{15}L_7$	4n-4	$C^{3}C[M-9]$
$Pd_{23}(CO)_{22}L_{10}$	4n-28	$C^{15}C[M-8]$
$Pd_{36}(CO)_{28}L_{14}$	4n-56	$C^{29}C[M-7]$
$Au_4Pd_{32}(CO)_{28}L_{14}$	4n-56	$C^{29}C[M-7]$
$Au_4Pd_{28}(CO)_{22}L_{16}$	4n-48	$C^{25}C[M-7]$
$Pd_{10}(CO)_{12}L_6$	4n-4	$C^{3}C[M-7]$
$Pd_{10}(CO)_{14}L_4$	4n-4	$C^{3}C[M-7]$
$Pd_{33}Ni_9(CO)_{41}L_6^{4-}$	4n-70	$C^{36}C[M-6]$
$Au_2Pd_{28}(CO)_{26}L_{10}$	4n-36	$C^{24}C[M-6]$
$HPd_{30}(CO)_{26}L_{10}^{-}$	4n-46	$C^{24}C[M-6]$
$Pd_{29}(CO)_{28}L_7^{2-}$	4n-44	$C^{23}C[M-6]$
$Pd_{23}(CO)_{20}L_{10}$	4n-32	$C^{17}C[M-6]$
$Pd_{30}(CO)_{26}L_{10}$	4n-48	$C^{25}C[M-5]$
$Pd_{16}(CO)_{13}L_9$	4n-20	$C^{11}C[M-5]$
$Tl_2Pd_{12}(CO)_9L_9^{2+}$	4n-16	$C^{9}C[M-5]$
$Pd_8(CO)_{10}L_4$	4n-4	$C^{3}C[M-5]$
$Pd_{23}(CO)_{20}L_8$	4n-36	$C^{19}C[M-4]$
$AuPd_{22}(CO)_{20}L_8$	4n-36	$C^{19}C[M-4]$
$Pd_{37}(CO)_{28}L_{12}$	4n-68	$C^{35}C[M-2]$
$Pd_{35}(CO)_{23}L_{15}$	4n-64	$C^{33}C[M-2]$
$Pd_{29}Ni_3(CO)_{22}L_{13}$	4n-58	$C^{30}C[M-2]$
$Pd_3(CO)_3L_3$	4n+0	$C^{1}C[M-2]$
$Os_{17}(CO)_{36}^{2-}$	4n-28	$C^{15}C[M-2]$
$Pd_{34}(CO)_{24}L_{12}$	4n-64	$C^{33}C[M-1]$
$Os_{20}(CO)_{40}^{2-}$	4n-38	$C^{20}C[M-0]$
$Pd_{54}(CO)_{40}L_{14}$	4n-108	$C^{55}C[M^{-1}]$
$Pd_{39}(CO)_{23}L_{16}$	4n-78	$C^{40}C[M^{-1}]$
$Pd_{52}(CO)_{36}L_{14}$	4n-108	$C^{55}C[M^{-3}]$
$Pd_{16}(CO)_7L_6$	4n-38	$C^{20}C[M^{-4}]$
$Pd_{59}(CO)_{32}L_{21}$	4n-130	$C^{66}C[M^{-7}]$
$Pd_{59}(CO)_{36}L_{14}$	4n-136	$C^{69}C[M^{-10}]$
$Pd_{69}(CO)_{36}L_{18}$	4n-168	$C^{85}C[M^{-16}]$
$Pd_{165}(CO)_{60}L_{30}$	4n-480	$C^{241}C[M^{-76}]$

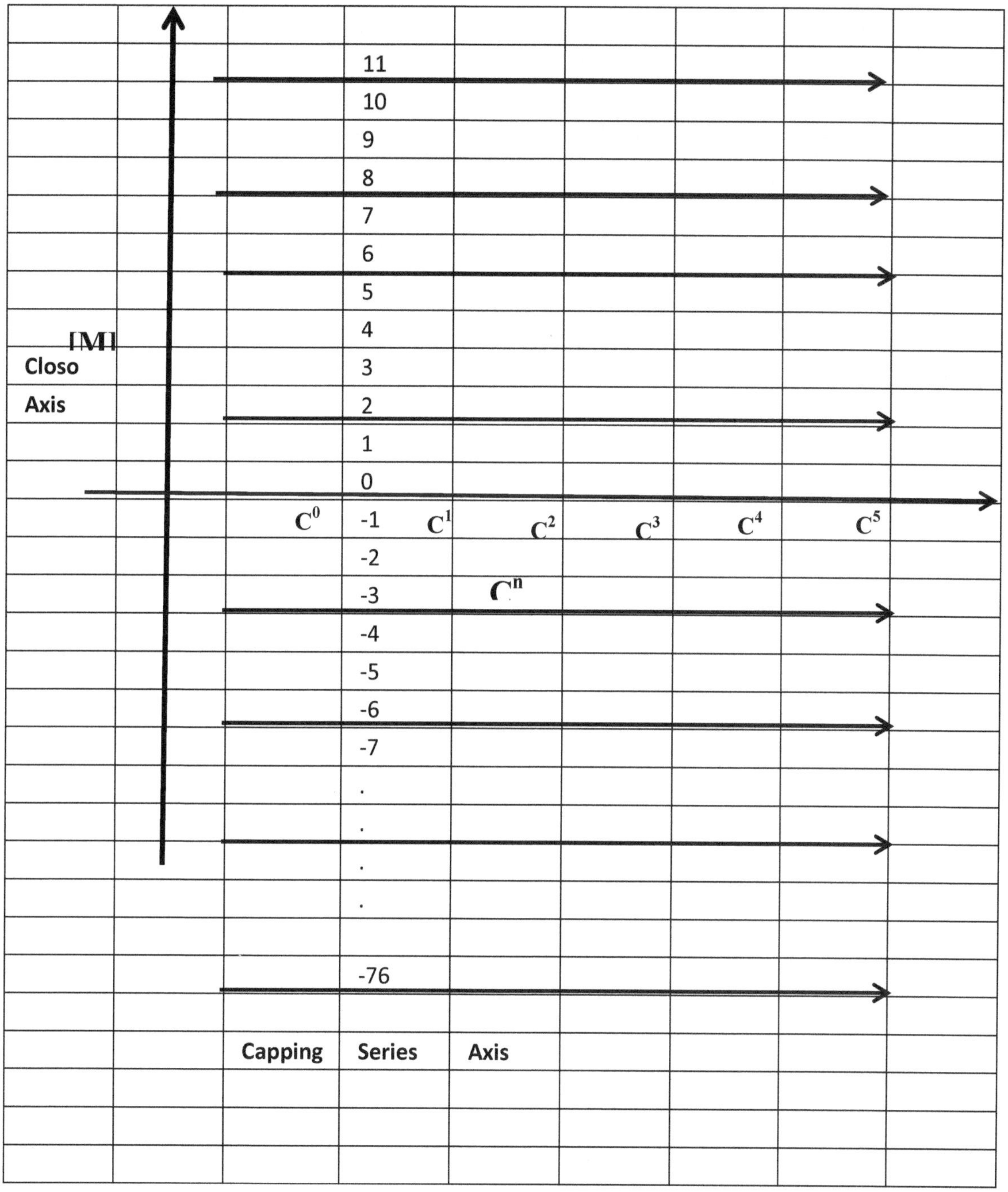

Figure 1. Underground High ways of Capping Carbonyl Clusters

Table 3. Palladium Capping Generating Function of Carbonyl Clusters

C^1	C^2	C^3	C^4	C^5	C^6	C^7	C^8	C^9	C^{10}
4n	4n-2	4n-4	4n-6	4n-8	4n-10	4n-12	4n-14	4n-16	4n-18
$Pd(CO)_2$	$Pd(CO)$	Pd							
$Pd_2(CO)_4$	$Pd_2(CO)_3$	$Pd_2(CO)_2$	$Pd_2(CO)$	Pd_2					
$Pd_3(CO)_6$	$Pd_3(CO)_5$	$Pd_3(CO)_4$	$Pd_3(CO)_3$	$Pd_3(CO)_2$	$Pd_3(CO)$	Pd_3			
$Pd_4(CO)_8$	$Pd_4(CO)_7$	$Pd_4(CO)_6$	$Pd_4(CO)_5$	$Pd_4(CO)_4$	$Pd_4(CO)_3$	$Pd_4(CO)_2$	$Pd_4(CO)$	Pd_4	
$Pd_5(CO)_{10}$	$Pd_5(CO)_9$	$Pd_5(CO)_8$	$Pd_5(CO)_7$	$Pd_5(CO)_6$	$Pd_5(CO)_5$	$Pd_5(CO)_4$	$Pd_5(CO)_3$	$Pd_5(CO)_2$	$Pd_5(CO)$
$Pd_6(CO)_{12}$	$Pd_6(CO)_{11}$	$Pd_6(CO)_{10}$	$Pd_6(CO)_9$	$Pd_6(CO)_8$	$Pd_6(CO)_7$	$Pd_6(CO)_6$	$Pd_6(CO)_5$	$Pd_6(CO)_4$	$Pd_6(CO)_3$
$Pd_7(CO)_{14}$	$Pd_7(CO)_{13}$	$Pd_7(CO)_{12}$	$Pd_7(CO)_{11}$	$Pd_7(CO)_{10}$	$Pd_7(CO)_9$	$Pd_7(CO)_8$	$Pd_7(CO)_7$	$Pd_7(CO)_6$	$Pd_7(CO)_5$
$Pd_8(CO)_{16}$	$Pd_8(CO)_{15}$	$Pd_8(CO)_{14}$	$Pd_8(CO)_{13}$	$Pd_8(CO)_{12}$	$Pd_8(CO)_{11}$	$Pd_8(CO)_{10}$	$Pd_8(CO)_9$	$Pd_8(CO)_8$	$Pd_8(CO)_7$
$Pd_9(CO)_{18}$	$Pd_9(CO)_{17}$	$Pd_9(CO)_{16}$	$Pd_9(CO)_{15}$	$Pd_9(CO)_{14}$	$Pd_9(CO)_{13}$	$Pd_9(CO)_{12}$	$Pd_9(CO)_{11}$	$Pd_9(CO)_{10}$	$Pd_9(CO)_9$
$Pd_{10}(CO)_{20}$	$Pd_{10}(CO)_{19}$	$Pd_{10}(CO)_{18}$	$Pd_{10}(CO)_{17}$	$Pd_{10}(CO)_{16}$	$Pd_{10}(CO)_{15}$	$Pd_{10}(CO)_{14}$	$Pd_{10}(CO)_{13}$	$Pd_{10}(CO)_{12}$	$Pd_{10}(CO)_{11}$
$Pd_{11}(CO)_{22}$	$Pd_{11}(CO)_{21}$	$Pd_{11}(CO)_{20}$	$Pd_{11}(CO)_{19}$	$Pd_{11}(CO)_{18}$	$Pd_{11}(CO)_{17}$	$Pd_{11}(CO)_{16}$	$Pd_{11}(CO)_{15}$	$Pd_{11}(CO)_{14}$	$Pd_{11}(CO)_{13}$
$Pd_{12}(CO)_{24}$	$Pd_{12}(CO)_{23}$	$Pd_{12}(CO)_{22}$	$Pd_{12}(CO)_{21}$	$Pd_{12}(CO)_{20}$	$Pd_{12}(CO)_{19}$	$Pd_{12}(CO)_{18}$	$Pd_{12}(CO)_{17}$	$Pd_{12}(CO)_{16}$	$Pd_{12}(CO)_{15}$
$Pd_{13}(CO)_{26}$	$Pd_{13}(CO)_{25}$	$Pd_{13}(CO)_{24}$	$Pd_{13}(CO)_{23}$	$Pd_{13}(CO)_{22}$	$Pd_{13}(CO)_{21}$	$Pd_{13}(CO)_{20}$	$Pd_{13}(CO)_{19}$	$Pd_{13}(CO)_{18}$	$Pd_{13}(CO)_{17}$
$Pd_{14}(CO)_{28}$	$Pd_{14}(CO)_{27}$	$Pd_{14}(CO)_{26}$	$Pd_{14}(CO)_{25}$	$Pd_{14}(CO)_{24}$	$Pd_{14}(CO)_{23}$	$Pd_{14}(CO)_{22}$	$Pd_{14}(CO)_{21}$	$Pd_{14}(CO)_{20}$	$Pd_{14}(CO)_{19}$
$Pd_{15}(CO)_{30}$	$Pd_{15}(CO)_{29}$	$Pd_{15}(CO)_{28}$	$Pd_{15}(CO)_{27}$	$Pd_{15}(CO)_{26}$	$Pd_{15}(CO)_{25}$	$Pd_{15}(CO)_{24}$	$Pd_{15}(CO)_{23}$	$Pd_{15}(CO)_{22}$	$Pd_{15}(CO)_{21}$
$Pd_{16}(CO)_{32}$	$Pd_{16}(CO)_{31}$	$Pd_{16}(CO)_{30}$	$Pd_{16}(CO)_{29}$	$Pd_{16}(CO)_{28}$	$Pd_{16}(CO)_{27}$	$Pd_{16}(CO)_{26}$	$Pd_{16}(CO)_{25}$	$Pd_{16}(CO)_{24}$	$Pd_{16}(CO)_{23}$
$Pd_{17}(CO)_{34}$	$Pd_{17}(CO)_{33}$	$Pd_{17}(CO)_{32}$	$Pd_{17}(CO)_{31}$	$Pd_{17}(CO)_{30}$	$Pd_{17}(CO)_{29}$	$Pd_{17}(CO)_{28}$	$Pd_{17}(CO)_{27}$	$Pd_{17}(CO)_{26}$	$Pd_{17}(CO)_{25}$
$Pd_{18}(CO)_{36}$	$Pd_{18}(CO)_{35}$	$Pd_{18}(CO)_{34}$	$Pd_{18}(CO)_{33}$	$Pd_{18}(CO)_{32}$	$Pd_{18}(CO)_{31}$	$Pd_{18}(CO)_{30}$	$Pd_{18}(CO)_{29}$	$Pd_{18}(CO)_{28}$	$Pd_{18}(CO)_{27}$
$Pd_{19}(CO)_{38}$	$Pd_{19}(CO)_{37}$	$Pd_{19}(CO)_{36}$	$Pd_{19}(CO)_{35}$	$Pd_{19}(CO)_{34}$	$Pd_{19}(CO)_{33}$	$Pd_{19}(CO)_{32}$	$Pd_{19}(CO)_{31}$	$Pd_{19}(CO)_{30}$	$Pd_{19}(CO)_{29}$
$Pd_{20}(CO)_{40}$	$Pd_{20}(CO)_{39}$	$Pd_{20}(CO)_{38}$	$Pd_{20}(CO)_{37}$	$Pd_{20}(CO)_{36}$	$Pd_{20}(CO)_{35}$	$Pd_{20}(CO)_{34}$	$Pd_{20}(CO)_{33}$	$Pd_{20}(CO)_{32}$	$Pd_{20}(CO)_{31}$

3. Conclusion

Palladium forms a wide range of capping carbonyl clusters. The revelation by series method indicate that palladium forms two types of capping clusters –those which undergo capping “above the ground” and those which do so “underground”. The difference between the above the ground and underground capping clusters is that the closo nucleus of the above the ground capping cluster is a positive function or has a positive value and when added to the capping function produces the molecular formula of the cluster with an enhanced molar mass. On the other hand the closo nucleus of the underground capping cluster is negative function and when subtracted from the capping cluster function gives rise to a molecular formula of the cluster with a reduced molar mass. The examples with $[M^{-q}]$ closo nuclear decapping clusters are given above. By using the capping generating series, some known palladium clusters can be generated as indicated in Table 3. A given molecular formula of a cluster of any size can be decomposed into mono-skeletal fragments from which the cluster series can be derived. The simple series such as bicapped, monocapped, closo, nido, arachno, hypho and klapo can readily be utilized to predict the borane-type of geometrical structures or related ones. The discovery of underground capping clusters has introduced a new dimension in the chemistry of carbonyl clusters. Each cap represents a class of series. For instance, C^1 indicates mono-capped series and C^{10} shows series with 10 caps. Clearly palladium portraying a series of C^{241} capping in $Pd_{165}(CO)_{60}L_{30}$ is perhaps one of the largest carbonyl cluster containing a large number of skeletal metal atoms. The series method used in the above examples indicate that it gives the same result of categorizing clusters as PSEPT method. A large variety of examples categorized by the series method used in this paper have been provided to indicate the viability and flexibility of the method as well as to make it more familiar to the reader. Since 14n series are isolobal to 4n series, it is much easier and faster to work with 4n series for categorization and structure prediction of clusters. This series method can easily be taught at both undergraduate and postgraduate levels.

References

Cotton, F. A., & Wilkinson, G. (1980). *Advanced Inorganic Chemistry, 4th Ed.* , John Wiley and Sons, New York.

Cox, D. N., & Mingos, D. M. P. (1981). Extended Hückel Molecular-orbital Calculations on Dodecahedral Metalloboranes which do not conform to the Polyhedral Skeletal Electron-pair Theory. *J. C. S., Dalton*, 1788-1797. http://dx.doi.org/10.1039/dt9810001788

Hoffmann, R. (1982). Building Bridges between Inorganic and Organic Chemistry. *Angew. Chem. Int. Ed. Engl., 21*,

711-724. http://dx.doi.org/10.1002/anie.198207113

Hughes, A. K., & Wade, K. (2000). Metal-metal and metal-ligand bond strengths in metal carbonyl clusters. *Coord. Chem. Rev., 197*, 191-229. http://dx.doi.org/10.1016/S0010-8545(99)00208-8

Jemmis, E. D. (2005). Building relationships between polyhedral boranes and elemental boron. *Inorg. Chem. 18*, 620-628.

Jemmis, E. D., & Balakrishnarajan, M. M. (2001). Polyhedral boranes and elemental boron. Direct structural relations and diverse electronic requirements. *J. Am. Chem. Soc., 123*, 4324-4330. http://dx.doi.org/10.1021/ja0026962

Jemmis, E. D., & Jayasree, E. G. (2003). Analogies between boron and carbon. *Acc. Chem. Res., 36*, 816-824. http://dx.doi.org/10.1021/ar0300266

Jemmis, E. D., & Pavankumar, P. N. V. (1984). Stability of Polyhedral Borane anions and Carboranes. *Proc. Indian Acad. Sci., 93*(3), 479-489.

Jemmis, E. D., & Prasad, D. L. V. K. (2008). Unknowns in the chemistry of Boron. *Current Science, 95(10),* 1277-1283.

Jemmis, E. D., Balakrishnarajan, M. M., & Pancharatna, P. D. (2001). Unifying electron counting rule for Macropolyhedral Boranes, Metallaboranes, and Metallocenes. *J. Am. Chem. Soc., 123*(18), 4313-4323. http://dx.doi.org/10.1021/ja003233z

Jemmis, E. D., Balakrishnarajan, M. M., & Pancharatna, P. D. (2002). Electronic Requirements for Macropolyhedral Boranes. *Chem. Rev. 102(1),* 93-144. http://dx.doi.org/10.1021/cr990356x

Jemmis, E. D., Jayasree, E. G., & Parameswaran, P. (2006). Hypercarbons in polyhedral structures. *Chem. Soc. Rev., 35*, 157-168. http://dx.doi.org/10.1039/B310618G

King, R. B. (2002). Spherical Chemical Bonding Models for three-diamensional aromaticity in deltahedral boranes. *PINSA, 68, A,* 473-486.

Kiremire, E. M. (2014). Numerical Sequence of Borane Series. *Orient. J. Chem., 30(3),* 1055. http://dx.doi.org/10.13005/ojc/300317

Kiremire, E. M. (2015a). Categorization and Structural Determination of Simple and More Complex Carbonyl Clusters of Rhenium and Osmium Using k values and the Cluster Table. *Orient. J. Chem. 31(1),* 293-302. http://dx.doi.org/10.13005/ojc/310133

Kiremire, E. M. (2015b). Classification of Transition Metal Carbonyl Clusters Using the 14n Rule Derived from Number Theory. *Orient. J. Chem. 31(2),* 605-618. http://dx.doi.org/10.13005/ojc/310201

Kiremire, E. M. (2015d). Isolobal Series of Chemical Fragments. *Orient. J. Chem.*, 31(Spl. Edn), 59-70.

Kiremire, E. M. R. (2015c). Capping and Decapping Series. *Int. J. Chem., 7(2),* 186-197. http://dx.doi.org/10.5539/ijc.v7n2p186

Lee, T., Zhenyang, L., & Mingos, D. M. P. (1989). Polyhedral Skeletal Electron Pair Theory of Bare Clusters. *Inorg. Chem., 28*, 2256-2261. http://dx.doi.org/10.1021/ic00311a004

Mednikov, E. G., & Dahl, L.F. (2010). Syntheses, structures and properties of primarily nanosized homo/heterometallic palladium CO/PR_3-ligated clusters. *Phil. Trans. R. Soc., A(2010), 368,* 1301-1332. http://dx.doi.org/10.1098/rsta.2009.0272

Mingos, D. M. P. (1972). A General Theory for Cluster and Ring Compounds of the Main Group and Transition Elements. *Nature(London), Phys. Sci., 236*, 99-102. http://dx.doi.org/10.1038/physci236099a0

Mingos, D. M. P. (1984), Polyhedral skeletal electron-pair approach. *Acc. Chem. Res., 17*, 311-319. http://dx.doi.org/10.1021/ar00105a003

Rossi, F., & Zanello, P. (2011). Electron Reservoir Activity of High-Nuclearity Transition Metal Carbonyl Clusters. *Portugaliae Electrochimica Acta, 29(5)*, 309-327. http://dx.doi.org/10.4152/pea.201105309

Wade, K. (1971). The structural significance of the number of skeletal bonding electron-pairs in carboranes, the higher boranes and borane ions and various transition metal carbonyl cluster compounds. *Chem. Commun.,* 792-793. http://dx.doi.org/10.1039/c29710000792

Wade, K. (1976). Structural and Bonding Patterns in Cluster Chemistry. *Adv. Inorg. Chem. Radiochem., 18*, 1-16. http://dx.doi.org/10.1016/s0065-2792(08)60027-8

Wales, D. J., (2005). Electronic Structure of Clusters. *Encyclopedia of Inorganic Chemistry, 2nd Ed., King R. B., Editor-in-Chief, John-Wiley and Sons, Ltd,* Chichester, *3*, 1506-1525.

Welch, A. J. (2013). The Significance of Wade's Rules. *Chem. Commun., 49*, 3615-3616. http://dx.doi.org/10.1039/c3cc00069a

12

Removal of Some Heavy Metals from Incinerator Bottom Ash Using Coconut (*Cocos Nucifera*) Husk

Doris F. Ogeleka[1], Harrison E. Avrukpere[1], Elias E. Elemike[1], Oghenekohwiroro Edjere[2], Iwekumo E. Agbozu[2]

[1]Department of Chemistry, Federal University of Petroleum Resources, Effurun, Delta State, Nigeria

[2]Department of Environmental Management and Toxicology, Federal University of Petroleum Resources, Effurun, Delta State, Nigeria

Correspondence: Doris F. Ogeleka, Department of Chemistry, Federal University of Petroleum Resources, Effurun, Delta State, Nigeria. E-mail: dorysafam@yahoo.com

Abstract

There has been a great limitation to the use of incinerator bottom ash (IBA) residue in the construction and manufacturing industries due to the high contents of heavy metals, which poses great threat to the environment and human health. This study was carried out to evaluate the efficiency of heavy metals (Fe, Zn, Cd and Pb) removal from domestic IBA filtrate using a biological material coconut (*Cocos nucifera*) husk as adsorbent and to remove the metals from the ash residue using the toxicity characteristic leaching procedure (TCLP). The mean percentage removal of the heavy metals for the incinerator bottom ash (IBA) filtrate varied from 75.84 - 99.42% for the two ash types. Similarly, the percentage removal of metals from the ash residues was between 92.34 and 99.97%. The data from this study showed that after the treatment, the heavy metals in the ash filtrate and residue were significantly reduced. Thus, it would be feasible to use the IBA residue as a partial replacement of cement in the construction and manufacturing industries for making monolithic blocks, interlocking bricks, concretes and road construction materials. This would help improve and provide good and effective waste management practice that would reduce the huge volume of ash residue from incineration combustion processes.

Keywords: incinerator bottom ash, waste management & utilization, adsorption, hazardous and domestic waste, TCLP, heavy metals

1. Introduction

Incineration is one of the major techniques for managing solid wastes in some developed and developing nations since it reduces the volume of waste by 80 to 90%. Municipal solid waste (MSW) consists of everyday items we use and dispose such as product packaging, furniture, clothing, bottles, food scraps, newspapers, appliances, paint, batteries etc. These wastes which come from homes, schools, hospitals, and businesses could cause detrimental effects on environmental media if not properly managed or disposed (EPA, 1980; El-Haggar, 2007). Depending on the municipality, solid waste could be managed using the four R's of waste management namely reduce, reuse, recycle and recover, however, incineration could be used for the final disposal of various types of waste. As waste generation rate increases, incineration processing costs increases, and available landfill space decreases, thus, the four R`s have become a central tenet in sustainable waste management efforts (Seadon, 2006; Davis, 2008; Suttibak & Nitivattananon, 2008; Tudor *et al.*, 2011). Some techniques commonly used for the treatment or management of waste in addition to incineration include: landfill/ engineered landfill, compositing, mechanical biological treatment (MBT), pyrolysis and gasification, detoxifying hazardous waste, thermal desorption unit (TDU), etc.

Generally, solid waste incineration produces two main types of ash, which can be grouped as incinerator bottom ash (IBA), which are solid waste that are not completely burned on the grate and incinerator fly ash (IFA), which are solid and condensable particulate matter which leaves the furnace suspended in the combustion gases and are subsequently collected in emission control devices. The waste (ash) generated from incineration, usually ends up in two ways; disposal in landfill or reuse as secondary raw materials. In most developed countries where land is scarce and environmental controls are enforced, environmental policies tend to reduce landfill disposals as much as possible (Lam *et al.*, 2010).

The ash residues produced from the burning of wastes could be used commercially as raw material in cosmetics production including soap making. It could also be used as partial replacement for cement in many manufacturing and construction industries in making noise barriers, capping layer on landfill sites and as aggregate in asphalt. However, this use is limited due to the presence of some contaminants in the ash, as a result of the chemical and physical characterization of the ash residues, which depend on the compositions of the raw waste, type of incinerator used and incineration process (Li *et al.*, 2004).The incineration bottom ash (IBA) is the most significant by-product from incineration as it accounts for 85 - 95% product resulting from combustion. It is important to note that the metals in the wastes are not destroyed during the incineration process and may pose harmful environmental consequences and health problems if not properly treated or managed (Ogwuegbu *et al.*, 2007). Much of the exposure to heavy metal pollution has been scientifically proven to be linked to causing free radical damage leading to heart attacks, cardiovascular disease, high blood pressure, strokes, cancer and many more. In environmental premises, bioavailability of these metals could lead to respiratory problem, growth impairment, reproduction defect as well as disruption of the ability of the organism to fight disease and function (Saleh, 1982).

Over the last few decades, several methods have been devised for the treatment and removal of heavy metals from contaminated materials. The commonly used procedures for removing metal ions from aqueous streams include; adsorption, ion exchange, biosorption, filter traps, electrodeposition and chemical precipitation (Rich & Cherry, 1987). The method adopted for this research is adsorption, which is a process through which a substance, originally present in one phase, is removed from that phase by accumulation at the interface between that phase and a separate (solid) phase (Jalal *et al.*, 2002; Agbozu & Emoruwa, 2014).The contaminants commonly found in the bottom ash which contribute adversely to the environment and human health and were considered in this study include zinc, iron, lead and cadmium. The objective of this study was to evaluate the removal of these contaminants from the incinerator bottom ash liquid extract (leachate) using coconut (*Cocos nucifera*) husk and from the ash residue using the toxicity characteristic leaching procedure (TCLP).

2. Materials and Method

2.1 Study Area

The selected point of sampling lies between 5°31'N and 5° 45'E, Warri, a City in the Niger Delta ecological zone of Delta State, Nigeria (Figure 1). It is an oil hub in southern Nigeria and experiences moderate rainfall and moderate humidity for most part of the year. The area is characterized by tropical equatorial climate with mean annual temperature of 32.8 °C and rainfall between 3000 – 4500 mm (Tamuno, 2000; Balouga, 2009). It is one of the major hubs of petroleum activities whose waste management practices involve the use of open dump sites for public utility while incineration and thermal desorption unit are used mainly by some specific Oil Exploration and Production (E&P) companies (Ekeh, 2005).

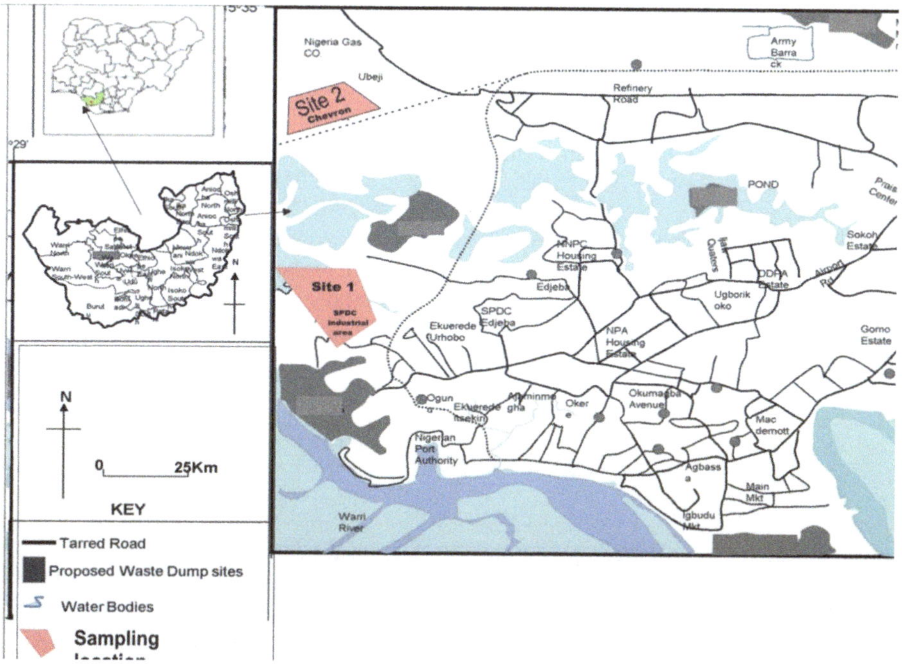

Figure 1. Map of sampling location in Warri, Delta state

2.2 Ash Sampling

The incinerator bottom ash (IBA) was collected from the waste management unit of two oil Exploration and Production (E&P) companies. Twenty four (24) samples were randomly collected in triplicate and stored for analysis. The waste types used in this study include: domestic waste ash (DW) - obtained through the incineration of domestic waste other than hazardous waste and hazardous waste ash (HW) - obtained through the incineration of medical waste, other biological and hazardous waste. These ash wastes would subsequently be referred to as ash A for domestic waste (DW) and ash B for hazardous waste (HW).

2.3 Bio-adsorbent Sampling and Treatment

Coconut (*Cocos nucifera*) husk was collected and sun dried for approximately seven (7) days and ground into fine particles and sieved to obtain 120μm (micrometer mesh). The particles were treated with 0.1 *M* HCl and re-introduced into an oven at a temperature of 30°C for 30 minutes and then preserved for further use.

2.4 Extraction and Determination of Metals in Ash Residue Using TCLP

Five (5) g of the crushed ash sample was accurately weighed into a sample extraction bottle and 96.5 mL of distilled water was added. This was covered with a watch glass and stirred vigorously for 5 minutes and the pH and electrical conductivity was determined. From the guidelines of EPA - SW-846 #1311 and DPR (EPA, 1992; DPR, 2011), extraction solution #2 was used for the metal extraction in the ash sample because the pH of the solution was greater than 5. If otherwise (i.e.< 5), extraction fluid #1 would have been used. Extraction fluid #1 was prepared by accurately measuring 5.7 mL glacial acetic acid, to which 500 mL of distilled water and 64.3 mL of 1 *M* NaOH was added and diluted to a volume of 1 litre. Extraction fluid #2 was prepared by accurately measuring 5.7 mL glacial acetic acid and added to distilled water make up to a volume of 1 litre.

For the metal extraction, 200 mL of extracting solution was added to 10 g of the crushed ash residue sample. This solution was shaken first for 2 minutes by hand to ensure saturation of the solid with the solution. Then the flask was shaken with an oscillator for 16 hours to ensure full saturation of the solid with the solution. The resultant slurry was centrifuged at 2000 rpm and the supernatant was carefully decanted through a glass funnel and the filtrate stored. This process was repeated three (3) times and the filtrate stored for the adsorption procedure while the concentration of heavy metals in the residues was determined using atomic absorption spectrophotometer.

2.5 Elution and Determination of Metals in Ash Filtrate

Ten (10) g powder of the pre-treated biological material - coconut (*Cocos nucifera*) husk was packed in a column stoppered at the tip with glass wool and 50 mL of the washed ash filtrate from the extraction of the residue introduced, allowed to drain through the packed column and repeated three times. The time taken for elution was recorded with a stop watch. Fifty (50) mL of the resultant eluate was digested using 5 mL concentrated nitric acid (AR). This reduced to the lowest volume possible (15 to 20 mL). Filtration was done after digestion to remove any insoluble material. The filtrate was then diluted to volume with distilled water in a 50 mL volumetric flask (APHA, 2005). The concentration of the heavy metals was analyzed using Atomic Absorption Spectrophotometer (AAS) by direct aspiration into a standardized computer interfaced with the instrument.

3. Results

The results obtained from the heavy metals analysis carried out on incinerator bottom ash are presented in Tables 1 and 2 with further illustrations in Figure 2-8. The tables indicate the concentration of heavy metals in the incinerator bottom ash filtrate before and after treatment with coconut husk as well as the extraction of metals from the ash residue.

Table 1. Mean percentage metal removal from ash filtrate

Parameters (mg/L)	Untreated	Treated	% removal	Untreated	Treated	% removal
	Ash A	Ash A	Ash A	Ash B	Ash B	Ash B
Iron	5.118 ± 0.05	1.236 ± 0.007	75.84 ± 0.03%	6.403 ± 0.09	1.132 ± 0.010	82.32 ± 0.06%
Zinc	3.152 ± 0.03	0.030 ± 0.005	99.05 ± 0.06%	4.106 ± 0.06	0.024 ± 0.002	99.42 ± 0.08%
Cadmium	1.046 ± 0.008	0.021 ± 0.004	97.99 ± 0.05%	0.092 ± 0.005	0.011 ± 0.001	88.04 ± 0.03%
Lead	< 0.001	< 0.001	-	< 0.001	< 0.001	-

The washing of the ash resulted in filtrate, which was bio-treated using coconut husk and resulted in a decrease in the metal concentrations (Table 1). There was significant difference at levels of $P<0.05$ between concentration obtained before and after adsorption (Figure 2-3). Although ash B contained slightly a higher magnitude of metals than ash A, this could be as a result of the constituent of the hazardous waste, which comprises mainly of medical waste (syringes, needles, blades, forceps, and other related bio-medical waste) which was not destroyed in the incinerated residues.

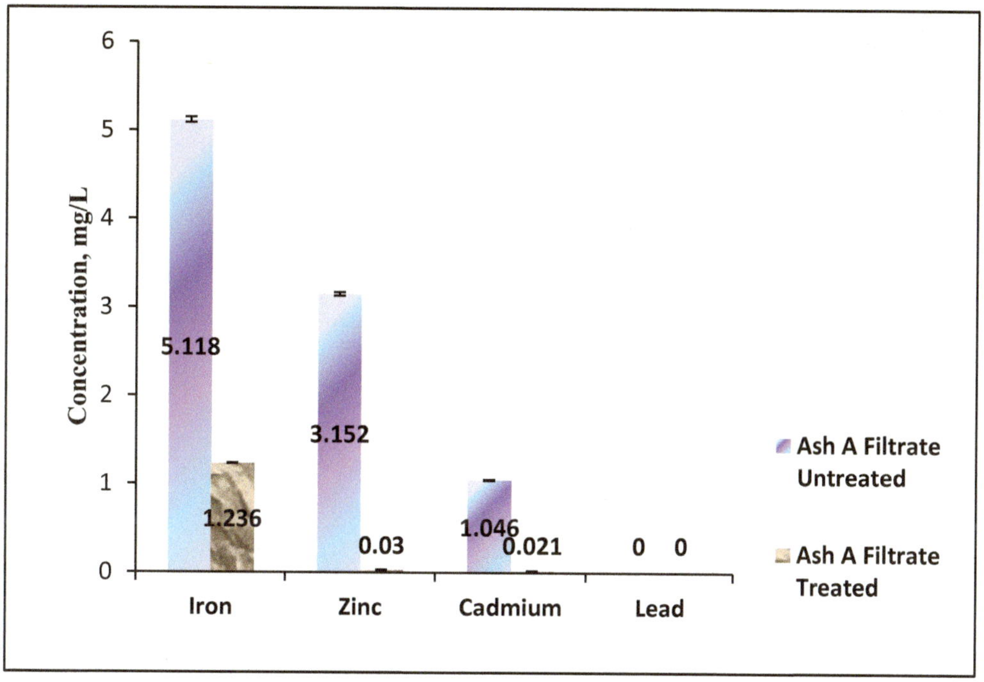

Figure 2. Mean ± SEM of Ash A filtrate before and after adsorption

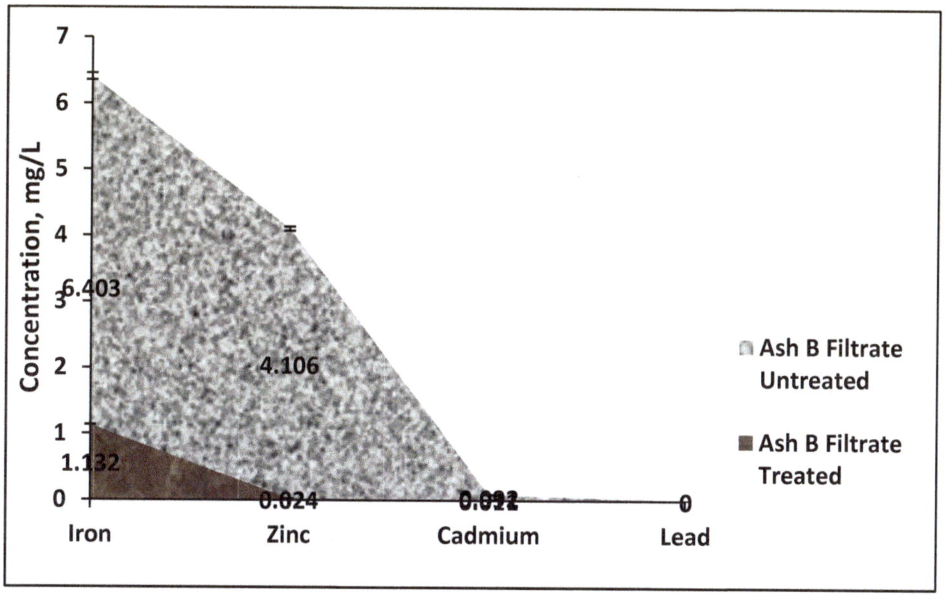

Figure 3. Mean ± SEM of Ash B filtrate before and after adsorption

The average percentage removal of heavy metals after treatment for the two ash samples resulted in a range of 75.84 - 99.05% and 82.32 – 99.42% for ash A and B respectively (Figure 4).

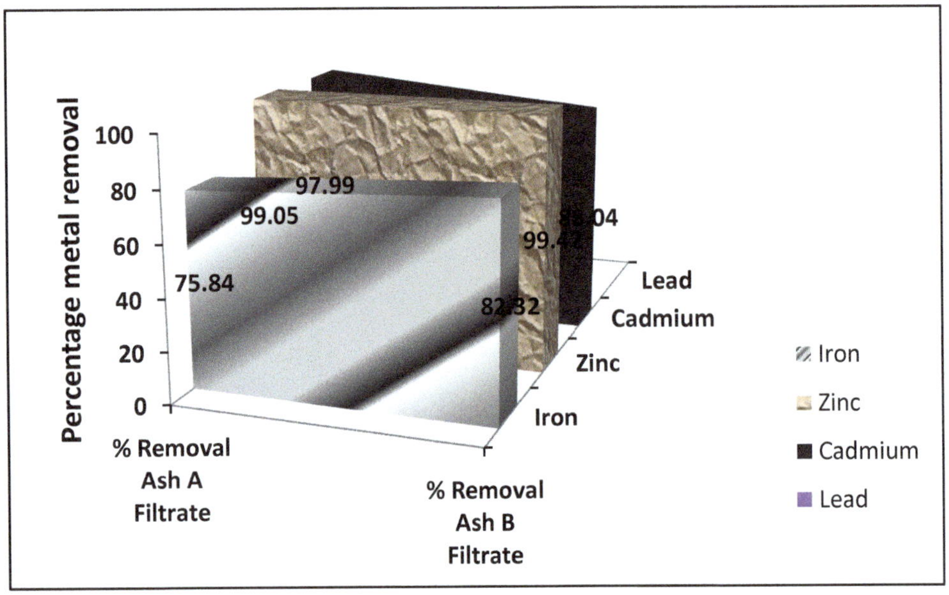

Figure 4. Mean % metal removal in Ash A and B filtrate

There was slight variation between the two ash samples in terms of percentage metal removal.

Table 2. Mean percentage metal removal from ash residue

Parameter, (mg/kg)	DPR Limits	Before extraction Ash A	After extraction Ash A	% removal Ash A	Before extraction Ash B	After extraction Ash B	% removal Ash B
Iron	N/A	175.340 ± 0.9	2.056 ± 0.05	98.83 ± 0.05%	181.280 ± 0.11	2.965 ± 0.06	98.36 ± 0.04%
Zinc	50.00	20.576 ± 0.34	1.576 ± 0.007	92.34 ± 0.04%	16.756 ± 0.56	0.524± 0.009	96.87 ± 0.06%
Cadmium	1.00	0.187 ± 0.005	< 0.001	99.97 ± 0.08%	0.237 ± 0.008	< 0.001	99.97 ± 0.09%
Lead	5.00	4.575 ± 0.08	0.001 ± 0.001	99.95 ± 0.07%	4.845 ± 0.06	0.001± 0.001	99.95 ± 0.08%

N/A = Not available

The ash residues contained some quantity of heavy metals before extraction, although the levels of zinc, cadmium, and lead were within the DPR stipulated limits of 50.00 mg/kg, 1.00 mg/kg, 5.00 mg/kg respectively. However, the concentrations in the ash residues were significantly reduced after extraction (Table 2). The levels of significance before and after extraction was at $P<0.05$ (Figures 5 and 6).

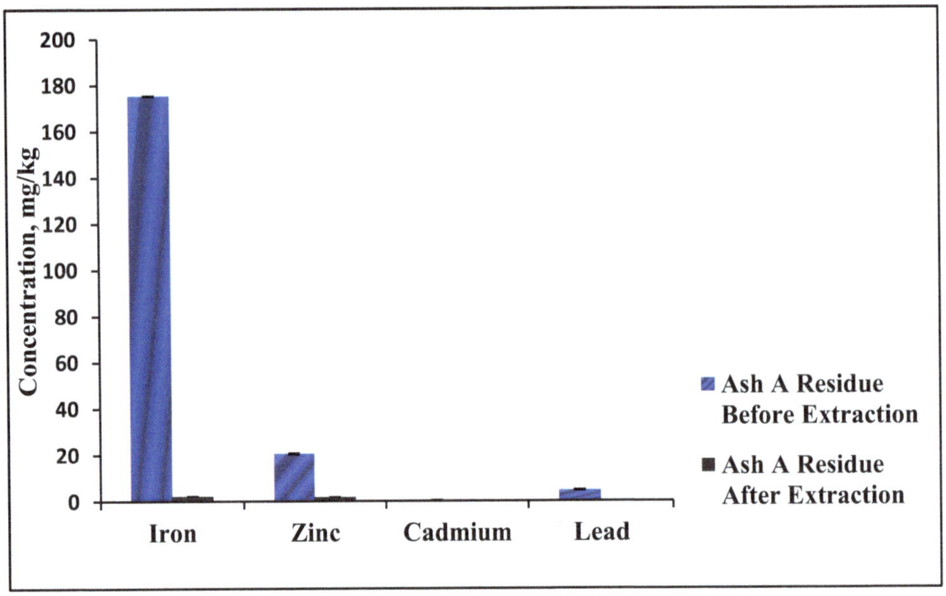

Figure 5. Mean ± SEM of Ash A residue before and after extraction

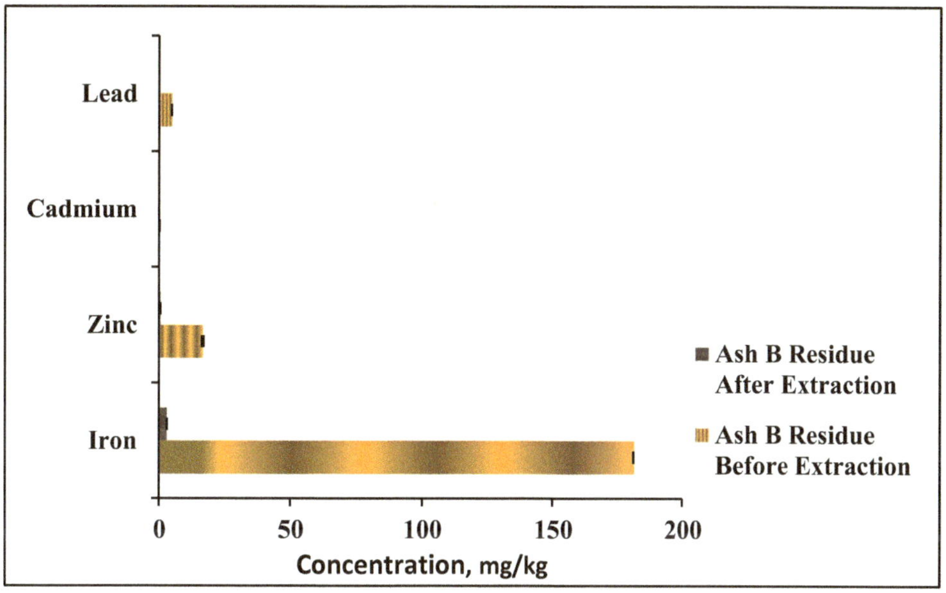

Figure 6. Mean ± SEM of Ash B residue before and after extraction

The percentage of the heavy metals removed after washing of ash A and B was 92.34–99.97 and 96.87–99.97 respectively (Figure 7).

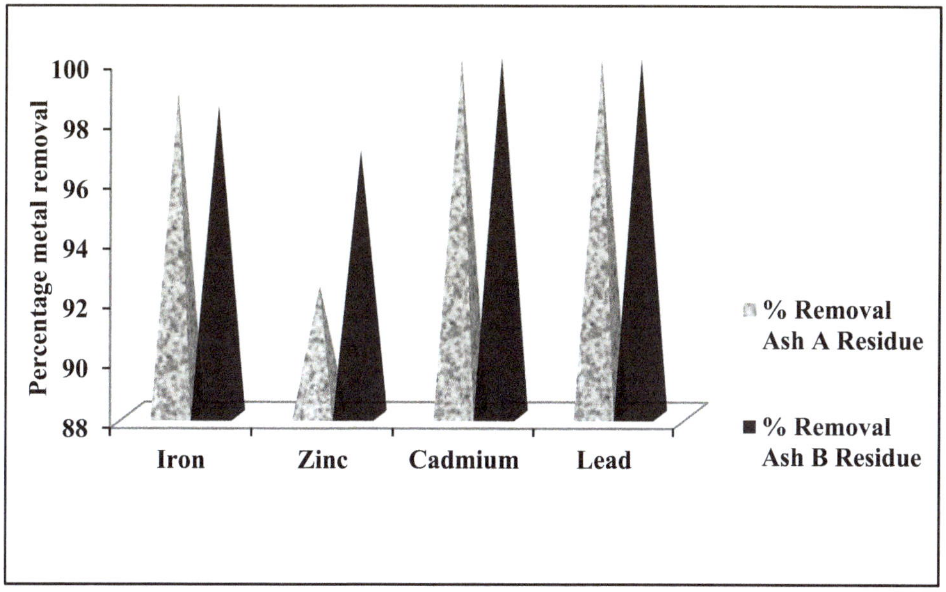

Figure 7. Mean % metal removal in Ash A and B residue

A significant % (> 90%) of metal was removed from the ash residue in the two ash types. However, the magnitude in variation between the two ash samples was slight except for zinc.

The pH and electrical conductivity values recorded for ash A and B was 10.76, 9.71 and 8.92 and 7.94 mS/cm respectively. The values for ash A was in higher magnitude than their corresponding DPR limits for pH (6.50-9.00) and electrical conductivity (8.00) (Figure 8).

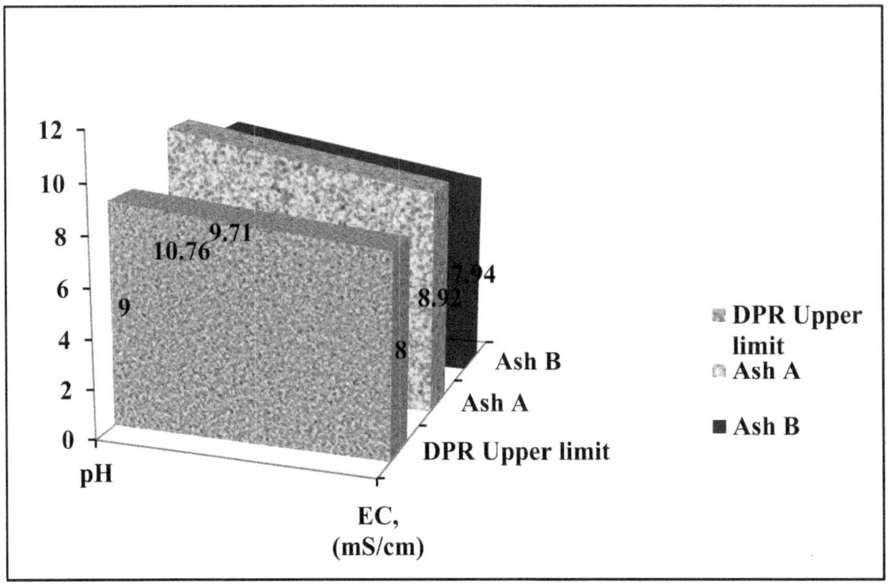

Figure 8. pH and Electrical conductivity in comparison with DPR limits

4. Discussion

4.1 Efficiency of Metal Removal from Ash Residue

The TCLP test using the EPA procedure of SW-846 method #1311 and the DPR protocol was to determine the mobility of metals and toxicants in the ash residue. The leaching test had been recognized as the primary and most widely used indicator for evaluating the retention capacity of toxicant in solid waste samples. The method is the process by which inorganic, organic contaminants or radionuclides are released from the solid phase into the liquid phase under the influence of mineral dissolution, desorption, complexation process as affected by the pH (EPA, 1992; DPR, 2011; Singh and Gadi, 2012).

The study evaluated the percentage of the metals that was leached out of the ash residue since the incinerator ash leftover/residue is considered toxic due to the fact that it still contained an appreciable amount of lead, cadmium, iron, zinc and other heavy metals. Thus, the binding capacity of substances in the solid matrix at high pH levels could be quite high and would not be expected to leach into underground or surface water except the pH of rain water or contact water of that of the environment falls below 3 pH units, which could cause dissolution of metals and allow them to be released from the solid matrix.

Moreover, if the incinerator ash is left untreated before being used as a raw material for construction and manufacture purposes, leaching would be highly detrimental to various forms of lives in the environment and human health due to the high heavy metal content that would be bio-available at very low pH (2–3) since the metals are not destroyed during the incineration process (Baby *et al.*, 2010). Meeting regulatory limits is a standard consideration for ensuring that the ash waste are not released at levels that can harm aquatic and terrestrial organisms as well as man being the end user or consumer of the organisms in the environment.

However, the heavy metals evaluated in this research showed relatively low concentrations after the TCLP treatment method and would not be expected to leach and affect the environment (surface, ground water, soil and sediment).That is the mobility of pollutant would be very minimal and may not likely impact environmental media if the ash residues are solidified/stabilized as bricks/concrete (Okieimen & Onyenkpa, 2000; Jalal *et al.*, 2002). The treatment would make the incinerator ash feasible to use for construction and manufacturing purposes as well as enhance proper and best waste management practice. However, since metals are bound effectively within the solidified waste matrix, the potential of converting a solidified ash waste to bricks and tiles for pavement, parking lots and paths or even for road construction would be highly exploited.

4.2 Efficiency of Metal Removal from Ash Filtrate

The susceptibility of pH could be linked to the nature of the ions in the filtrate and the adsorbent used. Similarly, the nature of the hydrogen ion concentration in solution is also as a result of the component that constitutes the ash sample, thus creating more sites for the adsorption process. Increase in concentration of the adsorbate brings about increase in competition of adsorbate molecules for few available binding sites on the surface of the adsorbent, hence, increase in the amount of metal ions removed. These results are in accordance with the work of Okieimen & Onyenkpa, (2000).

The percentage removal of heavy metal was due to the greater availability of the exchangeable sites and surface area of the adsorbent. The coconut husk was adequately treated to remove traces of heavy metals that would interfere or cause contamination of the process. The removal of metal ion from the ash filtrates was effective since the bio-material was able to release an appreciate level of the metals present in the solution. The percentage heavy metal removal obtained from treating the two ash samples with the biological material (coconut husk), indicated that it is a safe and cheap material for the treatment of the incinerator bottom ash filtrates / solutions (Abdel-Ghani & El-Chaghaby, 2009).

The use of the cellulosic biological waste material as an adsorbent in treatment of incinerator bottom ash filtrate and the TCLP method of metal removal from solid ash waste would go a long way in solving environmental pollution problems that seems to be intractable to the society. The results and findings of this study are in agreement with previous studies made on heavy metal removal using coconut husk as an adsorbent (Agbozu & Emoruwa, 2014).

5. Conclusion

We conclude that cellulosic biological waste (e.g. coconut husk) was a good adsorbent for the removal of heavy metals from incinerator bottom ash filtrate and solution. Similarly, the TCLP method was effective in the removal of some major heavy metal contaminants in the incinerator bottom ash (IBA) residues, thus, enhancing the chances of using it in construction industries for making monolithic blocks, interlocking bricks, concretes and other road construction materials. Thus, effective removal of the contaminants from the ash residue could also mean an environmentally safe and convenient use of the raw material in cosmetics production including soap making, road construction, as foundation material, in noise barriers, as capping layer on landfill sites and in some countries as an aggregate in asphalt etc.

References

Abdel-Ghani, N. T., & El-Chaghaby, G. A. (2009). Simultaneous removal of chromium, copper, cadmium and lead ions from aqueous solution by adsorption onto kaolin. *Int. J. Geotechnics Environ, 1*(2), 161-171.

Agbozu, I. E., & Emoruwa, F. O. (2014). Batch adsorption of heavy metals (Cu, Pb, Fe, Cr, and Cd) from aqueous solutions using coconut husk. *African Journal of Environmental Science and Technology, 8*(4), 240-246.

American Public Health Association APHA, America Water Works Association. AWWA, Water Environmental Federation WEF. (2005). Standard methods for the examination of water and wastewater. 21st ed.

Baby, J., Raj, J. S., Biby, E. T., Sankarganesh, P., Jeevitha, M. V., & Ajisha, S. U. (2010). Toxic effect of heavy metals on aquatic environment. *Int. J. Biol. Chem. Sci., 4*(4), 939-952.

Balouga, J. (2009). The Niger Delta: Defusing the time bomb. *International Association for Energy Economics, 1*, 8-11.

Davis, U. C. (2008). The 4 R's of waste reduction: Waste management. *Academic Journal, 40*, 13-17.

Department of Petroleum Resources (DPR) (2011).Environment Guidelines and Standards for the Petroleum Industry in Nigeria (EGASPIN) Revised Edition.

Ekeh, P. P. *(2005). Warri City and British colonial rule in Western Niger Delta. Urhobo Historical Society, 31.*

El-Haggar, S. M. (2007). Sustainable industrial design and waste management: Cradle-to-cradle for sustainable development. Oxford: Elsevier/Academic Press. 424.

Environmental Protection Agency (EPA). (1980). Environmental fact sheet, [EPA/530-SW-90-029c]: Characterization of municipal waste combustion ash, ash extracts, and leachates. Springfield, Va. 22161.

Environmental Protection Agency (EPA).(1992). Test methods for evaluating solid wastes (SW 846), physical and chemical methods. Method 1311, toxicity characteristic leaching procedure (TCLP), Environmental Protection Agency, Washington, DC, 1-38.

Jalal, R., Bhajourian, H., Asif, Y., Davarpanash, S. J., & Sepher, S. (2002). Removal and recovery of lead using nonliving biomass of algae. *J. Hazardous Mater, 24*, 421-430. http://dx.doi.org/10.1016/s0304-3894(02)00021-3

Lam, C. H. K., Ip, A. W. M., Barford, J. P., & McKay, G. (2010). Use of incineration MSW ash: A Review. *Sustainability, 2*, 1943-1968. http://dx.doi.org/10.3390/su2071943

Li, M., Xiang, J., Hu, S., Sun, L. S., Su, S., Li, P. S., & Sun, X. X. (2004). Characterization of solid residues from municipal solid waste incinerator. *Fuel, 83*, 1397-1405. http://dx.doi.org/10.1016/j.fuel.2004.01.005

Ogwuegbu, M. O. C., Egwurugwu, J. N., & Duruibe, J. O. (2007). Heavy metal pollution and human biotoxic effect. *Int. J. of Physics Sciences, 2*, 112.

Okieimen, F. E., & Onyenkpa, V. U. (2000). Binding of cadmium, copper, lead and nickel ions with melon (*Citrullus vulgaris*) seed husk. *Biol. Waste, 29*, 11-16. http://dx.doi.org/10.1016/0269-7483(89)90099-2

Rich, G., & Cherry, K. (1987). Hazardous waste treatment technologies. Pudram Publ. Co., 169.

Saleh, H. H. (1982). Fish liver as an indicator for aquatic environmental pollution. *Bull. Inst.Oceanogr. and Fish. 8*(1), 69-79.

Seadon, J. K. (2006). Integrated waste management: looking beyond the solid waste horizon. *Waste management, 26*(12), 1327-1336. http://dx.doi.org/10.1016/j.wasman.2006.04.009

Singh, N., & Gadi, R. (2012).Bioremediation of Ni(II) and Cu(II) from wastewater by the nonliving biomass of *Brevundimonas vesicularis .Journal of Environmental Chemistry and Ecotoxicology*, *4*(8), 137-142.

Suttibak, S., & Nitivattananon, V. (2008). Resources, conservation and recycling assessment of factors influencing the performance of solid waste recycling programs. *Conservation and Recycling*, *53*, 45-56. http://dx.doi.org/10.1016/j.resconrec.2008.09.004

Tamuno, T. (2000).The Niger Delta question, Port Harcourt: Riverside Communications. Tell Magazine. (2005). 18 April, 16-18.

Tudor, T., Robinson, G., Riley, M., Guilbert, S., & Barr, S. (2011). Challenges facing the sustainable consumption and waste management agendas. *Local Environment*, *16*(1), 51-66. http://dx.doi.org/10.1080/13549839.2010.548372

13

Toxicological Effects of Differently Polluted Dam Waters Spiked with Pesticides on Freshwater Snails *Lymnaea Natalensis*

Majaya Rachel Delfina[1], Donald Tapfuma[2], Sanele Mnkandla[1], Norah Basopo[1]

[1]Departments of Applied Biology and Biochemistry, National University of Science and Technology, P.O. Box AC 939, Ascot, Bulawayo, Zimbabwe

[2]Departments of Environmental Science and Health, National University of Science and Technology, P.O. Box AC 939, Ascot, Bulawayo, Zimbabwe

Correspondence: Norah Basopo, Departments of Applied Biology and Biochemistry, National University of Science and Technology, P.O. Box AC 939, Ascot, Bulawayo, Zimbabwe.

Abstract

Pesticides extensively used in agricultural fields to ensure high quality crop yields indirectly find their way to aquatic bodies where they affect aquatic biota. We investigated the effects of pesticides in different dam waters on esterase enzyme activity of the freshwater snail species *Lymnaea natalensis*. Groups of adult snails were exposed to 0.006 ppm chlorpyrifos and 0.003 ppm aldicarb in polluted water from Umguza dam and relatively pristine water from Hillside dam for 14 days. Carboxylesterase, acetylcholinesterase and arylesterase activities were measured. Both pesticides caused significant inhibition of esterase activity after the 14 day exposure period, with exposures to Umguza dam water showing higher inhibition as compared to exposures to Hillside dam water. Aldicarb and chlorpyrifos both showed a time-dependent inhibition of enzyme activity, the former causing a higher inhibitory effect as compared to the latter. Acetylcholinesterase was inhibited up to 80% following exposure to aldicarb while exposure to chlorpyrifos in Umguza water caused only 40% inhibition. Carboxylesterases were similarly inhibited with higher inhibition observed in snails exposed to Umguza dam water when compared to snails exposed to Hillside dam water, while arylesterases were inhibited in the range 80-90%, with an exception of chlorpyrifos spiked Hillside dam water which caused 45% inhibition. Contaminated Umguza dam water also appeared to enhance the effects of pesticides when compared to the relatively pristine Hillside dam water. Alteration of esterase activity can be used as an early warning signal indicating exposure to environmental pollutants. The results of this study therefore, highlight the adverse effects of pesticides on non-target aquatic organisms, evidenced by the inhibition of esterase activity.

Keywords: organophosphates, carbamates, esterases, snails, pollution

1. Introduction

Pesticides are applied widely to protect plants from diseases, weeds, fungal and insect damage. They usually come into contact with soil and aquatic systems, where they undergo a variety of transformations that provide a complex pattern of metabolites. Although the use of pesticides has resulted in increased crop production and other benefits, it has raised concerns about potential adverse effects on the environment and aquatic life (Aktar *et al.*, 2009).

Carbamate and organophosphate pesticides such as aldicarb and chlorpyrifos respectively, are widely used for agricultural and non-agricultural purposes (Maltby and Hills, 2008; Wang *et al.*, 2010). These pesticides however are environmentally toxic and have a great potential for unintended adverse effects through contamination of rivers, lakes and dams (Lamers *et al.*, 2011). Runoffs and soil erosion after heavy rains carry the applied pesticides to rivers, ponds and dams. The pesticides may also enter aquatic systems as non-point source pollutants through spray drifts or leaching (Tiryaki and Temur, 2010).

The presence of agricultural pollutants in aquatic bodies disturbs the ecosystem health, which necessitates the detection of biological changes due to exposure to pollutants (Kaviraj *et al*, 2014). This is achieved through the use of measurable indicators known as biomarkers. Biomarkers are change in biological parameters related to exposure to environmental chemicals (Kaviraj *et al*, 2014). Organisms, termed bioindicators, are used to measure biomarker responses following exposure. The organisms include, fish and mollusks which have good filtration capacities, are sensitive to damage following exposure to chronic or sublethal concentrations and exhibit ease of caging. These bioindicators provide good model systems for the investigation of stress induced damage, cell response and repair mechanisms employed to ameliorate damage (Valavanidis *et al.*, 2006).

The biomarkers used in this study are the esterase enzymes; acetylcholinesterase, carboxylesterase and arylesterase. Esterases are found in all living cells and they catalyse the hydrolysis of ester bonds (Vioque-Fernández, 2007). The activities of the esterases have been found to be constantly inhibited by organophosphates `and carbamates (Vioque-Fernández, 2007; Sayali *et al.*, 2013). Inhibition of acetylcholinesterases in aquatic organisms can be used as a warning signal of harmful effects of environmental pollutants such as pesticides on aquatic biota (Lionetto, 2013). Pesticides cause various toxicological effects on aquatic life. This was also highlighted by Anandhan *et al.*, (2012) who reported that in addition to neurotoxicity, organophosphates also caused immunotoxic effects in aquatic organisms. Acetylcholinesterase inhibition in fish exposed to pesticides has been associated with impaired homing behavior and anti-predator alarm responses in Chinook salmon (Scholz *et al.*, 2000) and decreased spontaneous swimming rates and feeding rates in coho salmon (Sandahl *et al.*, 2005). This study therefore assessed the effects of pesticide spiked dam waters on esterases of the freshwater snail, *L. natalensis* and the potential of using this enzyme system as biomarkers of exposure to pesticides.

2. Materials and Methods

2.1 Chemicals

All the pesticides, substrates and standards were purchased from Sigma Aldrich Chemical Company, Germany. All other laboratory reagents were of analytical grade.

2.2 Water Sampling

Water samples were collected from two Bulawayo dams: Hillside and Umguza, in the month of February during the agricultural and rainy season. Hillside dam, situated in the Hillside recreational conservancy (Sebata, 2015), was selected because it is located upstream of effluent discharge zones of the city of Bulawayo and is therefore considered relatively pristine. Umguza dam on the other hand is located downstream of effluent discharges from industrial, domestic and agricultural activities of and surrounding the city of Bulawayo (Siwela *et al.*, 2009; 2010; Dube *et al.*, 2010). Two sampling sites were selected at each dam. The physico-chemical water quality properties were determined in situ using the Eutech Cyberscan Instruments water proof series meters (pH: 300 Series, conductivity: 410 Series), before the water was drawn into 25 litre containers and transported to the laboratory. The waterproof Eutech Cyberscan pH 310 meter was used to measure pH while the Eutech Cyberscan CON 410 meter was used to measure conductivity and total dissolved solids (TDS).

2.3 Snail Breeding and Exposure

The snail species used in this study were the *Lymnaea natalensis.* The snails were bred in cement tanks containing tap water and were fed on fresh garden lettuce twice a week. Prior to the exposures, the snails were collected from the cement tanks and brought to the laboratory where they were placed in tap water for 3 days to acclimatize. Groups of adult snails (21) were separately exposed to 0.006 ppm chlorpyrifos and 0.003 ppm aldicarb for 14 days. The exposures were done in I L volume of collected dam water, at room temperature and in duplicate. Seven snails were removed from the exposure tanks at time intervals (day 3, 7 and 14) for homogenization and preparation of the post mitochondrial fraction for biochemical analysis. The sublethal concentrations used for exposures were a third dilution of the LC50 obtained following several screening trials.

2.4 Sample Preparation of the Post Mitochondrial Fraction (PMF)

After the exposure period, snails were sacrificed. The shells were carefully removed using forceps and the soft tissue isolated on ice. The whole body tissues were washed in distilled water, placed on filter paper to drain extra fluids, and weighed. Each pool of 7 snails was treated similarly. The snails were then homogenized in ice cold 0.1 M potassium phosphate buffer, pH 7. The volume used was 3x the weight of the snails. The homogenates were centrifuged at 10 000 x g for 15 minutes at 4°C and the supernatant (PMF) collected and stored at -80°C.

2.5 Protein Determination

Protein determination was carried out following a modified method of Lowry *et al.,* (1951) using bovine serum albumin (BSA) as a standard. Briefly, 5 ml of the alkaline solution (2.5 ml of 0.5 % $CuSO_4.5H_2O$ in 1% potassium sodium tartarate; 125 ml of 2 % NA_2CO_3 in 0.1N NaOH) was added to 0.5 ml of the test solution. The reaction mixture was then mixed thoroughly and allowed to stand at room temperature for 10 minutes. Following incubation, 0.5 ml of 1N folin-ciocalteau reagent was added, mixed rapidly and the reaction mixture further incubated for 30 minutes at room temperature. Absorbance was measured against an appropriate blank at 750 nm. The total protein concentration in the snail samples was obtained from the standard curve.

2.6 Assessment of Enzymatic Activity

2.6.1 Acetylcholinesterase

Determination of acetylcholinesterase activity was carried out following the method of Ellman *et al.* (1961) adapted for a microtitre plate reader as described by Kallander *et al.* (1997). Briefly, the following reagents were added to the microtitre plate:110 µl of 0.01 M Tri/HCl buffer pH 8.0, 20 µl of 3.2 mM 5.5 dithio-bis-(2-nitrobenzoic acid) (DTNB) and 50 µl of 1 mg/ml PMF. The mixture was incubated for 3 minutes before adding 20 µl of 10 mM acetylthiocholine iodide. The rate of production of a complex between thiocholine and DTNB was followed at 25°C for 5 minutes at 412 nm using the SpectraMax 340 pc plate reader. Each sample was assayed in quadruplicate.

2.6.2 Carboxylesterase

Carboxylesterase activity was measured using the substrate 4- nitrophenyl acetate following the method of Mackness *et al.* (1983). The reaction mixture contained 200 µl of reagent A (0.027 M Tris/HCl buffer, pH 7.6 and 0.0096 M 4-nitrophenyl acetate in a ratio of 2:1 respectively). The absorbance was measured at 405 nm for 5 minutes at 25 °C using the SpectraMax 340 pc plate reader. Each sample was measured in quadruplicate.

2.6.3 Arylesterase

Arylesterase activity was measured using phenylacetate as the substrate, following the method described by Lorentz *et al.* (1979). Briefly, 20 µl of PMF was diluted with 2 ml of activator solution (20 mM calcium chloride, 155 mM sodium chloride). From this solution 25 µl was pipetted and placed into a test tube containing 3.4 ml of substrate (0.516 mM 4 aminoantipyrine, 4.02 mM phenyl acetate, 50 mM Tris acetate buffer). The mixture was incubated for 20 minutes at 25 °C before adding 100 µl of 213 mM potassium ferrocyanide. The absorbance was measured at 492 nm. Each sample was assayed in quadruplicate.

2.7 Statistical Analysis

The results obtained were analyzed using one-way Analysis of Variance (ANOVA) statistical with Dunnet's Multiple Comparison test found in the GraphPad Prism 5 statistical analysis software program, as well as Microsoft excel. The significance of the results was ascertained at $*p<0.05$ and $**p<0.01$.

3. Results

3.1 Water Quality

Analysis of the water quality parameters revealed the pH values were 9.07 and 7.44 for Umguza and Hillside dams respectively (Table 1). Conductivity and total dissolved solids recorded in Umguza were higher compared to that of Hillside dam water (Table 1), possibly due to surface runoffs from the agricultural areas which probably caused an increase in the amount of ions present.

Table 1. Water quality analysis of Umguza and Hillside dam water.

	pH	Conductivity (µS)	Total dissolved solids (TDS in ppm)
Umguza dam	9.07	685	344
Hillside dam	7.44	187	94

3.2 Acetylcholinesterase Activity

Chlorpyrifos or aldicarb spiked Hillside dam water caused a steady increase in acetylcholinesterase enzyme inhibition. Day 3 showed inhibition above 25%, which increased to 40-50% at day 7 (Fig 1). At day 14, degree of inhibition increased to 70-80% (Fig 1). This revealed that the longer the exposures, the greater the inhibition of esterase activity. Umguza dam water spiked with aldicarb, caused ~28% enzyme inhibition at day 3 which increased to a high of ~ 80% at day 7, with no significant change thereafter (Fig 1). Chlorpyrifos on the other hand caused enzyme inhibition ranging between 30-40% throughout the exposure period (Fig 1).

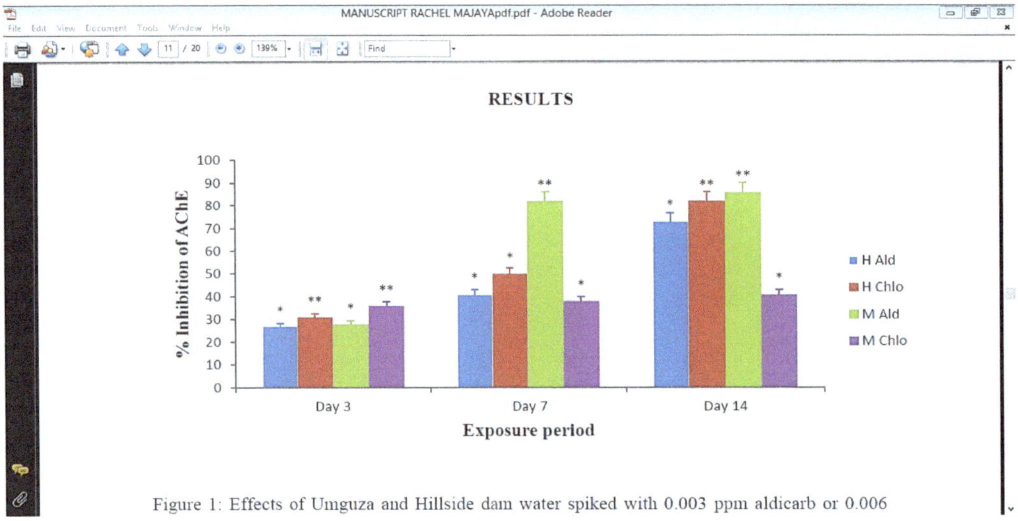

Figure 1. Effects of Umguza and Hillside dam water spiked with 0.003 ppm aldicarb or 0.006 ppm chlorpyrifos on acetylcholinesterase activity of the freshwater snail *L. natalensis.*

Acetylcholinesterase activity was measured using acetylcholine iodide as a substrate. H Ald = Hillside dam water with aldicarb; H Chlo = Hillside dam water with chlorpyrifos; M Ald = Umguza dam water with aldicarb; M Chlo = Umguza dam water with chlorpyrifos. Values represent the average of duplicate exposures and these are expressed as mean ± SD significantly different at *$p<0.05$ and **$p<0.01$.

3.3 Carboxylesterase Activity

Carboxylesterase activity following exposure to aldicarb spiked Hillside dam water was inhibited and the inhibition increased with increase in exposure duration in a nonlinear manner, with day 14 causing a highest inhibition of 80% (Fig 2). Similarly, exposure to chlorpyrifos caused a steady increase in inhibition, although the degree of inhibition was much lower than that of aldicarb (Fig 2). Umguza dam water spiked with aldicarb caused enzyme inhibition of above 60% at days 3 and 7 which later increased to ~90% at day 14 (Fig 2). Chlorpyrifos caused inhibition of up to ~70% at day 14 (Fig 2).

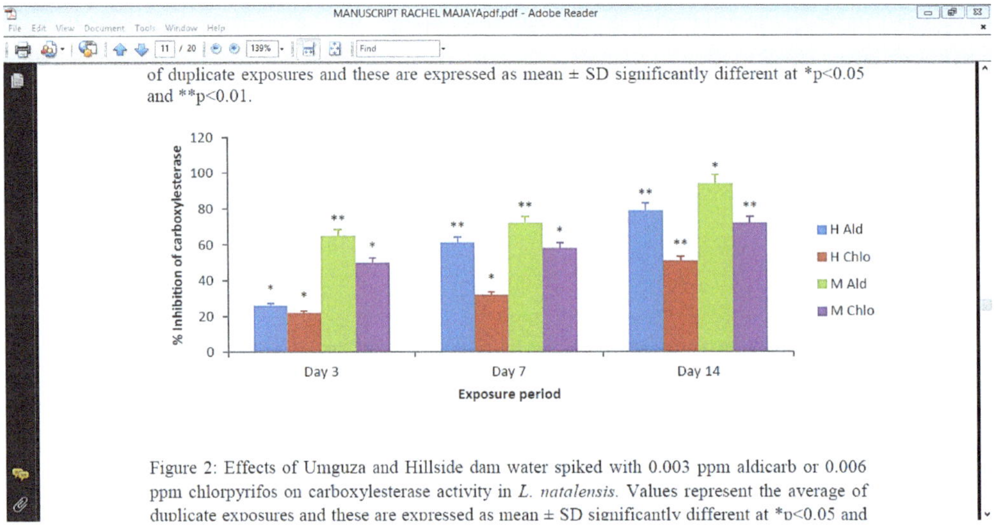

Figure 2. Effects of Umguza and Hillside dam water spiked with 0.003 ppm aldicarb or 0.006 ppm chlorpyrifos on carboxylesterase activity of the freshwater snail L. natalensis.

Carboxylesterase activity was measured using 4-nitrophenyl acetate as a substrate. H Ald = Hillside dam water with aldicarb; H Chlo = Hillside dam water with chlorpyrifos; M Ald = Umguza dam water with aldicarb; M Chlo = Umguza dam water with chlorpyrifos. Values represent the average of duplicate exposures and these are expressed as mean ± SD significantly different at *$p<0.05$ and **$p<0.01$.

3.4 Arylesterase Activity

Both dam waters spiked with pesticides inhibited arylesterase activity. Aldicarb caused a nonlinear increase in inhibition with the highest inhibition of ~ 80% at day 14 (Fig 3). Chlorpyrifos on the other hand caused a nonlinear dose dependent inhibition of arylesterase [above 45%] (Fig 3). Aldicarb spiked Umguza dam water caused enzyme inhibition of 80-90% throughout the exposure period while chlorpyrifos increased inhibition from ~60% at day 3 to ~80% at day 7 with insignificant change thereafter (Fig 3).

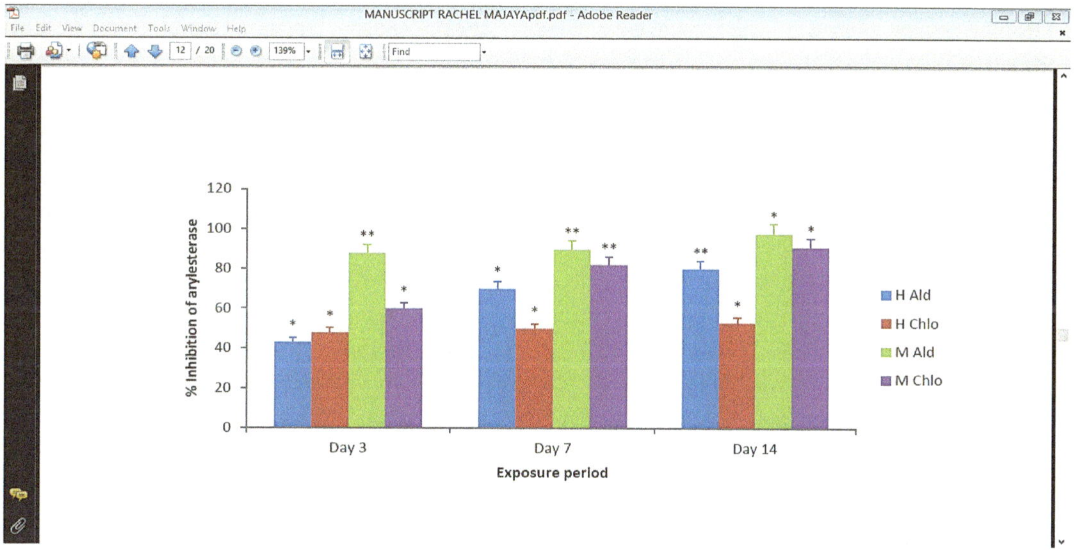

Figure 3. Effects of Umguza and Hillside dam water spiked with 0.003 ppm aldicarb or 0.006 ppm chlorpyrifos on arylesterase activity of the freshwater snail *L. natalensis.*

Arylesterase activity was measured using phenylacetate as a substrate. H Ald = Hillside dam water with aldicarb; H Chlo = Hillside dam water with chlorpyrifos; M Ald = Umguza dam water with aldicarb; M Chlo = Umguza dam water with chlorpyrifos. Values represent the average of duplicate exposures and these are expressed as mean ± SD significantly different at $*p<0.05$ and $**p<0.01$.

4. Discussion

Pesticides have an important use in the control of pests and weeds in agriculture and in the control of diseases in public health such as malaria. Unfortunately some of these pesticides are transported by air drifts when the pesticides are sprayed or washed off terrestrial surfaces as surface run-offs by rain into the aquatic systems. This, results in non-target species, which have physiologic or biochemical systems similar to those of target organisms, being affected (Tirello *et al.*, 2013; Lekhani, 2015).

The water quality analysis of the water from Umguza and Hillside dams used in this study showed that the pH, conductivity and TDS were comparable with values for surface waters obtianed in literature (*Chapman et al.,* 1996). According to Chapman *et al.* (1996), surface waters conductivity normally ranges between 10-1000 μS and pH ranges between 6-8.5 with an allowance of +/- 1 for water designated as a habitat for fish and other aquatic life and water for irrigation (Bhatnagar and Devi, 2013). Environmental Protection Agency's maximum contamination level for TDS is 500 ppm (US, EPA) and water samples in this study from both dams had values below that value. Total dissolved solids levels in Umguza dam water were however, higher than that from Hillside which may be attributed to the fact that Umguza dam is a confluence of numerous streams which may be contaminated by various anthropogenic pollutants (Dube *et al.*, 2010; Siwela *et al.*, 2010). Overall pesticide exposures in Umguza dam water caused higher inhibition of esterase activity compared to exposures in Hillside dam water (with the exception of AChE activity following exposure to chlorpyrifos spiked Umguza dam water). It may therefore be inferred that higher TDS levels in water enhance the effects of pollutants, however, further studies would have to be performed with dams with an equivalent or even higher TDS values to make this conclusion.

Esterase sensitivity to both pesticides was observed. Exposures to the carbamate aldicarb showed greater enzyme inhibition, compared to the organophosphate chlopyrifos. This may be due to the fact that carbamates induce toxicity faster than organophosphates, as they do not need to be bioactivated to be potent inhibitors (Barata *et al.*, 2004). Despite this, carbamate pesticides are susceptible to hydrolysis by carboxylesterases and arylesterases (Peterson and Talcott, 2013). In the present study, however, the two enzymes showed less capability of hydrolyzing pesticides. This may have been a result of the adaptability of the organism species to the chemical compounds. Some types of species are able to quickly hydrolyse / metabolise the compounds, while other species are resistant, failing to metabolise the compounds but rather become susceptible to inhibition by the toxicants (Wheelock *et al.*, 2005). Acetylcholinesterases are known to be specifically inhibited by carbamate pesticides (Apilux *et al.*, 2015). The present study showed higher inhibition caused by aldicarb with inhibition as high as 85% in Umguza dam water and 70% in Hillside dam water. Similar studies in literature have reported relatively high inhibitions of esterase activity due to carbamate exposure (Barata *et al.*, 2004); Apilux *et al.*, 2015). Kristoff *et al* (2006) reported inhibition of AChE activity of up to 65 % in the *Biomphalaria glabrata* snails, after exposure to a carbamate pesticide.

Some organophosphates, unlike carbamates, cannot directly inhibit acetylcholinesterases, but are metabolically activated by cytochrome P450 isoenzymes to form potent AChE inhibitors known as oxons (Dzul-Caamal *et al.*, 2012, Lionetto *et al.*, 2013). Exposures to chlorpyrifos in Hillside dam water showed increased inhibitions of up to 70% at day 14 while exposures in Umguza dam water only ranged between 30-40% throughout the exposure period. The observed difference in AChE inhibition may be due to the presence of chemical antagonists in the Umguza dam water, which may have reacted with the chlorpyrifos or its oxon form, subsequently producing a less toxic product. Antagonistic effects of pollutants were observed by Basopo *et al.* (2014) who showed inhibitions of esterase activity in snails exposed to pollutant mixtures being lower than in snails exposed to the individual chemicals.

Carboxylesterases are also targets of the oxon forms of organophosphates, however, their inhibitions do not result in the lethal cholinergic crisis observed in AChE inhibition (Ross *et al.*, 2010). Inhibition of carboxylesterases instead plays a protective role on the intrinsic target AChE as the carbooxylesterases scavenge the oxons generated within the organism, reducing the number of molecules available for inhibiting AChE (Barata *et al.*, 2004). In the present study, inhibition of carboxylesterases, following exposure to chlorpyrifos in both Hillside and Umguza dam water increased with prolonged exposure, suggesting enzyme increased affinity for the bioactivated chlorpyrifos. There is insufficient data, however, to conclude that CbE inhibition provided AChE protection. Further studies would have to be conducted as performed by Wheelock (2008) who, firstly, pretreated CbE with inhibitors and observed any potentiated toxicity of the organophosphate and secondly, considered any other protection contributed by other esterases such as butyrylcholinesterase.

Arylesterases are calcium dependent hydrolases that can detoxify active metabolites (oxons) of organophosphorus compounds (Aoki *et al.*, 2014). The results in the present study showed 50% of arylesterase inhibition throughout the exposure period in Hillside dam water. Arylesterases are high density lipoprotein associated enzymes and so any disturbance in the lipid profile may decrease its activity (Gabrowny *et al.*, 2007). This may account for the low activity observed. The increased inhibition (90% at day 14) seen following exposures in Umguza dam water may be due to the presence of substances such as metal chlorides, that are known to inhibit A esterases (Arias-Almeida and Rico-Martínez, 2011; Prasad *et al.*, 2009).

5. Conclusion

The present study showed that non-target organisms such as molluscs are potentially at risk to the toxic effects of pesticides. This is a cause of concern as the adverse effects of these agrochemicals may affect the aquatic ecosystem and biodiversity. Results from the i*n vitro* studies indicated that esterases from the exposed snails were consistently inhibited by exposure to pesticides and thus have a potential of being used as indicators of exposure to pesticide pollutants in aquatic ecosystems.

Acknowledgements

This research was supported by the International Program in Chemical Sciences (IPICS) Uppsala, Sweden.

References

Aktar, W., Sengupta, D., & Chowdhury, A. (2009). Impact of pesticides use in agriculture: their benefits and hazards. *Interdisciplinary Toxicology, 2*(1), 1-12. http://dx.doi.org/10.2478/v10102-009-0001-7

Anandhan, R., Bhuyan, G., Kavitha, V., & Selvam, A. (2012). (AChE) activity of two freshwater teleost *Channa striatus* and *Oreochromis mossambicus* in reference to Kedilam river, Cuddalore district, Tamil Nadu. *International Journal of Toxicology and Applied Pharmacology, 2*(4), 52-56.

Aoki, Y., Helzlsouer, K. J., & Strickland, P. T. (2014). Arylesterase phenotype specific positive association between arylesterase activity and cholinesterase specific activity in human serum. *International Journal of Environmental Research and Public Health, 11*, 1422-1443. http://dx.doi.org/10.3390/ijerph110201422

Apilux, A., Isarankura, Na-Ayudhya, C., Tantimongcolwat, T., & Prachayasittikul, V. (2015). Paper based acetylcholinesterase inhibition assay combining a wet system for organophosphate and carbamate pesticide detection. *Experimental and Clinical Sciences Journal, 14*, 307-319.

Arias-Almeida J. C., & Rico-Martínez, R. (2011). Inhibition of two enzyme systems in Euchlanis dilatata (Rotifera: Monogononta) as biomarker of effect of metals and pesticides. *Biomarkers 16*(1), 12-19. http://dx.doi.org/10.3109/1354750X.2010.520336

Barata, C., Solayan, A., & Porte, C. (2004). Role of B-esterases in assessing toxicity of organophosphorus (chlorpyrifos, malathion) and carbamate (carbofuran) pesticides to daphnia magna. *Aquatic Toxicology, 66*, 125-139. http://dx.doi.org/10.1016/j.aquatox.2003.07.004

Basopo, N., Mumbamarwo, L. T., Mnkandla, D., & Naik, Y. S. (2014). Pollutant mixtures as stressors of selected enzyme activities of the aquatic snail helisoma duryi. *Journal of Environmental Chemistry and Ecotoxicology, 6*, 27-37. http://dx.doi.org/10.5897/JECE2013.0319

Bhatnagar, A., & Devi, P. (2013). Water quality guidelines for the management of pond fish culture. *International Journal Environmental Sciences, 3*(6), 1980-2009.

Chapman, D. V., Unesco, O. W. H., & Programme, U. N. E. (1996). Water quality assessments: A guide to the use of biota, sediments, and water in environmental monitoring, E & FN Spon. http://dx.doi.org/10.4324/NOE0419216001

Dube, T., Makaka, C., & Sibanda, Z. (2010). An assessment of the effect of industrial and sewage effluent on aquatic invertebrates: A case study of a southern urban stream, Zimbabwe. *Journal of Sustainable Development, 3*(2), 210-214. http://dx.doi.org/10.5539/jsd.v3n2p210

Dzul-Caamal, R., Lilia Domínguez-López, M., García-Latorre, E., & Vega-López, A. (2012). Implications of cytochrome 450 isoenzymes, aryl-esterase and oxonase activity in the inhibition of the acetylcholinesterase of chirostoma jordani treated with phosphorothionate pesticides. *Ecotoxicology and Environmental Safety, 84*, 199-206. http://dx.doi.org/10.1016/j.ecoenv.2012.07.008

Ellman, G. L., Courtney, K. D., Andres, V., & Featherstone, R. M. (1961). A new and rapid colorimetric determination of acetylcholinesterase activity. *Biochemical pharmacology, 7*, 88-95. http://dx.doi.org/10.1016/0006-2952(61)90145-9

Gabrowny, K. H., Farag, E., & Sallam, A. (2007). Involvement of paraoxonase (pon) in oxidative stress induced by chlorpyrifos in albino rats. *Journal of Egyptian Society of Toxicology, 37*, 71-86.

Kallander, D., Fisher, S., & Lydy, M. (1997). Recovery following pulsed exposure to organophosphorus and carbamate insecticides in the midge, chironomus riparius. *Archives of environmental contamination and toxicology, 33*, 29-33. http://dx.doi.org/10.1007/s002449900219

Kaviraj, A., Unlu, E., Abhik, G. A., & El Nemr, A. (2014). Biomarkers of environmental pollutants. *BioMed Research International, 2014*, 1-2. http://dx.doi.org/10.1155/2014/806598

Kristoff, G., Guerrero, N. V., De D'angelo, A. M. P., & Cochón, A. C. (2006). Inhibition of cholinesterase activity by azinphos-methyl in two freshwater invertebrates: *Biomphalaria glabrata* and *Lumbriculus variegatus*. *Toxicology, 222*, 185-194. http://dx.doi.org/10.1016/j.tox.2006.02.018

Lamers, M., Anyusheva, M., La, N., Nyuyen, W., & Streck, T. (2011). Pesticide pollution in surface and groundwater by paddy rice cultivation: A case study from Northern Vietnam. *Clean-Soil, Air, Water, 39*(4), 356-361. http://dx.doi.org/10.1002/clen.201000268

Lekhani, L. (2015). How to reduce the impact of pesticide in aquatic environment. *Social Issues and Environmental Problems, 3*(9), 29-38.

Lionetto, M. G., Caricato, R., Calisi, A., Giordano, M. E., & Schettino, T. (2013). Acetylcholinesterase as a biomarker in environmental and occupational medicine: New insights and future perspectives. *BioMedical Research International, 2013*(8). http://dx.doi.org/10.1155/2013/321213

Lorentz, K., Flatter, B., & Augustin, E. (1979). Arylesterase in serum: Elaboration and clinical application of a fixed-incubation method. *Clinical chemistry, 25*, 1714-1720.

Lowry, O. H., Rosebrough, N. J., Farr, A. L., & Randall, R. J. (1951). Protein measurement with the folin phenol reagent. *Journal of Biological Chemistry, 193*, 265-275.

Mackness, M. I., Walker, C. H., Rowlands, D. G., & Price, N. R. (1983). Esterase activity in homogenates of three strains of the rust red flour beetle tribolium castaneum (herbst). *Comparative Biochemistry and Physiology Part C: Comparative Pharmacology, 74*, 65-68. http://dx.doi.org/10.1016/0742-8413(83)90150-0

Maltby, L., & Hills, L. (2008). Spray drift of pesticides and stream macroinvertebrates: Experimental evidence of impacts and effectiveness of mitigation measures. *Environmental Pollution, 156*(3), 1112-1120. http://dx.doi.org/10.1016/j.envpol.2008.04.013

Peterson, M. E., & Talcott, P. A. (2013). Small animal toxicology. W.B. Saunders Co., Philadelphia, PA.

Prasad, A. J. M., Kemparaju, K., Frank, E. A., & D'Souza, C. J. M. (2009). Purification of human serum paraoxonase: A simple and rapid method. *African Journal of Biochemistry Research, 3*(4), 125-129.

Ross, M. K., Streit, T. M., & Herring, K. L. (2010). Carboxylesterases: Dual roles in lipid and pesticide metabolism. *Journal of Pesticide Science, 35*, 257-264. http://dx.doi.org/10.1584/jpestics.R10-07

Sandahl, J. F., Baldwin, D. H., Jenkins, J. J., & Scholz, N. L. (2005). Comparative thresholds for acetylcholinesterase inhibition and behavioural impairment in coho salmon exposed to chlorpyrifos. *Environmental Toxicology and Chemistry, 24*, 136-145. http://dx.doi.org/10.1897/04-195R.1

Sayali, K., Sadichha, P., & Surekha, S. (2013). Microbial Esterases: An overview. *International Journal of Current Microbiology and Applied Science, 2*(7), 135-146.

Scholz, N. L., Truelove, N. K., French, B. L., Berejikan, B. A., Quinn, T. P., Casillas, E., & Collier, T. K. (2000). Diazinon disrupts anti-predator and homing behaviors in chinock salmon (Onc orhynchus, tshawytscha). *Canadian Journal of fisheries and Aquatic Sciences, 57*, 1911-1918. http://dx.doi.org/10.1139/f00-147

Sebata, S. (2015). Species composition and abundance of spider fauna of the Hillside Conservancy, Bulawayo, Zimbabwe. *Researcher, 7*(2), 44-49.

Siwela, A. H., Nyathi, C. B., & Naik, Y. S. (2009). Metal accumulation and antioxidant enzyme activity in *C. gariepinus*, catfish and *O. mossambicus*, tilapia collected from Mguza and Wright dams, Zimbabwe. *Bulletin of Environmental Contamination and Toxicology, 83*, 648-651. http://dx.doi.org/10.1007/s00128-009-9861-y

Siwela, A. H., Nyathi, C.B., & Naik, Y. S. (2010). A comparison of metal level and antioxidant enzymes in freshwater snails, *Lymnaea natalensis* exposed to sediment and water collected from Wright dam and lower Mguza Dam, Bulawayo, Zimbabwe. *Ecotoxicology and Environmental Safety, 73*(7), 1728-1732. http://dx.doi.org/10.1016/j.ecoenv.2010.08.001

Tirello, P., Pozzebon, A., & Duso, C. (2013). The effect of insecticides on non-target predatory mite Kampimodromus aberrans: laboratory studies. *Chemosphere, 93*(6), 1139-1144. http://dx.doi.org/10.1016/j.chemosphere.2013.06.046

Tiryaki, O., & Temur, C. (2010). The fate of pesticides in the environment. *Journal of Biological and Environmental Science, 4*(10), 29-38.

United States Environmental Protection Agency (US EPA). National Primary Drinking Water Regulations. www.epa.gov/safewater/ contaminants/index.html [Accessed 29.02.16].

Valavanidis, A., Vlahogianni, T., Dassenakis, M., & Scoullos, M. (2006). Molecular biomarkers of oxidative stress in aquatic organisms in relation to toxic environmental pollutants. *Ecotoxicology and environmental safety, 64*, 178-189.

Vioque-Fernadez, A., de Almeida, E. A., & López-Barea, J. (2007). Esterases as pesticide biomarkers in crayfish (*Procambarus clarkii,* Crustacea): tissue distribution, sensitivity to model compounds and recovery from inactivation.

Wang, C., Lu, G., & Cui, J. (2010). Responses of ache and gst activities to insecticide coexposure in *carassius auratus*. *Environmental Toxicology, 27*, 50-57. http://dx.doi.org/10.1002/tox.20612

Wheelock, C. E., Phillips, B. M., Anderson, A. S., Miller, J. L., Miller, M. J., & Hammock, B. D. (2008). Applications of carboxylesterase activity in environmental monitoring and toxicity identification evaluations. *Reviews of Environmental Contamination and Toxicology, 195*, 117-178. http://dx.doi.org/10.1007/978-0-387-77030-7_5

Wheelock, C. E., Shan, G., & Ottea, J. A. (2005). Overview of carboxylesterases and their role in metabolism of insecticides. *Journal of Pesticide Science, 30*, 75–83. http://dx.doi.org/10.1584/jpestics.30.75

14

Scalable Route to Chiral Phosphoramidites

Sergiy G. Krasutsky[1], Sheila H. Jacobo[1]

[1]Chemical Development Department, Albany Molecular Research Inc, United States

Correspondence: Sergiy G. Krasutsky, Chemical Development Department, AMRI, 3 University Place, Rensselaer, NY, 12144, United States. E-mail: Sergiy.Krasutsky@amriglobal.com.

Abstract

The chiral copper-catalyzed conjugate addition is a well-developed methodology to create asymmetric C-C bonds. Many efforts have been dedicated to design efficient systems and identifying highly efficient ligands. Among the most proficient ligands, the ones possessing a bis-phenol or a bis-naphthol moiety are the most prominent. Herein we describe a scalable convenient route that can afford phosphoramidites in excellent yield.

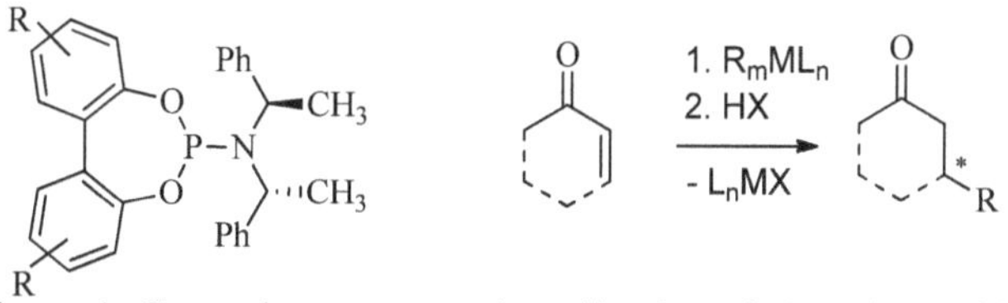

Figure 1. General structure and application of phosphoramidites

Keywords: chiral phosphoramidite, bis-phenol, Asymmetric conjugate addition

1. Introduction

The engagement of chiral phosphoramidites as ligands in enantioselective copper-catalyzed conjugate carbon-carbon bond formation using organozinc reagents has become an essential tool in the hands of a synthetic chemist (Ferringa, 2000). A series of chiral ligands have been prepared, particularly the preparation of ligand **3** has been described in the literature (Vuagnoux-d'Augustin et al; 2007) and is shown in Scheme 1. This ligand was used in the copper mediated asymmetry-inducing step in a production of taxadienone (Mendoza et al., 2012). However, during the process optimization route of taxadienone (Krasutsky et al., 2015) a substantial quantity of this intermediate was required and thus it was highly desirable to develop a method for preparation of ligand **3** that minimized the purification efforts as well as improved the overall yield of the product. The process described within this paper eliminated the silica gel chromatography step and provided a more convenient and efficient approach to chiral phosphoramidites.

2. Results and Discussion

Since bis-phenol **2** was not commercially available, a convenient route to make it was also desired. Although compound **2** is known in the literature (Neelamegam et al., 2010) a simple purification method was essential to obtain high quality material. Complete conversion of **1** to **2** was observed when 1.2 eq. of persulfate was used. The reported purification procedure using silica column purification did not purge closely eluting impurities. Instead, we found that the product can be crystallized from MTBE/heptanes in 60% yield.

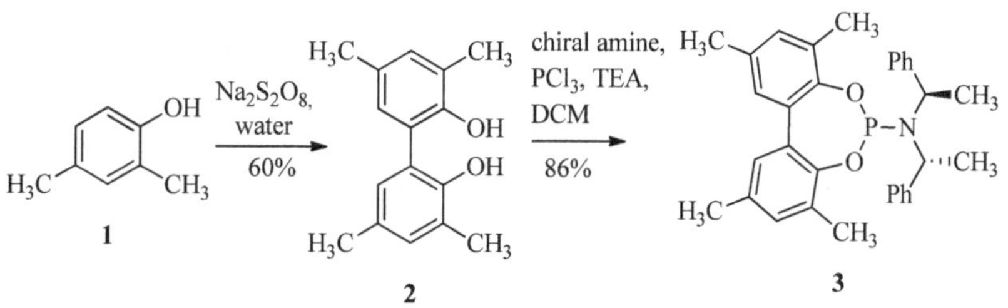

Scheme 1. Preparation of bis-phenol **2** and corresponding phosphoramidite ligand **3**.

The second reaction, formation of phosphoramidite **3**, was reported to proceed in THF (Vuagnoux-d'Augustin et al;

2007). However, the reaction was plagued by the precipitation of salts and incomplete conversion. As a result, the decision was made to run the reaction in dichloromethane instead of THF. This change allowed the reaction to exist as a homogeneous solution. The order of addition was also investigated. When phosphorus trichloride was reacted with the bis-phenol **2** prior to the chiral amine addition, the reaction required substantial additional time due to the decreased reactivity of bis-phenoxy phosphino-chloride intermediate. Conversely, when the amine was added prior to the addition of bis-phenol **2**, the desired product was formed in 2 hours and was isolated in 86% yield (the optimal stoichiometric ratio was found to be chiral amine:PCl3:bis-phenol 2 0.95:1:1**)**. An alumina plug was used for the purification. The reaction was scaled up to 10 g of the chiral amine to give a quantitative yield of phosphoramidite **3**. Encouraged with this result, we extended this strategy to prepare bis-naphthyl phosphoramidite **5** (Scheme 2). This catalyst was found to be useful for iridium catalyzed amination of allylic carbonates (Leitner et al., 2004).

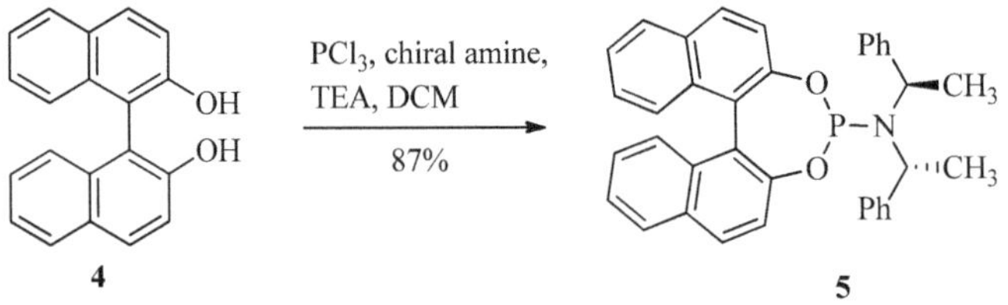

4 **5**

Scheme 2. Preparation of bis-naphthyl phosphoramidite **5**.

The reaction proceeded as expected and the ligand **5** was obtained in quantitative yield. This result demonstrated the overall usefulness of this methodology.

3. Experimental

Chemical materials, solvents, and reagents were reagent grade and were used without further purification unless otherwise noted. Reaction progress was monitored using analytical thin-layer chromatography (TLC) on 0.25-mm Merck F-254 silica gel glass plates. Visualization was achieved by UV light (254 nm). ^{1}H NMR, ^{13}C NMR and ^{31}P spectra were recorded with a Bruker AVANCE 500 MHz spectrometer; chemical shifts were reported in parts per million (δ) using TMS as the internal standard with coupling constants (*J*) reported in hertz (Hz).

Synthesis of bis-phenol 2: 5-L, three-neck round bottomed flask equipped with a mechanical overhead stirrer, thermocouple, N_2 inlet, and addition funnel was sparged with N_2 and charged with iron(II) sulfate heptahydrate (15.0 g, 54.0 mmol, 0.05 equiv.), 2,4-dimethylphenol (131 g, 1072 mmol, 1.00 equiv.), and water (1.0 L). A solution of sodium persulfate (306 g, 1.29 mol, 1.20 equiv.) in water (1.1 L) was added slowly over 4 hours at 15–25 °C. No significant exotherm was observed. The reaction was stirred overnight. After this time, the reaction was diluted with MTBE/heptanes (1:1 ((v/v), 800 mL). The contents of the flask were rested to allow phase separation. The aqueous layer was separated, extracted with MTBE/heptanes (1:1 (v/v), 2 × 300 mL). The combined organic layers were concentrated, and passed through a silica plug (300 g, 20% MTBE in heptanes). The fractions containing the product were concentrated. The obtained residue was dissolved in MTBE (150 mL) and heptanes (500 mL) were added. The solution was then concentrated down to 300 mL and heptanes (500 mL) was added. The solution was stirred overnight affording slurry, and the product was collected by filtration. The filter cake was dried under high vacuum at 35 °C to afford bis-phenol **2** [77.80 g, 60% yield] as a white solid. IR (KBr, cm^{-1}): ῡ = 3520, 2920, 1465, 1210, 1185, 1115, 860, 784. ^{1}H NMR (500 MHz, $CDCl_3$): δ 7.00 (m, 2H), 6.86 (m, 2H), 5.06 (s, 2H), 2.28 (s, 6 H), 2.27 (s, 6 H). ^{13}C NMR (125 MHz, $CDCl_3$): δ 149.2, 132.0, 130.0, 128.5, 125.2, 122.2, 20.4, 16.2. MS (ESI+): M/z (%) = 243 (100) $[M+H]^+$. CHNS analysis of **2**, $C_{16}H_{18}O_2$ (242): calculated C 79.31, H 7.49; found C 79.21, H 7.35.

Synthesis of ligand 3: A 2-L, three-neck round bottomed flask equipped with a mechanical overhead stirrer, thermocouple, N_2 inlet, and addition funnel was sparged with N_2, charged with phosphorus trichloride (6.09 g, 44.4 mmol, 1.00 equiv.), dichloromethane (300 mL), and cooled to 0 °C. The mixture was stirred for 30 min. Triethylamine (22.5 g, 222 mmol, 5.00 equiv.) was added keeping the internal temperature in 0–10 °C range. A solution of (+)-bis((*R*)-1-phenylethyl)amine (10.0 g, 44.4 mmol, 1.00 equiv.) in dichloromethane (100 mL) was added slowly over 20 min at 0–5 °C. The reaction mixture was allowed to warm up to room temperature over 2 hours. A solution of bis-phenol **2** (10.8 g, 44.4 mmol, 1.00 equiv.) in dichloromethane (100 mL) was then added. The reaction was stirred for 2 h, concentrated, and diluted with heptanes (500 mL). The salts were filtered and washed with heptanes (~1 L). The combined organic solution was passed through neutral alumina (200 g). The fractions containing the product were concentrated to afford ligand **3** [19.0 g, 86% yield] as a white solid. IR (KBr, cm^{-1}): = 3516, 2927, 1493, 1463, 1282, 1214, 1185, 1110, 1064, 962, 861, 787. ^{1}H NMR (500 MHz, $CDCl_3$): δ 7.00 (m, 2H), 6.86 (m, 2H), 5.06 (s, 2H), 2.28 (s, 6 H), 2.27 (s, 6 H). ^{13}C NMR (125 MHz, $CDCl_3$): δ 147.9 ($J_{C\text{-}P}$ = 7.6 Hz), 147.1 ($J_{C\text{-}P}$ = 2.2 Hz), 143.4, 133.3, 132.5, 131.3, 131.2, 131.0, 130.9, 130.2, 130.1, 130.0, 129.3, 128.1, 128.0, 127.9, 127.6, 126.5, 52.5, 52.4, 20.8, 20.7, 17.3, 16.3. ^{31}P

NMR (202 MHz, $CDCl_3$): δ 142.3. MS (ESI+): M/z (%) = 496 (100) $[M+H]^+$. CHNS analysis of **3**, $C_{32}H_{34}NO_2P$ (495): calculated. C 77.55, H 6.91; N 2.83; found C 77.49, H 6.26, N 2.86.

4. Conclusion

Herein we described a useful procedure to prepare chiral phosphoramidites in excellent yield. The process described within this paper eliminated the silica gel chromatography step and provided a more convenient and efficient procedure to chiral phosphoramidites.

References

Ferringa, B. L. (2000). Phosphoramidites: Marvellous ligands in catalytic asymmetric conjugate addition. *Accounts of Chemical Research*, *33*, 346-353. http://dx.doi.org/10.1021/ar990084k

Krasutsky, S. G., Jacobo, S. H., Tweedie, S. R., Krishnamoorthy, R., & Filatov A. S. (2015). Route Optimization and Synthesis of Taxadienone. *Organic Process Research and Development*, *19*(1), 284-289. *http://dx.doi.org/10.1021/op500314c*

Leitner, A., Shu, C., & Hartwig, J. F. (2004). Editing the stereochemical elements in an iridium catalyst for enantioselective allylic amination. *Proceedings of the National Academy of Sciences of the United States of America, 101*(16), 5830-5833. http://dx.doi.org/10.1073/pnas.0307967101

Mendoza, A., Ishihara, Y., & Baran, P. S. (2012). Scalable enantioselective total synthesis of taxanes. *Nature Chemistry*, *4*, 21-25. http://dx.doi.org/10.1038/nchem.1196

Neelamegam, R., Palatnik, M. T., Fraser-Rini, J., Slifstein, M., Abi-Dargham, A., & Easwaramoorthy, B. (2010). An unprecedented oxidative trimerization of sesamol catalyzed by laccases. *Tetrahedron Letters*, *53*(26), 2497-2499. http://dx.doi.org/10.1016/j.tetlet.2010.02.173

Vuagnoux-d'Augustin, M., & Alexakis, A. (2007). Copper-Catalyzed Asymmetric Conjugate Addition of Trialkylaluminium Reagents to Trisubstituted Enones: Construction of Chiral Quaternary Centers. *Chemistry A European Journal*, *13*(34), 9647-9662. http://dx.doi.org/10.1002/chem.200701001

Appendix A

The NMR spectra for the prepared compounds

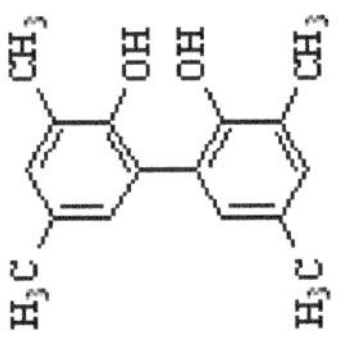

2

^1H NMR Spectrum of **2** ($CDCl_3$, 500 MHz)

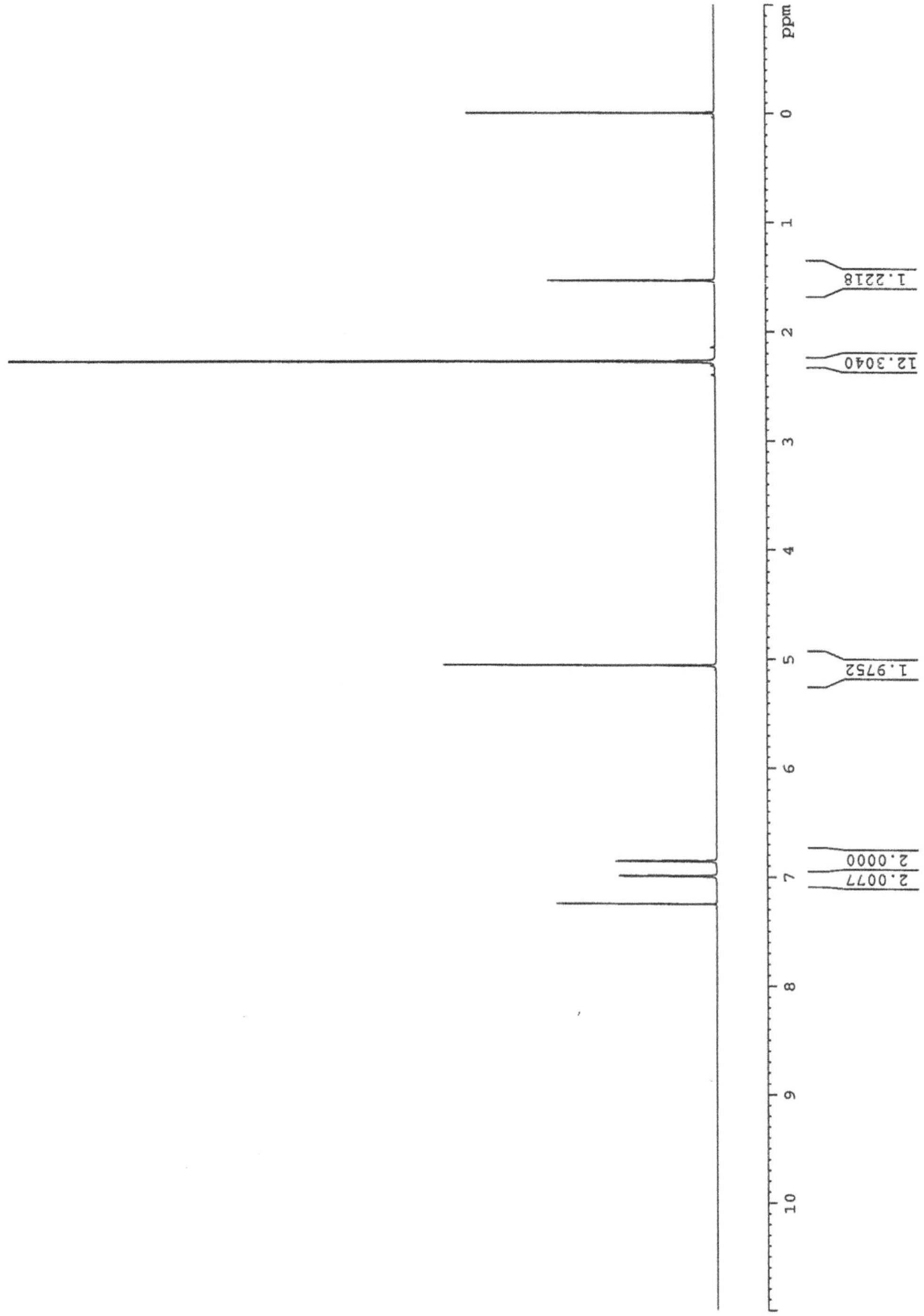

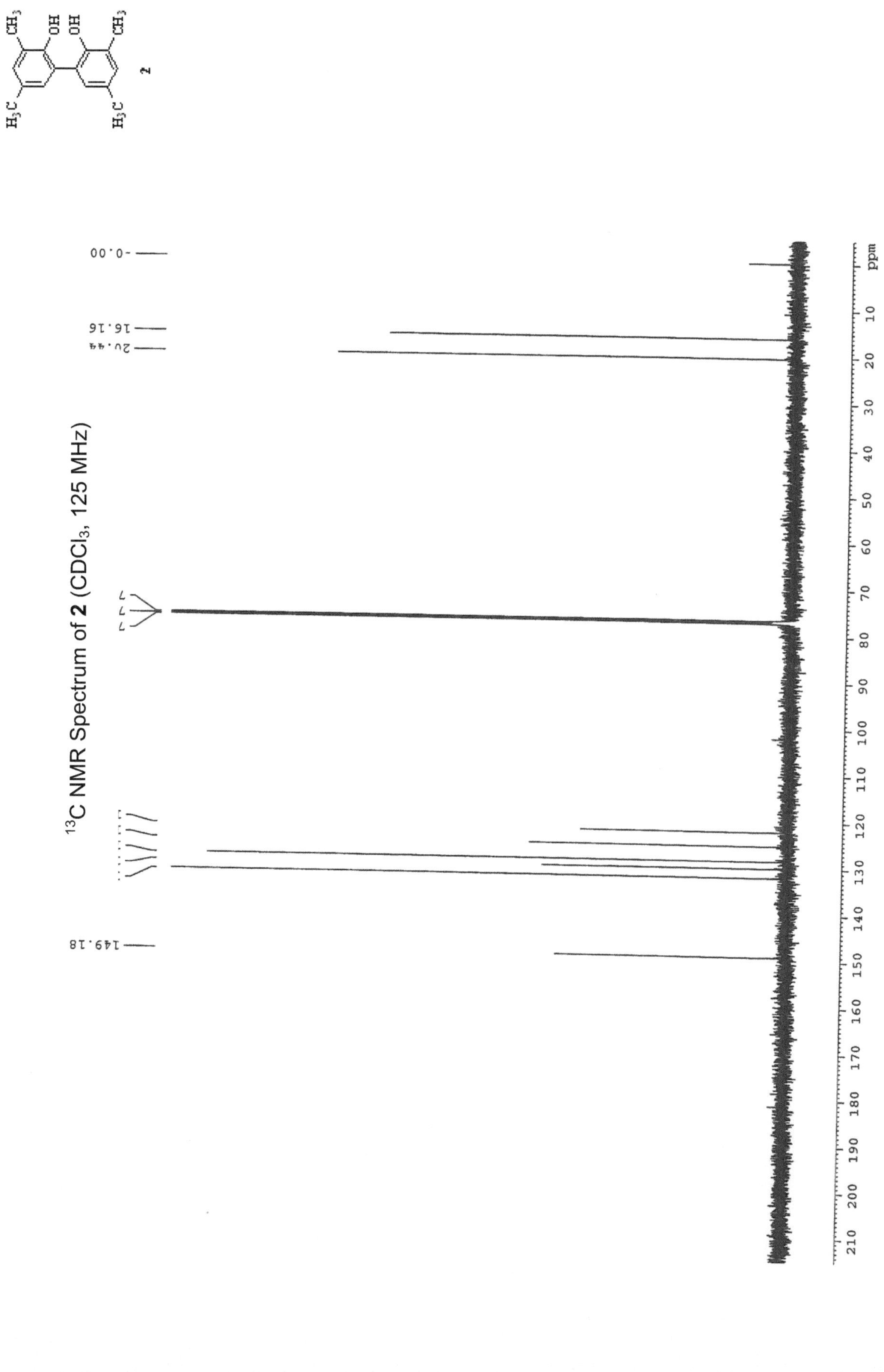

2

^{13}C NMR Spectrum of **2** ($CDCl_3$, 125 MHz)

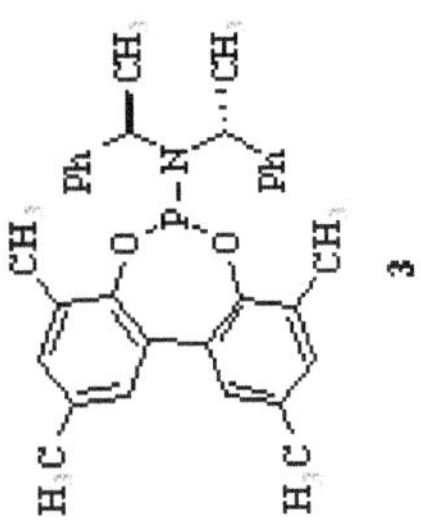

3

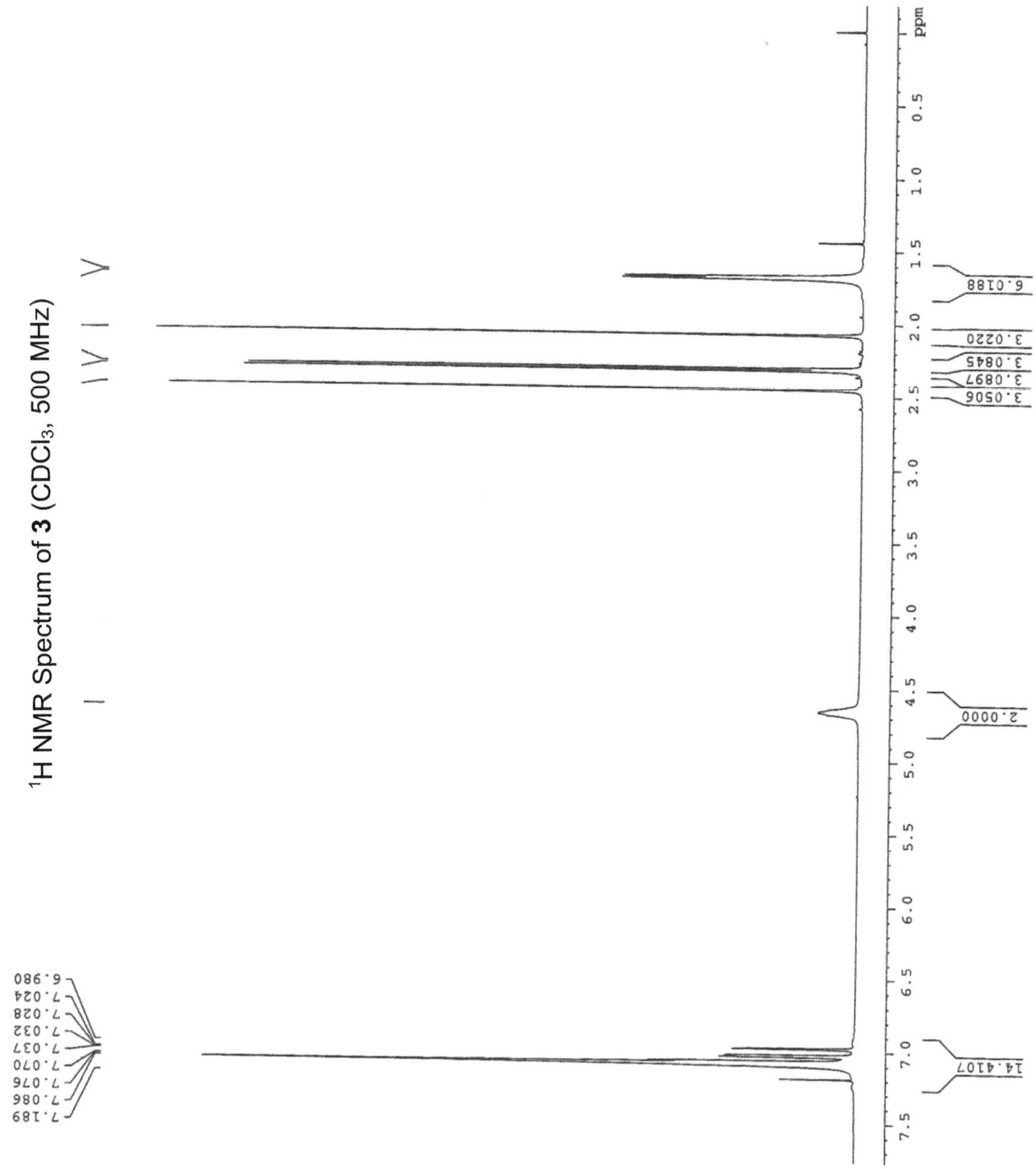

^1H NMR Spectrum of **3** ($CDCl_3$, 500 MHz)

^{13}C NMR Spectrum of **3** ($CDCl_3$, 125 MHz)

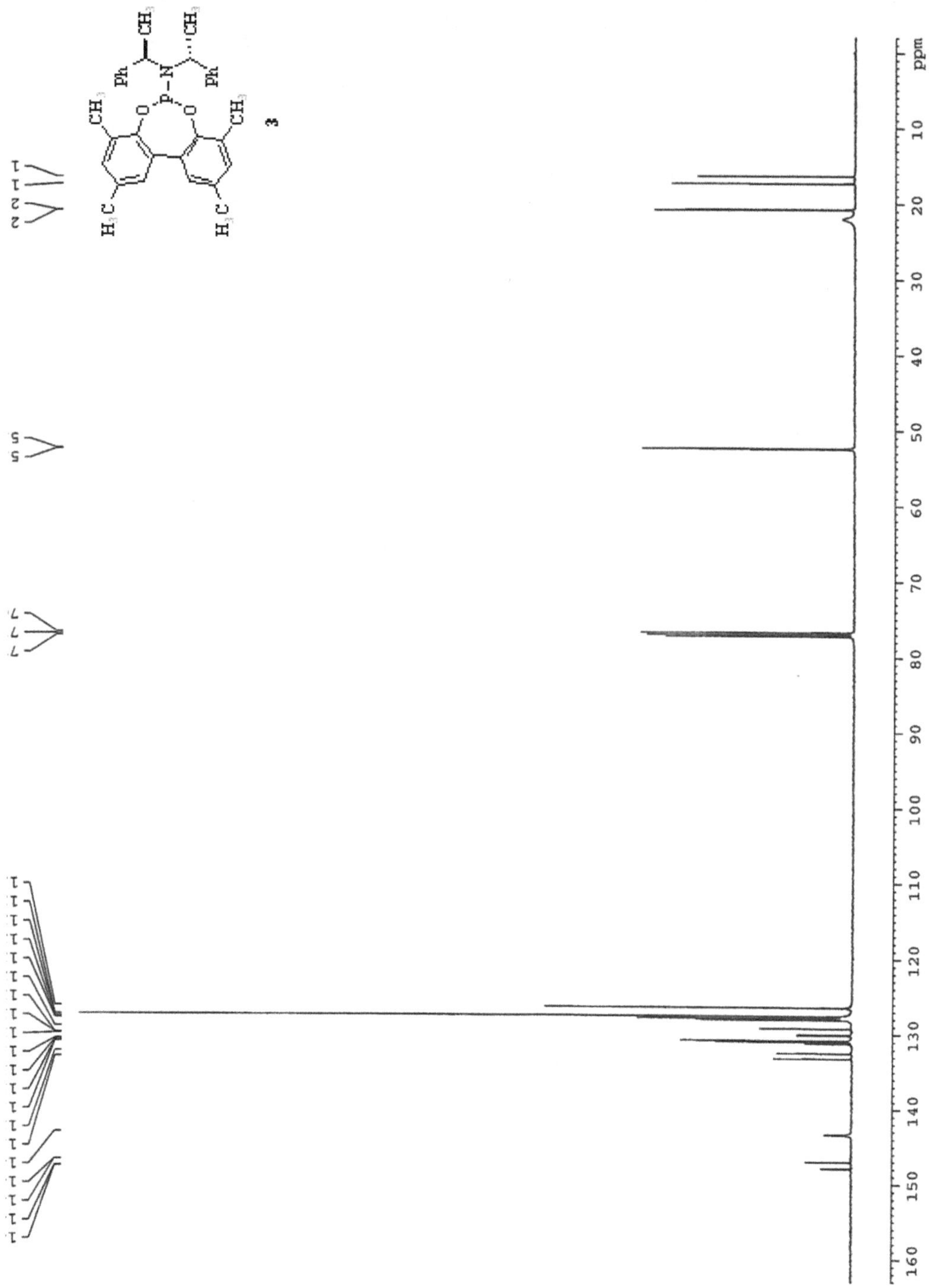

^{31}P NMR Spectrum of **3** ($CDCl_3$, 202 MHz)

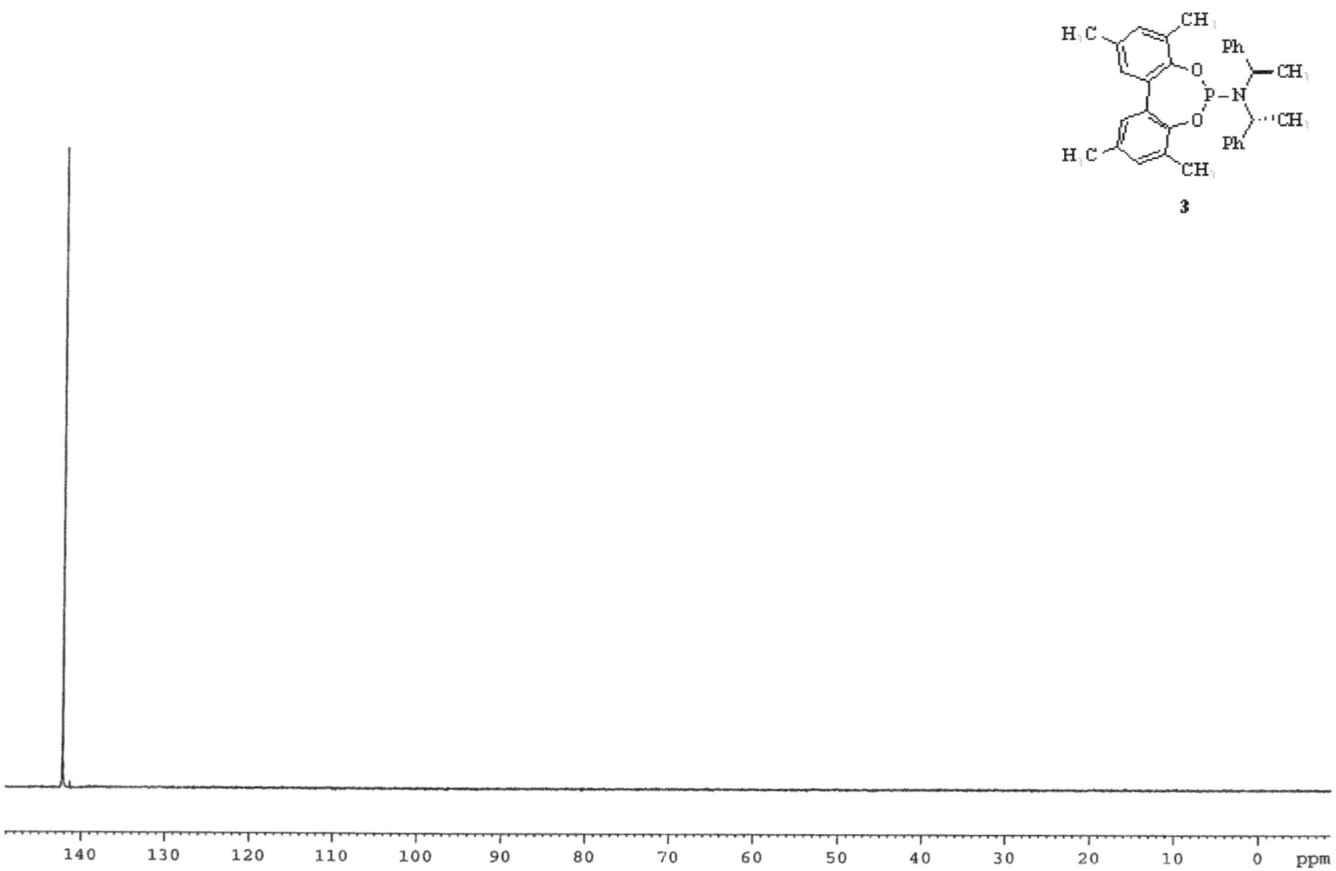

15

Determination of Uranium and Thorium in the Industrial Phosphoric Acid Using X-Ray Fluorescence Spectrometry with Wavelength Dispersive (WDXRF)

Mohamed Guitouni

Correspondence: Mohamed Guitouni, X-ray fluorescence Laboratory, National Institute of Research and Physico-Chemical Analysis, Technological Pole Sidi Thabet 2020, Aryanah, Tunisia. E-mail: guitmed@yahoo.fr

Abstract

The method of determination for uranium and thorium in the industrial phosphoric acid by wavelength-dispersive X-ray fluorescence spectrometry (WDXRF) responds positively to all the validation tests and can be adopted for the dosage for these elements. It is clear that the procedure of the proposed method is simple and requires less time to complete the analysis. The concentration for uranium and thorium in the Tunisian phosphoric acid is in the order of 22 ppm for uranium with a coefficient of variation of 0.77% and of about 4 ppm for thorium with a coefficient of variation of 1.37%.

Keywords: WDXRF, radioactive, uranium, thorium, analysis, validation

1. Introduction

Industrial Phosphoric acid is one of the most important products of the chemical industry, it is used in the manufacture of phosphate fertilizers; TSP: $Ca(H_2PO_4)_2$ and DAP: $(NH_4)_2HPO_4$, feed livestock ($CaHPO_4.2H_2O$) and sequestering phosphate (sodium hexametaphosphate). The production of phosphoric acid is based on the chemical treatment of the rock phosphate with sulfuric acid, this process can be described by the following overall reaction of (Slack, 1968; Waggaman, 1969):

$$Ca_{10}\ (PO_4)_6F_2 + 10\ H_2SO4 + 20\ H_2O \rightarrow 10\ (CaSO_4\ 2\ H_2O) + 2\ HF + 6\ H_3PO_4 \qquad (1)$$

The sulfuric attack dissolved at the same time as apatite, almost all existing impurities in the soil. These consist of heavy metals, radionuclides and other natural elements (Becker, 1989). Some of these impurities can be recycled such as (U, Th, Y, Sc, rare earth ...) and could be recovered, and others that are penalizing for further use of acid such as (Cd , Cr, Pb, Mg ...), they must be eliminated or minimized.

In this work we focus on uranium and thorium and particularly to techniques for determining these elements. The methods currently used are:

- Gravimetry which consists in the precipitation of the uranyl ion with 2-hydroxy-1-naphthaldehyde, uranium is determined as U_3O_8 after calcining the precipitate at 850 ° C. (Lopez et al., 1989) or the precipitation of the UO_2^{2+}ion, in the pH range 3–5, by a solution of *p*-aminobenzoic acid and pyridine in water. An amorphous yellow precipitate forms, which is easily filtered and has the composition $UO_2\ (H_2N\text{-}C_6H_4\text{-}COO)_2$. 2Py.2$H_2O$. which is converted by calcinationto U_3O_8 (Ripan and Sacelean, 1965).The uranium (VI) is precipitated quantitatively in the range of pH 5.0 to 7.5 with N-benzoyl-o-tolylhydroxylamine in the form of $UO_2(C_{14}H_{12}O_2N)_2$ (Mrinal and Majumdar, 1970).The thorium, forms an insoluble complex with vanillin in the range of pH 4.0 to 6.2. Thorium (IV) can be determined gravimetrically and separated from cerium, uranium (VI) and trivalent rare earth. The precipitate obtained is converted into oxide ThO_2 by calcination and then weighed (Jain and Singh, 1962). Thorium (IV) make a precipitate with benzenephosphonic acid in the form $Th(C_6H_5PO_3)_2{\cdot}3H_2O$ at pH around 0.5. The precipitate containing three molecules of H_2O and it is stable between 140 ° C and 180 ° C. between 240 ° C and 300 ° C, the precipitate contains only two molecules of H_2O. $Th(HPO_4)_2$ is obtained anhydrous above 800 ° C (Charles banks and Rodney Davis, 1955).

-Volumetry is used for dosing the uranium (VI) by carrying out a reduction using ferrous sulfate in phosphoric medium. The excess of Fe^{2+} is determined by the dichromate (Al Ammar and Basheer, 1993).

-The inductively coupled plasma coupled to mass spectrometry (ICP-MS) (Aydin and Soylak, 2007): the separation and concentration of thorium (IV) and uranium (VI) by solid phase extraction on Duolite XAD761 (resin adsorption). The chelates of thorium (IV) and uranium (VI) with 9-phenyl-3-fluorone formed and adsorbed on the XAD761 Duolite.

Thorium (IV) and uranium (VI) are quantitatively eluted with HCl 2 mol L^{-1} and determined by inductively coupled plasma coupled to mass spectrometry (ICP-MS).

Another method to extract the uranium present in an aqueous phase; we have the extraction of the uranium using the D2EHPA and Synergistic D2EHPA-TOPO Mixture (Ali et al., 2012; Beltrami et al., 2013; Beltrami et al., 2014; Orabi et al., 2015), using a synergistic DOPPA–TOPO mixture (Krea and khalaf, 2013), using of DNPPA and bidentateoctyl(phenyl) CMPO (Mondal et al., 2014) and the adsorption method of the uranium (Hussein and Morsy, 2011; Morsy and Hussein, 2012; Ahmed et al., 2016)

- The spectrophotometric method; ion UO_2^{2+} make a chelate with meloxicam by the following reaction: 2 (meloxicam H) + $UO_2^{2+} \rightarrow UO_2(\text{meloxicam})_2 + 2H^+$, the product is a yellow complex with an absorbance passes through a maximum at 398 nm (Lutfullah et al., 2008). Thorium, extract from a hydrochloric solution using the acid di (2-ethylhexyl) orthophosphoric, where the color is developed in the organic solution by adding an Arsenazo III solution and isopropanol, is determined by spectrophotometry with a maximum absorbance at 660 nm (Cerrai and Ghersini, 1967).

- The high resolution gamma spectrometry is used to determine the concentration of uranium and thorium in the surface soils, where soil samples are ground and sieved then stored for four weeks before analysis, it is carried for 18 hours (Tzortzis and Tsertos, 2004).

- The thin layer chromatography (Hodisan at al., 1998) is also used for the determination of uranium and thorium in presence of other metal ions using the iso-acid propyldithiophosphoric (PRDTP-i) in mobile phase as a complexing agent to differentiate between the species studied by modification of their retention time.

-The determination of the uranium isotope and thorium isotope is possible by coupling the ion chromatography and Inductively Coupled Plasma coupled to mass spectrometry IC-ICP-MS (Benkhedda et al., 2005; Zheng and Yamada, 2007; Rathore, 2008; Goodall and Lythgose, 1999; Betti, 1997), but the IC-ICP-MS has limitations due to the structure of the sample (capacity of the column, the acidity ...).

-The determination of thorium is possible by Energy dispersive X-ray fluorescence (EDXRF) (Mirashi et al., 2010), but we have problems of resolution and selectivity with (EDXRF).

-Other techniques such as the fluorimetry (Rathore, 2008; Sanjukta et al., 2008) The neutron activation analysis (Shinotsuka and Ebihara, 1997; Byrne and Benedik, 1988) and the alpha spectrometry (Hou and Roos, 2008; Saidou et al., 2008; Benedik et al., 2009; Jia et al., 2008) can be used for the determination of uranium and thorium, but these techniques require very long preparation of the sample (pre-concentration of the sample separation ...).

Although a variety of methods for the determination of uranium and thorium in different types of samples have been developed, there are several disadvantages for every method;

-The Precipitation methods: uranium and thorium have the same chemical properties, so they are difficult to separate, they precipitate together. We must then search the optimal conditions for selectivity. In addition the sensitivity of these gravimetric methods depends on a certain number of factors, the principal one being the interference, it is caused by the structure and concentration of the matrix elements (interference with the rare earth essentially).

-For colorimetric methods there are the stability problems for complexes and the problems of selectivity for the two elements (uranium and thorium).

- The inductively coupled plasma coupled to mass spectrometry (ICP-MS): the disadvantages of this method are the salinity of the matrix of samples processed and the isobaric interferences.

-The neutron activation analysis and the alpha spectrometry require a long process (counting and purification) and the use of a radioactive tracer.

- The gamma spectrometry, despite that it allows direct determination of uranium and thorium through their decay products, requires a long time to establish an equilibrium radiochemical decay products.

These methods require the use of many reagents and require multiple steps in preparations which causes a large error in measurement, in addition these methods require a very long time. Therefore it would be useful to have a simple, fast and accurate method for the determination of uranium and thorium in industrial phosphoric acid. The X-ray fluorescence spectrometry (XRF) is always an acceptable technique due to its acceptable precision and accuracy associated with its lower cost compared to the techniques mentioned above.We propose to perform the analysis of uranium and thorium in the industrial phosphoric acid by the wavelength-dispersive X-ray fluorescence spectrometry (WDXRF) with the use of the intensity of the line Lα (Lα1 and Lα2) for the determination of uranium and thorium. The X-ray fluorescence was used to determine traces of uranium and thorium in zircon, forming a tablet with 100 mg of zircon (Voldet, 1972), the author has determined the values in the range of 100 to 2000 ppm of uranium and 100 to 300 ppm of thorium in rocks zircon of Greenland. XRF is also used for the determination of thorium in the range of 0.1 to 1.2% of the plutonium in (Woltermanet al., 1974).

2. Experimental

2.1 Materials

2.1.1 Spectrometer

An X-ray fluorescence spectrometer (Magix 3kW, PW2403, PANalytical) was used for the determination of the uranium; it is a sequential spectrometer with wavelength-dispersive with a channel of measure based on a single goniometer covering the complete range of measure. It is equipped with an X-ray tube which is the X-ray source, the anode of the X-ray tube is in rhodium.

2.1.2 Scintillation Detector

It is constituted by a crystal of iodide of sodium in which the atoms of thallium are distributed in a homogeneous way (NaI; Tl), by a photocathode and by a tube photomultiplier.

It works by converting the X-ray in light which is then measured with a photomultiplier.

2.2 Sample Preparation

The samples of industrial phosphoric acid are analyzed directly without any prior preparation. Indeed, the industrial phosphoric acid is put in a cup (special sample holder for liquid) whose bottom is a polymer film and it is subjected directly to incident X-rays from the X-ray tube. In our case, it comes to an inverted optical. The X-ray tube is located under the sample. If the film broken during the analysis of the liquid, it can damage these parts. It is therefore necessary to adopt a sufficiently resistant film and make a compromise with the X-ray absorption and the presence of additives in the film (which allow the film to more resistant to some products). This problem does not arise in the case of direct optical.

For this problem, we used a film for liquid analysis by XRF which does not absorb X-rays and which has no effect or interference matrix which may influence our results.

This film is *X-Ray film MYLAR®, diameter 63.5mm.* We used a mass of 3 g with 0.1 mg of accuracy for industrial phosphoric acid sample. The analysis is made under helium (900 hPa pressure) to prevent boiling under the vacuum.

2.3 Measurement Conditions

Measurement conditions of uranium and thorium in industrial phosphoric acid using WDXRF aresummarized as follows;

•Atmosphere: Helium (900 hPa pressure)
•Power: 60kV, 50 mA
•Crystal:LiF220
•Collimator: 150microns
• X-ray: U-Lα1 line of 13.631 keV energy and Th- Lα1 line of 12.986keV energy.

The WDXRF analysis of uranium and thorium in industrial phosphoric acid is shown in Figure 1.

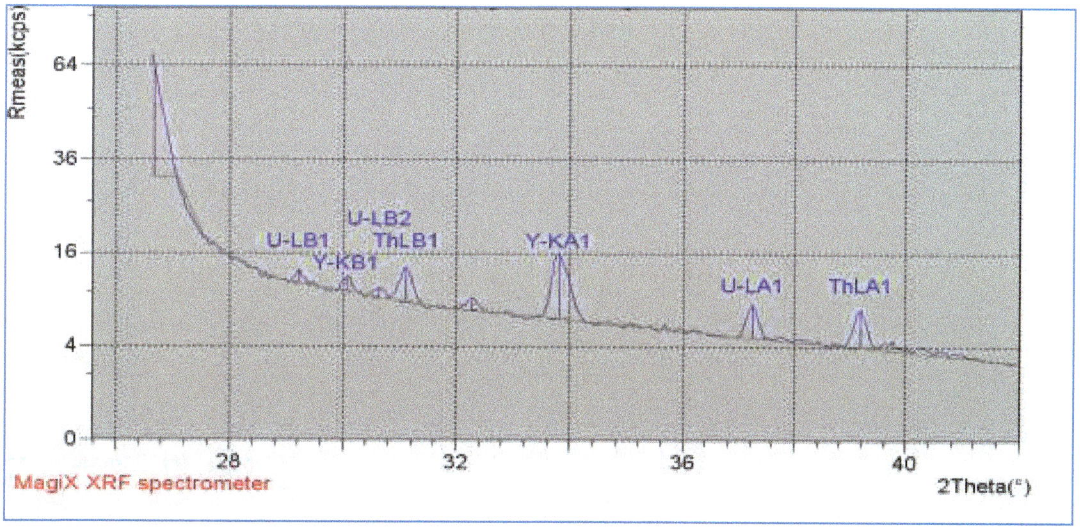

Figure 1. Detectionof uranium and thorium peak using the crystal LiF220

- The experimental angle of the line of uranium: U-Lα1 (2θ °) [noted U-LA1]: 37.247°

- The experimental angle of the line of thorium: Th- Lα1 (2θ °) [noted Th-LA1] : 39.175°

3. Validation

The proposed method was validated by linearity, detection limit and quantification limit, specificity, fidelity (repeatability and reproducibility) and accuracy (AFNOR NF XPT 90-210, 1999; AFNOR NF ISO 5725-1, 1994).

3.1 Results and Discussion

3.1.1 Linearity, Detection Limit and Quantification Limit

The linearity of the proposed method was investigated by repeating the analysis (*n* = 6) at six concentration levels of 0, 20, 40, 60, 80 and 100 ppm for uranium and 0, 10, 20, 30, 40 and 50 ppm for thorium.

The intensity obtained at each concentration was plotted against the initial concentration of uranium and thorium and the linear regression equation was evaluated in the statistical treatment of calibration data. The characteristics of the regression were calculated using Microsoft Excel™. We have the following curves in figure 2 and in figure 3:

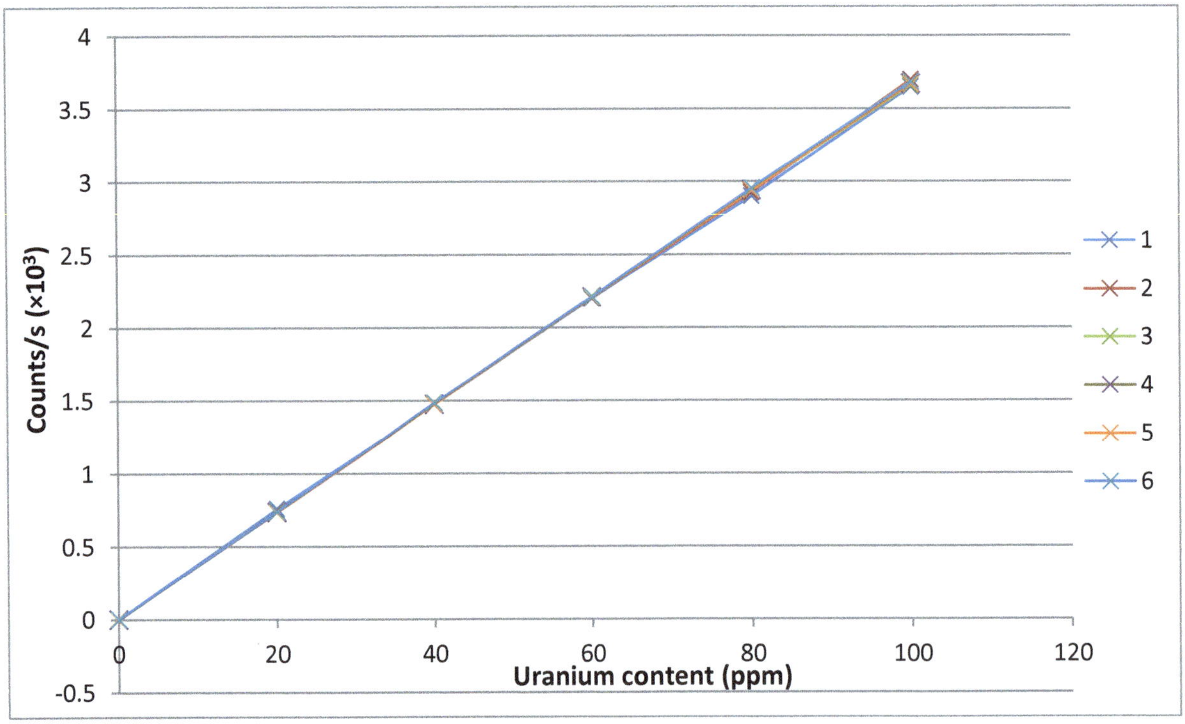

Figure 2. Calibration of the method for the determination of uranium

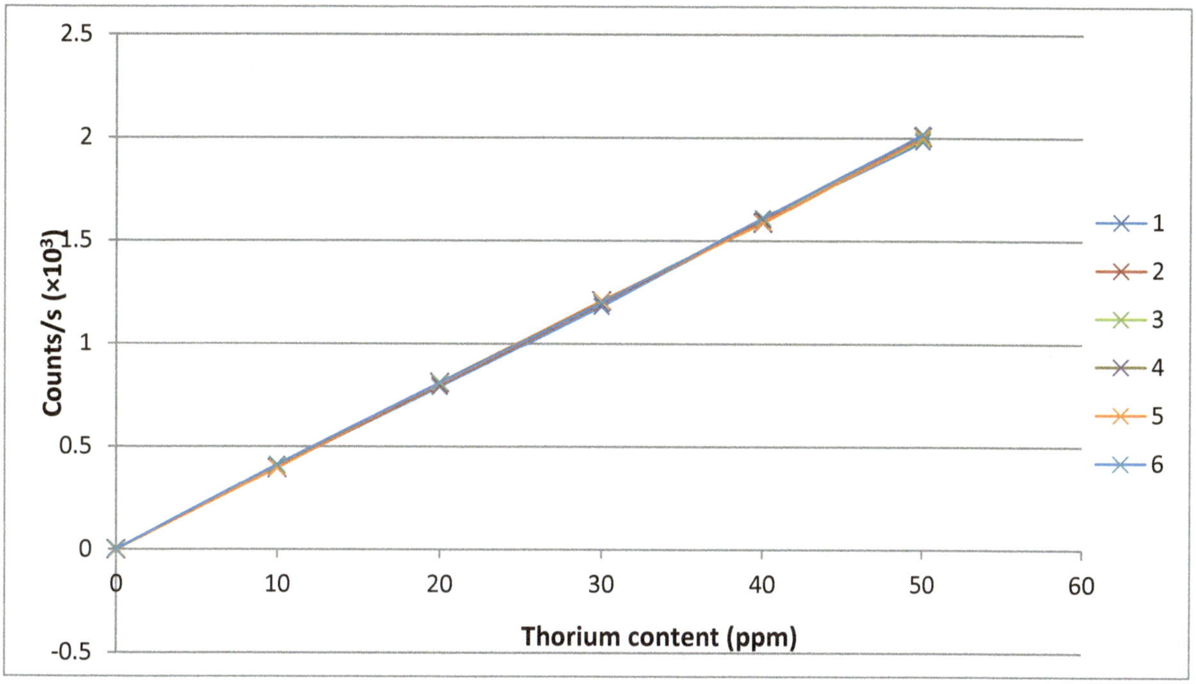

Figure 3. Calibration of the method for the determination of thorium

The measurement collected to perform the linearity can be used to calculate the sensitivity (b_1) and its standard deviation S(b_1) and the blank value b_0 and its standard deviation S(b_0).

These standard deviations (S(b_1) and S(b_0)) define as follows the limits of detection and quantification, based on the two definitions:

$$\text{The limit of detection:LD} = \frac{b_0+3.S(b_0)}{b_1} \tag{2}$$

$$\text{The limit of quantification: LQ} = \frac{b_0+10.S(b_0)}{b_1} \tag{3}$$

$$b_1 = \frac{SPE(x,y)}{SCE(x)} \tag{4}$$

$$b_0 = \bar{\bar{y}} - b_1.\bar{\bar{x}} \tag{5}$$

$$S(b_1) = \sqrt{\frac{s^2(e)}{SCE(x)}} \quad \text{standard deviation of the sensitivity} b_1 \tag{6}$$

$$S(b_0) = \sqrt{s^2(e)\left(\frac{1}{np} + \frac{\bar{\bar{x}}^2}{SCE(x)}\right)} \text{standard deviation of the blank} b_0 \tag{7}$$

$s^2(e)$: experimental variance of the regression.

$SCE(x)$: Sum of squared deviations for the variable x.

$SPE(x,y)$: Sum of the products of deviations for the variable x and y.

The results of the study linearity test are summarized in the following Table 1:

Table 1. Statistical results of the calibration

Designation	Observed value for uranium	Observed value for thorium
Number of levels(p)	6	6
Total number of measurements	36	36
sensitivityb_1	0.037 ppm	0.040 ppm
blank value b_0	0.006 ppm	0.002 ppm
Equation of linear regression	Y=0.037X+0.006	Y=0.040X+0.002
Correlation coefficient	0.9999	0.9998
Standard deviation of sensitivity S(b_1)	5.918×10^{-5}	8.235×10^{-5}
Standard deviation of the blank value S(b_0)	0.003	0.002
Detection limit (DL)	0.483 ppm	0.235 ppm
Quantification limit (QL)	1.168 ppm	0.671 ppm

The results of the test of evaluation of the linearity showed that the model of regression is considered acceptable for both for uranium and thorium and that the domain of linearity is validated.

3.2 The Specificity

To perform this test, we made 16 (p=16) standard additions on selected samples to cover the scope of the method. The regression line that passes from the points from the following equation is calculated:

$$r_i = C_0 + C_1.v_i \quad (8)$$

C_0 the intercept of the regression line.

C_1 the slope of the regression line.

v_i added content of uranium or thorium in the samples.

r_i observed content of uranium or thorium = content measured after addition - content measured before addition.

We carryon a graph the values of v_i additions and observed values r_i and the straight recovery calculated and the line of slope1 (equation y = x). The curves are shown in figure 4 for uranium and figure 5 for thorium.

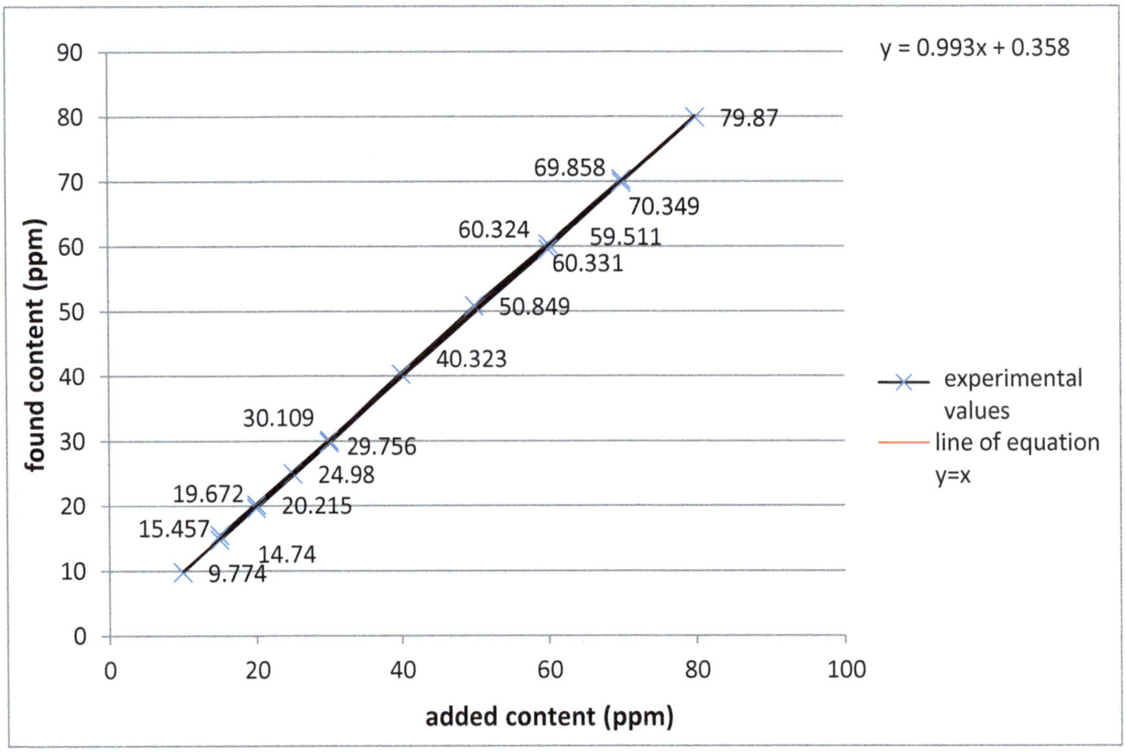

Figure 4. Study of the specificity of the method for the determination of uranium

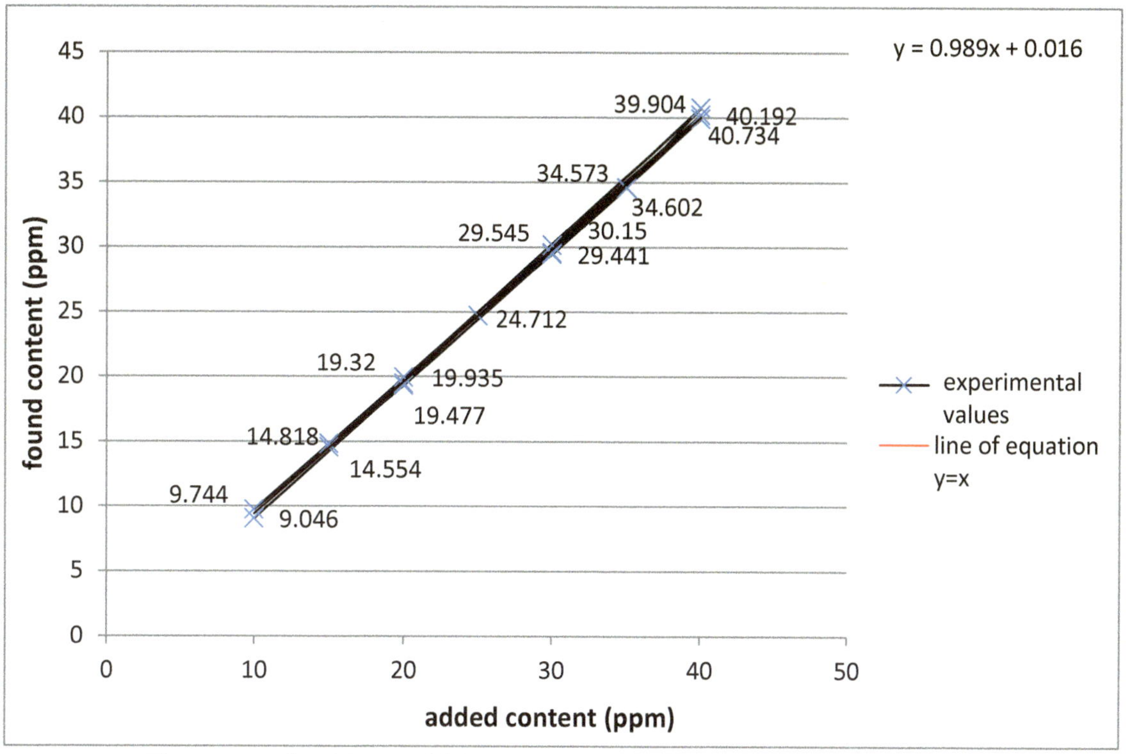

Figure 5. Study of the specificity of the method for the determination of thorium

3.3 Interpretation

To check if the slope of the regression line is equal to1, a test is performed by calculating thet_{obs}which follows aStudent distribution with$p$ -2degrees of freedom.t_{obs}it is a criterion calculated to check the slope as it is described in the validation standard.

$$t_{obs} = \frac{|C_1 - 1|}{\sqrt{\frac{s^2(e)}{SCE(v)}}} = \frac{|C_1 - 1|}{s(C_1)} \tag{9}$$

$s^2(e)$: experimental variance of the regression.

$SCE(v)$: Sum of squared deviations for the variable v.

$s(C_1)$:standard deviation of the slopeC_1

a) If $t_{obs}$is less than or equal to the critical value read from the Student table to the risk1 -α/2 with ν= p- 2= 14degrees of freedom, it is possible to conclude that the absence of interference and that the specificity is acceptable.

b) If $t_{obs}$is greater than the critical value ,the method is not specific. In general, the risk of error α is taken to 1%.

t_{crit} value of a variable of Student to the risk1 -α/2 withν= p- 2degrees of freedom, the risk of error α is taken to 1% (t_{crit} = 2.977 from the Student table).

- We obtain $t_{obs\ for\ uranium}$ = 0.788 and $t_{obs\ for\ thorium}$ = 2.046

To check if the intercept is equal to 0, a test is performed by calculating the t'_{obs}which follows a Student distribution withp -2 degrees of freedom. t'_{obs} it is a criterion calculated to check the intercept as it is described in the validation standard.

$$t'_{obs} = \frac{|C_0|}{s(C_0)} \tag{10}$$

$s(C_0)$:standard deviation of the interceptC_0

a) If $t'_{obs}$is less than or equal to the critical value read from the Student table to the risk1 -α/2 withp -2 degrees of freedom, it is possible to conclude that the intercept is not different from 0.

b) If t'_{obs} is greater than the critical value, it is possible to conclude that the intercept is not equal to 0 and that the method is not specific. In general, the risk of error α is taken to 1%.

- We obtain $t'_{\text{obs for uranium}} = 0.849$ and $t'_{\text{obs for thorium}} = 0.085$
    - $t_{obs} < t_{crit}$ it is possible to conclude that the absence of interference and that the specificity is acceptable for uranium and for thorium.
    - $t'_{obs} < t_{crit}$ it is possible to conclude that the intercept is not different from 0 for uranium and for thorium.
- The slope of the regression line for uranium and for thorium is equal to1 and the intercept of the regression line for these two elements is equal to 0. It is thus possible to conclude that the absence of interference in the defined scope and the specificity for these two elements is acceptable.

3.4 The Fidelity (Repeatability and Reproducibility)

3.4.1 Repeatability

In the case of the study of the repeatability of the method of determination of uranium and thorium, 10 industrial phosphoric acid samples covering the scope were analyzed in triplicate repeatability conditions.

The internal repeatability variance of the alternative method $S_r^2(x)$ is equal to the sum $SCE_r(x)$divided by the number of degrees of freedom appropriate.

$$\mathrm{S_r^2}(x) = \frac{\mathrm{SCE_r}(x)}{\mathrm{N}-p} \quad (11)$$

With:

$$SCE_r(x) = \sum_{i=1}^{p} \sum_{j=1}^{n} (x_{ij} - \bar{x}_i)^2 \text{ Sum of squared intra-sample differences} \quad (12)$$

$$\bar{x}_i = \frac{\sum_{j=1}^{n_j} x_{ij}}{n_i} \text{Average of the sample } i \quad (13)$$

$$N = \sum_{j=1}^{p} n_j \text{Total number of measurements} \quad (14)$$

$$\mathrm{S_r}(x) = \sqrt{S_r^2(\mathrm{x})} \text{ standard deviation of repeatability intra laboratory} \quad (15)$$

We can calculate the coefficient of variation of repeatability:

$$CV_r(\%) = \frac{S_r(x)}{\bar{\bar{x}}} \times 100 \quad (16)$$

We obtain a value of standard deviation of repeatability intra-laboratory $\mathrm{S_r}(x)$:

$\mathrm{S_r}(x)_{\text{for uranium}} = 0.147$ and $\mathrm{S_r}(x)_{\text{for thorium}} = 0.042$

and an average coefficient of variation of repeatability intra-laboratory CV_r: $CV_{r\ for\ uranium} = 0.67$ % and $CV_{r\ for\ thorium} = 1.02$ %.

3.4.2 Checking the Stability of Fidelity

It is performed by means of the test of Cochran to the risk error α = 1%. This test is to ensure if the fidelity is constant throughout the application domain.

$$C_{obs} = \frac{S_{\max}^2(x)}{\sum_{\mathrm{i}=1}^{\mathrm{p}} S_{\mathrm{i}}^2(x)} \quad (17)$$

$S_{\max}^2(x)$the maximumexperimental variance.

$\sum_{\mathrm{i}=1}^{\mathrm{p}} S_{\mathrm{i}}^2(x)$the Sum ofexperimental variances.

C_{obs}this is a ratio calculated from the equation cited above to verify the stability of fidelity.

C_{crit}value of a variable of Cochran to the risk α=1% with *n* number of distributionand number of sample *p*.

We compare $C_{obs}$with C_{crit}which is the critical value of the table of Cochran at the risk of error 1 %.The

Statistical results of the verification of the stability of fidelity are presented in Table 2.

Table 2. Statistical results of the verification of the stability of fidelity

Designation	Observed value for uranium	Observed value for thorium
$S^2_{\max}(x)$	0.066	0.005
C_{obs}	0.108	0.097
Number of samples (p)	10	10
Number of replicates(n)	3	3
$C_{crit}(1\%)$	0.536	0.536

C_{obs} for uranium and for thorium is lower than the value of the table at the risk of error 1 %. It is possible to conclude that the fidelity in terms of repeatability intra-laboratory is constant in all the field of application for these two elements. The fidelity in terms of repeatability is acceptable.

3.4.3 Reproducibility

This test is used to calculate the intra-laboratory reproducibility of the method studied, that is to say his fidelity when repetitions are made by several operators or longer time intervals with respect to the method. A sample of industrial phosphoric acid was analyzed for 10 different days (p=10) with 3 replicates (n=3).The results are summarized in Table 3 for uranium and in Table 4 for thorium.

Table 3. Results of reproducibility test for uranium

Day	Replicates			Average	Variance	$S^2_r(x)$	$S^2_L(x)$	$S_R(x)$	$CV_R(\%)$
	1	2	3						
1	22.106	22.091	22.213	22.137	0.004				
2	22.189	22.205	22.218	22.204	0.001				
3	21.897	21.943	22.186	22.009	0.024				
4	22.435	22.376	22.528	22.446	0.006				
5	22.075	21.796	22.137	22.003	0.033	0.013	0.016	0.171	0.771
6	21.994	22.184	22.159	22.112	0.011				
7	22.221	22.307	22.213	22.247	0.003				
8	22.059	22.124	22.267	22.150	0.011				
9	21.987	22.379	22.224	22.197	0.039				
10	22.428	22.376	22.331	22.378	0.002				

Table 4. Results of reproducibility test for thorium

Day	Replicates			Average	Variance	$S^2_r(x)$	$S^2_L(x)$	$S_R(x)$	$CV_R(\%)$
	1	2	3						
1	4.03	4.152	4.094	4.092	0.004				
2	3.985	4.026	4.100	4.037	0.003				
3	4.082	4.117	4.142	4.113	0.001				
4	4.080	3.994	3.975	4.016	0.003				
5	4.128	4.147	4.109	4.128	0.001	0.002	0.001	0.056	1.371
6	4.044	4.105	4.086	4.078	0.001				
7	4.021	3.966	4.087	4.025	0.003				
8	4.093	4.140	4.115	4.116	0.001				
9	3.977	4.076	4.059	4.037	0.003				
10	4.124	4.033	4.083	4.080	0.002				

To calculate the variance of internal reproducibility, we rely on these formulas:

$$N' = N - \frac{\sum_{i=1}^{p} n_i^2}{N} \text{ Corrected average number of repetitions} \quad (18)$$

$$S^2_L(x) = \frac{(p-1)\left(\frac{SCE_L(x)}{p-1} - S^2_r(x)\right)}{N'} \text{ Variance inter-sample.} \quad (19)$$

$$S^2_R(x) = S^2_L(x) + S^2_r(x) \text{ Variance of internal reproducibility.} \quad (20)$$

$$CV_R(\%) = \frac{S_R(x)}{\bar{\bar{x}}} \times 100 \text{ Coefficient of internal reproducibility.} \quad (21)$$

n_i: repetition number per day (n=3).

N : total number of measurements.
$SCE_L(x)$: Sum of squared inter-sample differences .
$S_r^2(x)$: variance internal repeatability.
$S_R(x)$: standard deviation of internal reproducibility.

- We obtain a value of coefficient of variation of reproducibility intra-laboratory CV_R:

$CV_{R\ for\ uranium}$ = 0.77 % and $CV_{R\ for\ thorium}$ = 1.37 %.

- Fidelity in terms of internal reproducibility is acceptable.

3.4.4 The Accuracy

In the case of the study of the accuracy of the method of determination of uranium and thorium in industrial phosphoric acid, 10 repetitions were performed on a NIST (National Institute of Standards and Technology) standards for uranium and thorium (the content of U is equal to 20 ppm and the content of Th is equal 10 ppm to). We obtained an average concentration of 19,865 ppm for uranium with a variance of 0.205 and 10.083 ppm for thorium with a variance of 0.095.

To evaluate the accuracy from a reference standard:

- Repeat by the same operator determining the analyte n test taken.
- Calculate the differencebetween the averages of$n$determination ($\bar{x}$) and the certifiedconcentration ofuranium and thorium in the reference standard (x_{SR}).
- Calculatethe standard deviation ofmeasurements obtained $S_R(x)$.
- Calculate $$t_{cal} = \frac{|\bar{x} - x_{SR}|}{\frac{S_R(x)}{\sqrt{n}}} \qquad (22)$$

t_{cal}this is a ratio calculated from the equation cited above to verify the accuracy.
t_{cal} =2.083 for uranium and for thorium ist_{cal} =2.747

- Obtain $t_{crit} = t_{V,1-\frac{\alpha}{2}}$from the Student table. (t_{crit}= 3.355)
- Compare t_{cal}tot_{crit}

The decision rule is as follows:

- If $t_{cal} > t_{crit}$: error accuracy is significant, it is possible to conclude that the method is not justified.
- If $t_{cal} \leq t_{crit}$: error accuracy is not significant, it is possible to conclude that the method is justified.

- The value of t_{cal} for uranium and for thorium is less than the value of the Student table. It is possible to conclude that the accuracy error is not significant. The method is considered justified.
- After evaluating the criteria of linearity, specificity, limit of detection and quantification, accuracy and fidelity, the method is acceptable in the scope defined.

4. Conclusion

The present work has developed a method for the determination of uranium and thorium in the industrial phosphoric acid obtained with wet process. This method has been validated on Tunisian phosphoric acid used as a liquid concentrate without dilution. It offers several advantages in particular the speed of measurement, accuracy and low cost of operation. The proposed method has the advantage of not using reagents for the preparation of samples for analysis, the samples were analyzed directly without dilution or preparation which increases the accuracy of the analyzes. Moreover the relatively lower cost of the sensor for the wavelength dispersive X-ray fluorescence (WDXRF) offers a major advantage compared to the energy dispersive analyzer (EDXRF) spectrometry, which has been frequently employed for determination of trace elements in phosphoric acid and the (WDXRF) analyzer has many advantages regarding resolution and selectivity over the (EDXRF). The application of the proposed method for the determination of uranium and thorium in industrial phosphoric acid shows that it is contains more uranium than thorium(of about22ppm of uranium and 4 ppm of thorium).

Acknowledgment

This research was supported by the National Institute of Research and Physico-Chemical Analysis, (INRAP),Tunisia.

References

AFNOR NF ISO 5725-1. (1994). Application of statistics. Accuracy (trueness and precision) of measurement methods and results. Part 1: general principles and definitions.

AFNOR NF XP T 90-210. (1999). Water quality-Protocol for the evaluation of an alternative quantitative physico-chemical analysis method against a reference method.

Ahmed, H., Orabi, Enass, M., El-Sheikh, W. H., Saleh, A. O., Youssef, M. Y., & El-Kady, Z. M. S. (2016). Potentiality of uranium adsorption from wet phosphoric acid using amine-impregnated cellulose, *Journal of Radiation Research and Applied Sciences, 9*, 193-206. http://dx.doi.org/10.1016/j.jrras.2015.12.003

Al Ammar, A. S., & Basheer, H. M. (1993). Further development in the high-precision volumetric method for the determination of uranium in nuclear-grade uranium compounds. *Journal of Radioanalytical and Nuclear Chemistry, 171*(2), 435–441. http://dx.doi.org/10.1007/BF02219867

Ali, H. F., Ali, M. M., Taha, M. H., & Ahmed, F. A. (2012). Uranium extraction mechanism from analytical grade phosphoric acid using D2EHPA and synergistic D2EHPA-TOPO mixture. *IJNESE, 2*(2), 57-61.

Aydin, F. A., & Soylak, M. (2007). Solid phase extraction and preconcentration of uranium(VI) and thorium(IV) on Duolite XAD761 prior to their inductively coupled plasma mass spectrometric determination. *Talanta, 72*(1), 187-192. http://dx.doi.org/10.1016/j.talanta.2006.10.013

BECKER, P. (1989). Phosphates and phosphoric acid: raw materials, Technology and economics of the wet process, Fertilizer Science Technology Ser (2nd ed.), 6, Marcel Dekker Inc, New York.

Beltrami, D., Chagnes, A., Haddad, M., Varnek, A., Mokhtari, H., & Courtaud, B. (2013). Recovery of uranium (VI) from concentrated phosphoric acid by mixtures of new bis (1,3dialkyloxypropan-2-yl) phosphoric acids and tri-n-octylphosphine oxide. *Hydrometallurgy, 140*, 28-33. http://dx.doi.org/10.1016/j.hydromet.2013.08.008

Beltrami, D., Chagnes, A., Haddad, M., Laureano, H., Mokhtari, H., & Courtaud, B. (2014). Studies on the selectivity of cetrimide for uranium extraction from wet process phosphoric acid. *Hydrometallurgy*, 207-214.

Benedik, L., Vasile, M., Spasova, Y., & Watjen, U. (2009). Sequential determination of 210Po and uranium radioisotopes in drinking water by alpha-particle spectrometry. *Applied Radiation and Isotopes, 67*(5), 770–775. http://dx.doi.org/10.1016/j.apradiso.2009.01.033

Benkhedda, K., Epov, V., Evans, N., & Douglas, R. (2005). Flow-injection technique for determination of uranium and thorium isotopes in urine by inductively coupled plasma mass spectrometry.*Analytical and Bioanalytical Chemistry, 381*(8), 1596-1603. http://dx.doi.org/10.1007/s00216-005-3147-y

Betti, M. (1997). Use of ion chromatography for the determination of fission products and actinides in nuclear applications. Journal of Chromatography *A*, 789(1-2), 369–379. http://dx.doi.org/10.1016/S0021-9673(97)00784-X

Byrne, A. R., & Benedik, L. (1988). Determination of uranium at trace levels by radiochemical neutron-activation analysis employing radioisotopic yield evaluation. *Talanta, 35*(3), 161–166. http://dx.doi.org/10.1016/0039-9140(88)80058-4

Cerrai, E., & Ghersini, G. (1967). Spectrophotometric determination of thorium with arsenazo III in the organic phase after extraction with di-(2-ethylhexyl) orthophosphoric acid. *Analytica. Chimica. Acta., 37*, 295–309. http://dx.doi.org/10.1016/S0003-2670(01)80677-2

Charles, B. V., & Rodney, D. J. (1955). Analytical applications of the reaction of thorium with benzenephosphonic acid. *Analytica Chimica Act., 12*, 418–431.

Goodall, P., & Lythgoe, C. (1999). Rapid separation of uranium and plutonium by extraction chromatography for determination by thermal ionisation mass Spectrometry. *Analyst, 124*, 263-269. http://dx.doi.org/10.1039/a809219b

Hussein, A., & Morsy, A. (2012). Uranium recovery from wet-process phosphoric acid by a commercial ceramic product. *Arabian Journal of Chemistry.* http://dx.doi.org/10.1016/j.arabjc.2012.09.007.

Hodisan, T., Curtui, M., Cobzac, S., Cimpoiu, C., & Haiduc, I. (1998). The limit of detection improvement in TLC determination of uranium and thorium in the presence of other metal ions. *Journal of Radioanalytical and Nuclear Chemistry, 238*(1), 179–183. http://dx.doi.org/10.1007/BF02385377

Hou, X., & Roos, P. (2008). Critical comparison of radiometric and mass spectrometric methods for the determination of radionuclides in environmental, biological and nuclear waste samples. *Analytica Chimica Acta, 608*(2), 105–139. http://dx.doi.org/10.1016/j.aca.2007.12.012

Jain, B. D., & Singh, J. J. (1962). Separation and gravimetric determination of thorium and cerium (IV) with vanillin. *Analytica Chimica Acta, 27*, 359–362. http://dx.doi.org/10.1016/S0003-2670(00)88516-5

Jia, G., Torri, G.,Ocone, R., Di Lullo, A., De Angelis, A., & Boschetto, R. (2008). Determination of thorium isotopes in mineral and environmental water and soil samples by α-spectrometry and the fate of thorium in water. *Applied Radiationand Isotopes, 66*(10), 1478–1487. http://dx.doi.org/10.1016/j.apradiso.2008.03.015

Krea, M., & Khalaf, H. (2000). Liquid-liquid extraction of uranium and lanthanides from phosphoric acid using a synergistic DOPPA-TOPO mixture. *Hydrometallurgy, 58*(3), 215-225. http://dx.doi.org/10.1016/S0304-386X(00)00129-8

Lopez-de-Alba, P. L., Gonzalez, S., & Gomez, L. J. (1989). Gravimetric determination of uranyl ion with 2-hydroxy-1-naphthaldehyde/2H-1N/. *Journal of Radioanalytical and Nuclear Chemistry, 136*(3), 203–210. http://dx.doi.org/10.1007/BF02164120

Lutfullah, N. A. M., Rahman, N., & Najmul, H. A. S. (2008). Optimized and validated spectrophotometric method for the determination of uranium(VI) via complexation with meloxicam. *Journal of Hazardous Materials, 155*(1-2), 261–268. http://dx.doi.org/10.1016/j.jhazmat.2007.11.055

Mirashi, N. N., Sangita, D., Sanjay, K. S., Satyajeet, C., Misra, N. L., & Aggarwal, S. K. (2010). Energy dispersive X-Ray fluorescence determination of thorium in phosphoric acid solutions.*SpectrochimicaActa Part B, 65*, 579–582. http://dx.doi.org/10.1016/j.sab.2010.05.006

Mondal, S., Singh, D., Anitha, M., Sharma, J., Hubli, R., & Singh, H. (2014). New synergistic solvent mixture of DNPPA and bidentateoctyl(phenyl) CMPO for enhanced extraction of uranium (VI) from phosphoric acid medium. *Hydrometallurgy,* 147-148, 95-102. http://dx.doi.org/10.1016/j.hydromet.2014.04.023

Morsy, A., & Hussein, A. (2011). Adsorption of uranium from crude phosphoric acid using activated carbon, *Journal of Radioanalytical and Nuclear Chemistry, 288*, 341-346. http://dx.doi.org/10.1007/s10967-011-0980-7

Mrinal, K. D., & Majumdar, A. K. (1970). N-benzoyl-o-tolylhydroxylamine as a gravimetric reagent for uranium (VI).*Analytica Chimica Acta, 50*, 243–247.

Orabi, A., El-Sheikh, E., Mowafy, A., Abdel-Khalekb, M., & El Kady, M. (2015). Studies on the selectivity of cetrimide for uranium extraction from wet process phosphoric acid, *International Journal of Mineral Processing, 137*, 26-33. http://dx.doi.org/10.1016/j.minpro.2015.02.012

Rathore, D. P. S. (2008). Advances in technologies for the measurement of uranium in diverse matrices. *Talanta, 77*(1), 9–20. http://dx.doi.org/10.1016/j.talanta.2008.06.019

Ripan, R., & Sacelean,V. (1965). Une nouvelle méthode pour le dosage gravimétrique de l'uranium. *Talanta, 12*, 69–72.

Saidou, B, F., Laedermann, J. P., Kwato N. M. G., & Froidevaux, P. (2008). A comparison of alpha and gamma spectrometry for environmental natural radioactivity surveys. *Applied Radiation and Isotopes, 66*(2), 215–222. http://dx.doi.org/10.1016/j.apradiso.2007.07.034

Sanjukta, K. A., Niyoti, S. S., Pandey, S., Sounderajan, S., & Enkateswaran, G.V. (2008). Direct determination of uranium in seawater by laser fluorimetry.*Talanta, 77*(1), 422–426. http://dx.doi.org/10.1016/j.talanta.2008.07.002

Shinotsuka, K., & Ebihara, M. (1997). Precise determination of rare earth elements, thorium and uranium in chondritic meteorites by inductively coupled plasma mass spectrometry-a comparative study with radiochemical neutron activation analysis. *Analytica Chimica Acta, 338*(3), 237-246. http://dx.doi.org/10.1016/S0003-2670(96)00528-4

Slack, A.V. (1968). Phosphoric Acid, Marcel Dekker, New York.

Tzortzis, M., & Tsertos, H. (2004). Determination of thorium, uranium and potassium elemental concentrations in surface soils in Cyprus. *Journal of Environmental Radioactivity, 77*(3), 325–338. http://dx.doi.org/10.1016/j.jenvrad.2004.03.014

Voldet, P. (1972). Dosage de traces d'uranium, de thorium et de plomb par fluorescence-x. étudesur des "zircons" synthétiques et application à quelques zircons naturels du Groenland. *Analytica Chimica Acta, 62*, 297–304.

Waggaman, W. H. (1969). Phosphoric acid and phosphatic fertilizers, Hafner Publishing Company, New York & London.

Woltermann, H. A., Eckstein, R. R., Redding, P. L., & Tomes, S. A. (1974). Determination of thorium in plutonium by X-ray spectrometry.*Journal of Nuclear Materials, 54*(1), 117-120. http://dx.doi.org/10.1016/0022-3115(74)90082-8

Zheng, J., & Yamada, M. (2006).Determination of U isotope ratios in sediments using ICP-QMS after sample cleanup with anion-exchange and extraction chromatography. *Talanta, 68*(3), 932–939. http://dx.doi.org/10.1016/j.talanta.2005.06.065

16

Physicochemical Characterization of Moroccan Phosphate and Coke of a Solid Combustible

Ikbal Bouatba[1], M'barek Benchanaa[1], Mustapha El Hammioui[1], Latifa Bilali[1]

[1]Laboratoire de Recherche, Reactivite des Materiaux et Optimisation des Procedes: REMATOP, Faculte des Sciences Semlalia, Universite Cadi Ayyad, Avenue Prince Moulay Abdellah, B.P. 2390, 40000 Marrakech, Maroc

Correspondence: M'barek Benchanaa, Laboratoire de Recherche, Reactivite des Materiaux et Optimisation des Procedes: REMATOP, Faculte des Sciences Semlalia, Universite Cadi Ayyad, Avenue Prince Moulay Abdellah, B.P. 2390, 40000 Marrakech, Maroc. E-mail: mbarek_benchanaa@yahoo.fr

Abstract

This paper concerns the study of the phosphate ore and a coke of solid combustible. The determination of their physicochemical characteristics and the study of their structural, textural and morphological aspects before and after thermal treatment are important and necessary for understanding of the reaction mechanisms and of the modelling of the processes of combustion and calcination.

The characterization of these materials was accomplished according to various experimental techniques such as: X-ray diffraction; infrared spectroscopy, scanning electron microscopy, porosimetry and the BET method. The results obtained show that the X-ray diffraction pattern of the coke presents only the crystalline phase of the graphite. The diffraction pattern of the raw phosphate as well as the crystallographic parameters at 25°C
(a = 9.36 Å, c = 6.90 Å) are consistent with those given in the literature and indeed correspond to those of the crystal lattice of pure fluorapatite. The specific surface of natural phosphate is in the range of 13.36 $m^2.g^{-1}$ and that of the coke is 0.361 $m^2.g^{-1}$. The total pore volume of the phosphate ore measured by the condensation of nitrogen adsorbed was 0.178 mL and the mean opening diameter of the pores is 0.029 m. The real density of the phosphate (2.96 $g.cm^{-3}$) and that of the coke (1.36 $g.cm^{-3}$), were determined using an ultra pycnometer. Morphological observation of the grain surface shows fairly homogenous areas and a porous structure in the phosphate ore. The coke grains show on the contrary a difference in the grain size and a non-porous structure.

Keywords: phosphate ore, coke, characterisation, calcination, combustion, energy, industrial process

1. Introduction

Morocco has important reserves of phosphate ores. According to the chemical and granulometric characteristics of each phosphate type, the mineral extracted is subjected to treatments which are principally washing, drying, calcination and dry enrichment. All these processing operations are energy consuming.

In this context, we have initiated a research project that aims first to contribute to the energy optimization of industrial processes, particularly related to the calcination of phosphate ores, and second, to develop techniques for combustion, using coke as a solid fuel. Many authors have focused on the study of complex phenomena occurring during the combustion of coke (Dubuisson et al., 1995; Salvador et al., 2003; Slaoui & Bounahmidi, 2004). Others analysed the characterisation of the petroleum coke (Heintz, 1996; Suelves et al., 2002; Commandre & Salvador, 2005) and its application (Méndez et al., 2006; Murthy et al., 2014).

The present work is part of an extensive research and development program on energy optimization of the calcination process of the Moroccan phosphate ore using coke a solid fuel. We focused as a first step on the characterization of phosphate ore samples and coke samples used in the calcination of Moroccan phosphates. The main objective is to determine the physicochemical and mineralogical characteristics as well as the structural features, textural and morphological aspects of these samples before starting the study of their behavior in terms of combustion and calcination.

2. Materials and Experimental Techniques

The experimental study is based on two samples: the coke of a solid combustible and bright natural phosphate. These samples were provided by the CERPHOS (Center for studies and research on phosphates of the OCP group).

2.1 Sample Preparation

The homogeneity of the material and its particle size must be well controlled. For this, we subjected the samples to preparation operations such as:

- Coning and quartering

Sampling is an operation which splits a large sample into several smaller samples that are virtually identical. This operation can be accomplished using a splitter or manually through a series of divisions-mixtures.

- Granulometric separation

This operation consists of the separation of the desired elements from unwanted particles by sieving into three fractions: $_1$< 100 µm, 100 µm < $_2$< 400 µm and $_3$> 400 µm. Only the fraction $_2$ was kept, because it represents a great part of the sample (nearly 80%).

2.2 Materials Characterisation Techniques

The characterization of the samples was achieved using different techniques and experimental processes such as: X-ray diffraction, infrared spectroscopy, scanning electron microscopy; porosimetry and the BET method.

3. Results & Discussion

3.1 Radiocrystallographic Analyses

The use of X-ray diffraction permits determination of the structure of the samples. It allows, at first, to control the different phases of the studied material, and subsequently the determination of its crystal parameters.

Figures 1 and 2 present the The diffraction pattern performed on powdered samples, obtained by grinding the raw samples (phosphate and cokeof a solid combustible) at 25°C. The diffraction pattern of the raw phosphate sample (Figure 1) shows various lines. Some can be attributed to the carbonate fluorapatite ($(CaF)Ca_4(PO_4)_3$) contained in the ore and others correspond to the carbonate $CaCO_3$ phase.

The values of the crystallographic parameters of the phosphate ore analysed are: a = 9.36 Å and c = 6.90 Å at 25°C. The X-ray diffraction pattern of the coke (Figure 2) shows a single line corresponding to the crystalline phase of graphite.

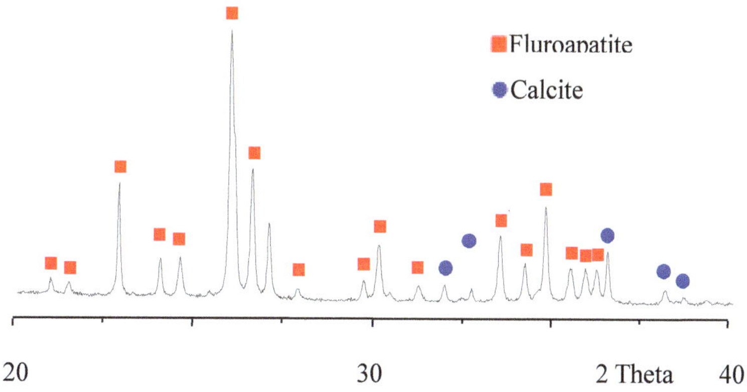

Figure 1. X-ray diffraction pattern of the raw phosphate.

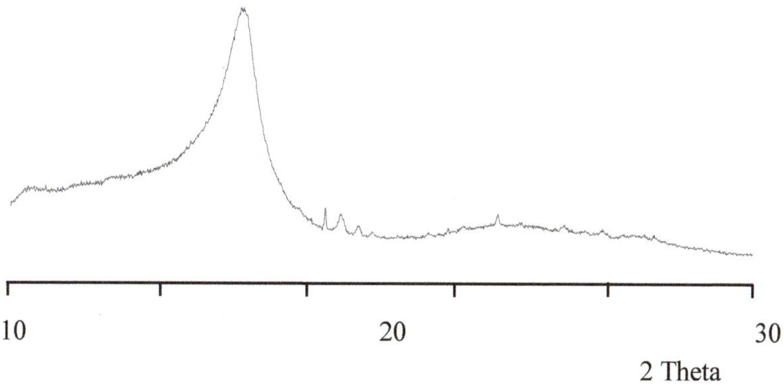

Figure 2. X-ray diffraction pattern of the coke of a solid combustible.

3.2 Infrared Spectroscopy

The presence of apatite and carbonate phases in bright phosphates are confirmedby infrared spectroscopy technique (Figure 3).

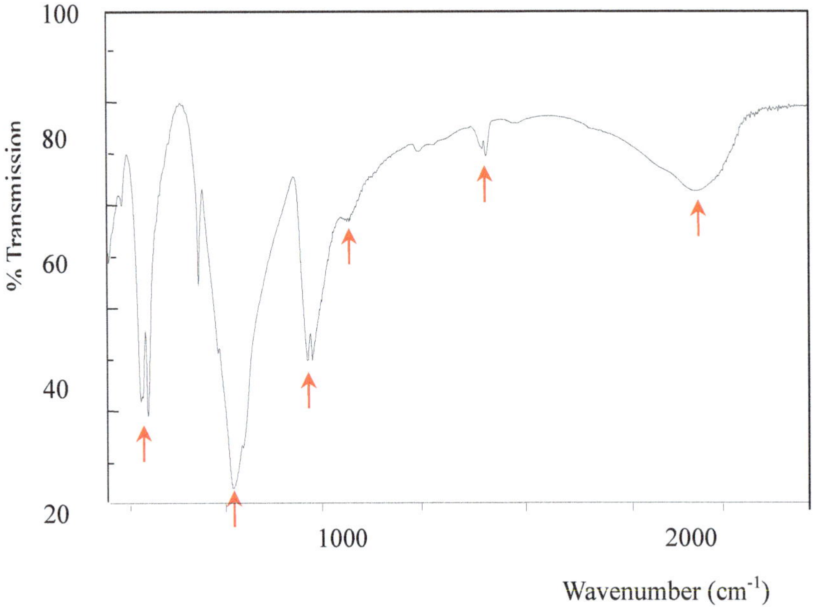

Figure 3. Infrared spectrum of natural phosphate before treatment.

The analysis of the infrared spectrum shows that:

- between 4000 and 3000 cm^{-1}, a large peak is observed around 3480 cm^{-1}corresponds to the water absorption band. It is a combination of antisymmetric stretching mode, symmetric stretching mode and an overtone of the bendingmode (Bernath, 2002 ; Tennyson et al., 2014),
- the 2360 cm^{-1} peak corresponds to the atmospheric CO_2,
- the 1644 cm^{-1} peak corresponds to the third vibration mode of adsorbed water molecule, known as angular deformity,
- the 1425 cm^{-1} doublet isfrom carbonate ν_3 and the 873 cm^{-1}peak is from carbonate ν_2. (Knubovets et al., 1997; Marshall et al., 1999).
- the1190-976 cm^{-1}peaks are from phosphate ν_3, 961 cm^{-1}shouldisfrom phosphate ν_1 and peaks around 660-520 cm^{-1} are from phosphate ν_4 (Pleshko et al., 1991; Gadaleta et al. 1996; Rehman & Bonfield, 1997).

3.3 Specific Area Measurement

The specific surface areas of the samples were measured by the BET method which consists of the adsorption of an inert gas at its liquefaction temperature on the solid. The specific surface of the solid is then calculated from conventional formulas proposed by Brunauer, Emmet & Teller (Brunauer et al., 1938).

The study of the adsorption-desorption isotherm is used to determine the distribution of the pore sizes. The isothermal adsorption-desorption of nitrogen on Youssoufia phosphate is reported in Figure 4.

The nitrogen adsorption-desorption isotherm on natural phosphate shows that it is of type II according to BDDT classification (Brunauer et al., 1940). It is characterized by a very gradual increase in the adsorbed amount based on the pressure equilibrium. This type of adsorption is a multimolecular adsorption characteristic, obtained during the adsorption on non-porous surfaces or macroporous adsorbents.

We also note the presence of a hysteresis between the desorption and the adsorption curves. This phenomenon is related primarily to the phase difference between adsorption and desorption of nitrogen in the solid pores. According to the IUPAC classification (Sing, 1985), the observed hysteresis is an H_3 type. It is observed in the case of an adsorbent that forms aggregates and can be attributed to capillary condensation taking place in a non rigid construction, but it is not a defined mesoporosity. Using this method, we have estimated for the phosphate mineral studied a specific surface area of about 13.37 m².g^{-1}.

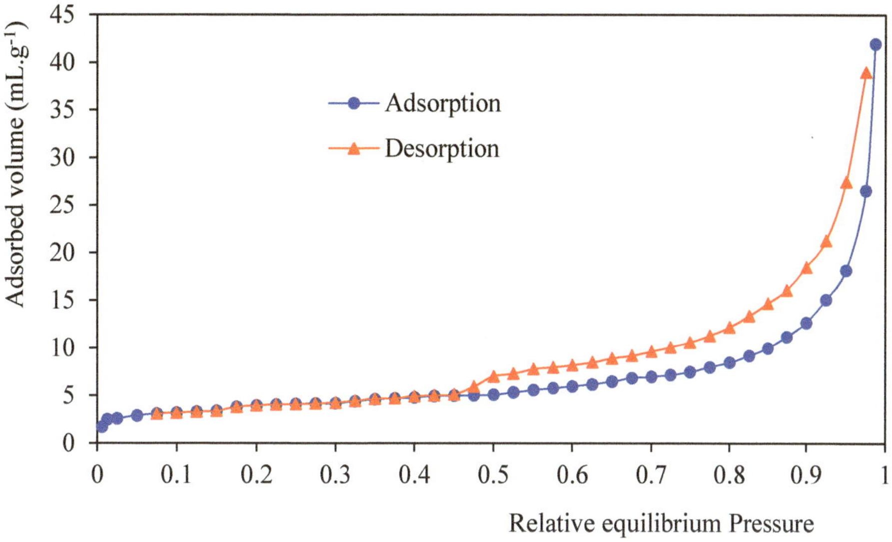

Figure 4. Natural Phosphate BET Isotherm.

3.4 Analyze of Phosphate Porosity

In any calcination process, it is very useful to know the distribution of pores on the material surface. The best method used to study the pore distribution is the BHJ method (Barrett et al., 1951). This technique is used to determine the ratio (volume/area) of the pore distribution taking into account the change in the thickness of the adsorbed layer and the condensation in the pores of the solid. As a function of pore diameter of the natural phosphate, we report the evolutions of the cumulative volume (Figure 5), the volume distribution of the pore size (Figure 6) and the cumulative surface pores (Figure 7).

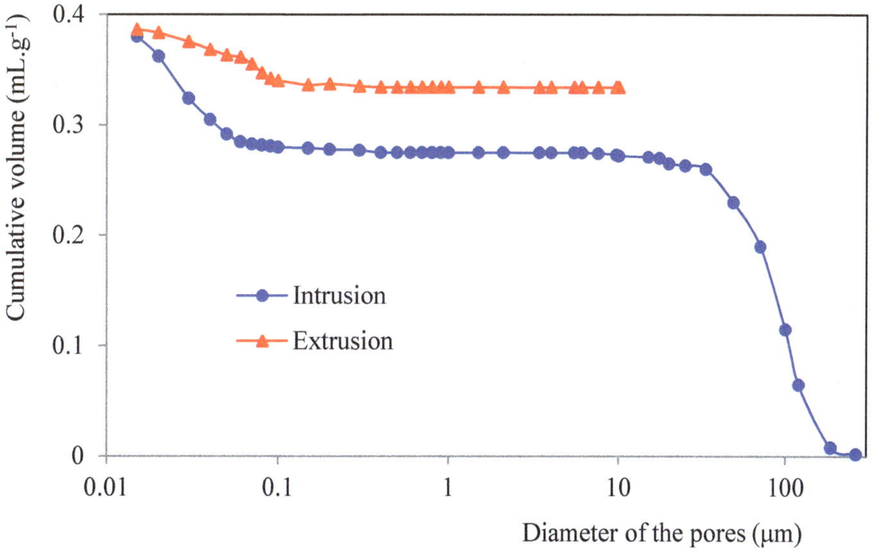

Figure 5. Cumulative volume versus the pore diameterin the natural phosphate.

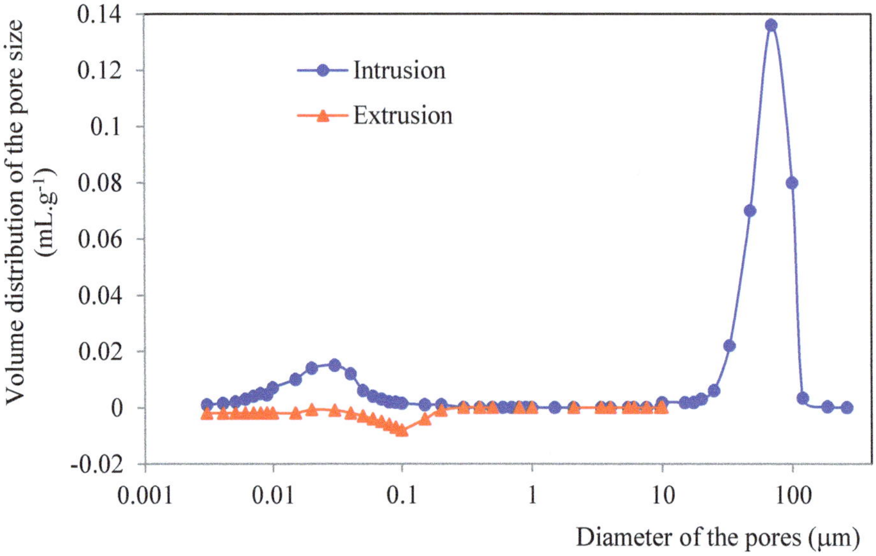

Figure 6. Volume distribution of the pore size in the natural phosphate.

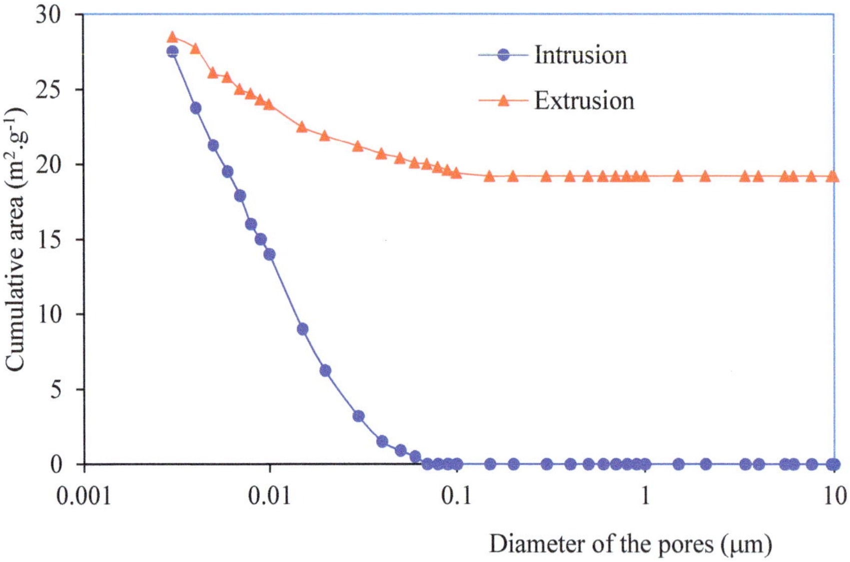

Figure 7. Cumulative surface versus the pore diameter in the natural phosphate.

We can also see the evolutions of the cumulative volume and volume distribution as functions of the equilibrium pressure (Figures 8 and 9).

The analysis of these curves show that:

- for a pore size between 30 µm and 320 µm, there is a progressive gas adsorption until the recovery of the monolayer,
- for a pore size between 0.1 µm and 31 µm, there is a saturation stage indicating no adsorption despite the increase in the equilibrium pressure. This is interpreted by the fact that the mesopores, in which capillary condensation is occurred, are no longer available for adsorption.
- when the pore size is between 31 µ m and 0.1 µ m, there is a saturation level that indicates a lack of adsorption despite the increased pressure balance. This is interpreted by the fact that the mesopores under which took place capillary condensation is no longer available for adsorption,
- for a pore size less than 0.1 µm, the adsorption is produced at so high pressures and the pores are not large enough and do not substantially contribute to the specific surface area.

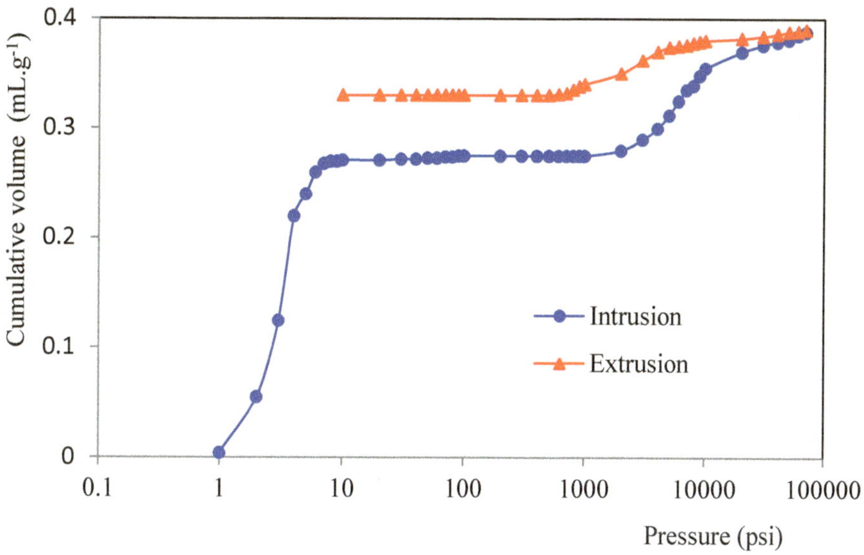

Figure 8. Change in cumulative volume as a function of the relative equilibrium pressure.

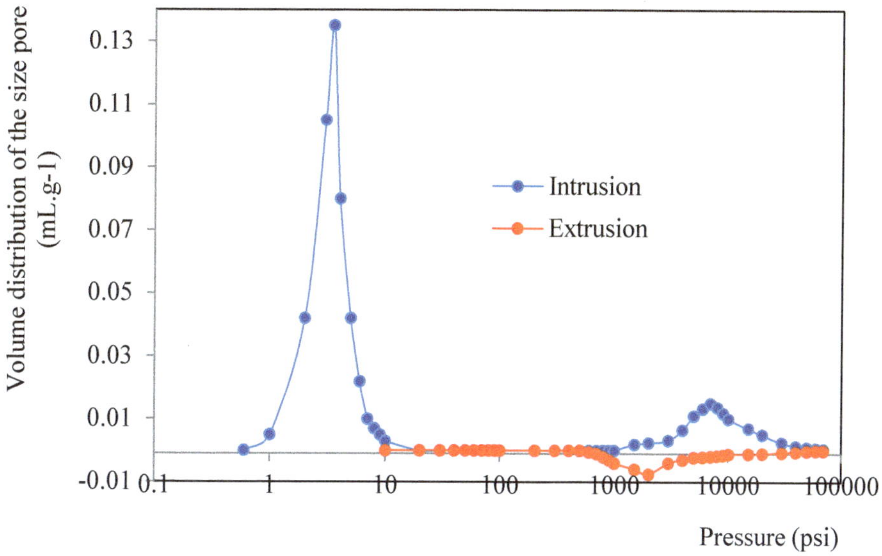

Figure 9. Volume distribution of the poresize as a function of the relative equilibrium pressure in the natural phosphate.

From these results, we can deduce the total pore volume of the natural phosphate measured by the condensation of adsorbed nitrogen: $V_T = 111\ mm^3.g^{-1}$ and the mean pore opening diameter is: $d = 0.029$ µm.

The adsorption-desorption isotherm of nitrogen on coke of a solid combustible is illustrated in Figure 10. According to the BDDT classification (Brunauer et al., 1940), it shows that it is a type III and it is different from the adsorption isotherms of type II at low pressure. This change at the beginning of the adsorption isotherm is due to the weaknesses of the adsorbent/adsorbate interactions. Using this method, we found out that the specific surface of the coke of a solid combustible is approximately $0.36\ m^2.g^{-1}$.

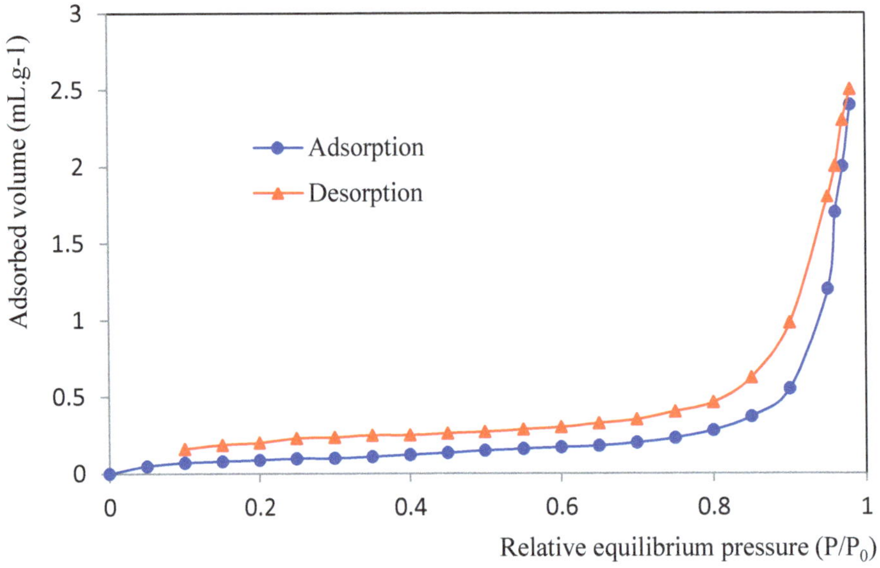

Figure 10. Coke BET Isotherm.

3.5 Scanning Electron Microscopy

This technique allows the observation of the sample morphology before treatment. The morphological observation of phosphate grains before treatment shows a fairly homogeneous zones and a porous structure. The morphology observation of the grain surface of coke of a solid combustible shows that there are different sizes of grains, heterogeneous areas and non-porous structure and anisotropic.

Figure 11. Morphological observation the surface of natural phosphate before treatment

(a) 100 μm, (b) 10μm, (c) 5μm, (d) 1μm.

Figure 12. Morphological observation the surface of coke of a solid combustible before treatment

(a) 100 µm, (b) 10 µm, (c) 5 µm, (d) 1 µm.

4. Conclusion

We determine the physicochemical and morphological characteristics of the phosphate and coke of a solid combustible in order to know the properties of these materials. This will help when it comes to analyse its behaviours during the phosphate thermal treatment process as well as during the different reaction mechanisms involved in the process of combustion of coke of a solid combustible and phosphate calcinations.

References

Barrett, E. P., Joyner, L. G., & Halenda, P. P. (1951). The determination of pore volume and area distributions in porous substances. I. Computation from nitrogen isotherms. *Journal of the American Chemical Society*, *73*(1), 373-380. http://dx.doi.org/10.1021/ja01145a126

Bernath, P. F. (2002). The spectroscopy of water vapour: Experiment, theory and applications. *Physical Chemistry Chemical Physics*, *4*, 1501-1509. http://dx.doi.org/10.1039/b200372d

Brunauer, S., Deming, L. S., Deming, W. E., & Teller, E. (1940). On a theory of the van der Waals adsorption of gases. *Journal of the American Chemical Society*, *62*(7), 1723-1732. http://dx.doi.org/10.1021/ja01864a025

Brunauer, S., Emmett, P. H., & Teller, E. (1938). Adsorption of gases in multimolecular layers. *Journal of the American Chemical Society*, *60*(2), 309-319. http://dx.doi.org/10.1021/ja01269a023

Commandre, J. M., & Salvador, S. (2005). Lack of correlation between the properties of a petroleum coke and its behaviour during combustion. *Fuel Processing Technology*, *86*(7), 795-808. http://dx.doi.org/10.1016/j.fuproc.2004.08.001

Dubuisson, C., Charrette, A., & Kocaefe, D. (1995). Kinetic of the calcination of petroleum coke in neutral and oxidizing atmospheres. *The Canadian Journal of Chemical Engineering*, *73*(1), 110-119. http://dx.doi.org/10.1002/cjce.5450730112

Gadaleta, S. J., Paschalis, E. P., Betts, F., Mendelsohn, R., & Boskey, A. L. (1996). Fourier transform infrared spectroscopy of the solution-mediated conversion of amorphous calcium phosphate to hydroxyapatite: New correlationsbetween X-ray diffraction and infrared data. *Calcif Tissue Int.*, *58*(1), 9-16. PMID: 8825233

Heintz, E. A. (1996). The characterization of petroleum coke. *Carbon*, *34*(6), 699-709. http://dx.doi.org/10.1016/0008-6223(96)00037-1

Knubovets, R., Nathan, Y., Shoval, S., & Rabinowitz, J. (1997). Thermal transformations in phosphorites. *Journal of Thermal Analysis*, *50*(1-2), 229-239. http://dx.doi.org/10.1007/BF01979564

Marshall, C. P., Rose, H. R., Lee, G. S. H., Mur, G. L., & Wilson, M. A. (1999). Structure of organic matter in conodonts with different colour alteration indexes. *Organic Geochemistry*, *30*(10), 1339-1352. http://dx.doi.org/10.1016/S0146-6380(99)00106-0

Mendez, M. O. A., Lisboa, A. C. L., Coutinho, A. R., & Otani, C. (2006).Activated petroleum coke for natural gas storage. *The Journal of the Brazilain Chemical Society*, *17*(6), 1144-1150. http://dx.doi.org/10.1590/S0103-50532006000600011

Murthy, B. N., Sawarkar, A. N., Deshmukh, N. A., Mathew, T., & Joshi, J. B. (2014). Petroleum coke gasification: A review. *The Canadian Journal of Chemical Engineering*, *92*(3), 441-468. http://dx.doi.org/10.1002/cjce.21908

Pleshko, N., Boskey, A., & Mendelsohn, R. (1991). Novel infrared spectroscopic method for the determination of crystallinity of hydroxyapatite minerals. *Biophysical Journal*, *60*(4), 786-793. http://dx.doi.org/10.1016/S0006-3495(91)82113-0

Rehman, I., & Bonfield, W. (1997). Characterization of hydroxyapatite and carbonated apatite by photo acoustic FTIR spectroscopy. *Journal of Materials Sciences: Materials in Medecine, 8*, 1-4. http://dx.doi.org/10.1023/A:1018570213546

Salvador, S., Commandre, J. M., & Stanmore, B. R. (2003). Reactionrates fort he oxidation of highly sulphurised petroleum cokes: the influence of thermogravimetric conditions and some coke properties. *Fuel*, *82*(6), 715-720. http://dx.doi.org/10.1016/S0016-2361(02)00363-0

Sing, K. S. W. (1985). Reporting physisorption data for gas/solid systems with special reference to the determination of surface area and porosity. *Pure and Applied Chemistry*, *57*(4), 603-619. http://dx.doi.org/10.1351/pac198557040603

Slaoui, S., & Bounahmidi, T. (2004). Experimental study and modeling of coke combustion kinetics. *Comptes Rendus Chimie*, *7*(5), 547-557. http://dx.doi.org/10.1016/j.crci.2004.01.010

Suelves, I., Lazaro, M. J., Diez M. A., & Moliner R. (2002). Characterization of chars obtained from co-pyrolysis of coal and petroleum residues. *Energy & Fuels*, *16*(4), 878-886. http://dx.doi.org/10.1021/ef010264m

Tennyson, J., Bernath, P. F., Brown, L. R., Campargue, A., Csaszar, A. G., Daumont, L., Gamache, R. R., Hodges, J. T., Naumenko, O. V., Polyansky, O. L., Rothman, L. S., Vandaele, A. C., & Zobov, N. F. (2014). A database of water transitions fromexperiment and theory (IUPAC Technical Report). *Pure and Applied Chemistry*, *86*, 71-83. http://dx.doi.org/10.1515/pac-2014-5012

17

Thermal Properties and Crystal Structure of $BaC_4H_4O_6$ Single Crystals

Takanori Fukami[1], Seiya Hiyajyo[1], Shuta Tahara[1], Chitoshi Yasuda[1]

[1]Department of Physics and Earth Sciences, Faculty of Science, University of the Ryukyus, Japan

Correspondence: Takanori Fukami, Department of Physics and Earth Sciences, Faculty of Science, University of the Ryukyus, Okinawa 903-0213, Japan. E-mail: fukami@sci.u-ryukyu.ac.jp

Abstract

Single crystals of barium L-tartrate, $BaC_4H_4O_6$, were grown at 308 K by a gel method using silica gel as the growth medium. Differential scanning calorimetry, thermogravimetric-differential thermal analysis, and X-ray diffraction measurements were performed on the single crystals. The space group symmetry (orthorhombic, $P2_12_12_1$) and structural parameters were determined at room temperature. The crystal structure consisted of BaO_9 polyhedra, $C_4H_4O_6$ molecules, and zig-zag hydrogen-bonded chains along the *a*- and *c*-axes linked by O–H···O and C–H···O hydrogen bonds between adjacent molecules. Weight losses due to thermal decomposition of $BaC_4H_4O_6$ occurred in the temperature range of 450–1530 K. We suggest that the evolution of $2H_2$, 2CO, CO, $(1/2)O_2$, and O_2 gases caused the observed weight losses and that the decomposition product, barium monocarbide BaC, formed a residue in the vessel.

Keywords: $BaC_4H_4O_6$, BaC, thermal decomposition, TG-DTA, X-ray diffraction

1. Introduction

Tartaric acid (chemical formula: $C_4H_6O_6$; systematic name: 2,3-dihydroxybutanedioic acid) is found in grapes, currants, gooseberries, oranges, apples, and in most acidulous fruits. It has two chiral carbon atoms in its structure, which provides for four possible different forms of chiral, racemic, and achiral isomers: L(+)-tartaric, D(–)-tartaric, racemic (DL-) tartaric, and meso-tartaric acid (Bootsma & Schoone, 1967; Fukami, Tahara, Yasuda, & Nakasone, 2016b; Song, Teng, Dong, Ma, & Sun, 2006). The most common form in nature is L-tartaric acid, while meso-tartaric acid is man-made and does not occur in nature. Solutions of L- and D-tartaric acid rotate the plane of polarized light to the left and right, respectively, whereas of DL- and meso-tartaric acid do not induce rotation of plane-polarized light. The properties of optically active molecules derived from tartaric acid were discovered by Biot (Lowry, 1923).

Many tartrate compounds are formed by the reaction of tartaric acid with compounds containing positive ions (two monovalent cations or one divalent cation) (Boese & Heinemann, 1993; Buschmann & Luger, 1985; Fukami, Hiyajyo, Tahara, & Yasuda, 2016a; Fukami et al., 2016c; Hawthorne, Borys, & Ferguson, 1982; Labutina, Marychev, Portnov, Somov, & Chuprunov, 2011; Silgo, Platas, Pérez, López, & Torres, 1999; Weil, 2015). These compounds are used in numerous industrial applications, for example, as transducers and in linear and non-linear mechanical devices because of their excellent dielectric, ferroelectric, piezoelectric, and nonlinear optical properties (Firdous et al., 2010; Kader et al., 1991; Torres et al., 2002). Several types of tartaric acid crystals, such as potassium hydrogen tartrate ($KHC_4H_4O_6$), and calcium tartrate ($CaC_4H_4O_6$), develop naturally in bottled wine and are the major component of the harmless sediment found in wine (Boese & Heinemann, 1993; Buschmann & Luger, 1985; Derewenda, 2008; Hawthorne et al., 1982). The crystalline tartrate sediment as a by-product has to be removed from the wine after yeast fermentation of the grape juice. More than 150 years ago, Louis Pasteur first separated the two enantiomers of sodium ammonium tartrate ($NaNH_4C_4H_4O_6$) by utilizing the asymmetric habit of their crystals (Gal, 2013; Pasteur, 1848). In addition, he discovered the change in optical rotation induced by the different structures of each enantiomer in water solution. The discovery of enantiomers has played an important role in advancing the scientific understanding of molecular chirality.

Silgo et al. (1999) established the crystal structure of barium L-tartrate $BaC_4H_4O_6$ by means of single-crystal X-ray diffraction at room temperature. The crystal structure was orthorhombic with space group $P2_12_12_1$, and consisted of BaO_9 polyhedra, $C_4H_4O_6$ molecules, and O–H···O hydrogen bonds. The positions of hydrogen atoms were calculated based on geometric criteria (O–H and C–H distances of 0.82 and 0.98 Å, respectively), and the atoms were included in the refinement at the fixed positions with constant isotropic thermal parameters. Shah and Patel (2010) reported the stretching modes of O–H, C–H, C–O, C–C, and Ba–O in $BaC_4H_4O_6$ based on FTIR and X-ray diffraction measurements. Moreover, they confirmed that the crystal structure was essentially the same as that reported by Silgo et al (1999).

The purpose of this study is to determine the exact crystal structure of $BaC_4H_4O_6$, including the positions of all hydrogen atoms, at room temperature using X-ray diffraction measurements. We report the thermal properties of this compound, as determined by means of differential scanning calorimetry (DSC) and thermogravimetric-differential thermal analysis (TG-DTA).

2. Experimental

2.1 Crystal Growth

Light yellow block crystals of $BaC_4H_4O_6$ were grown in silica gel medium at 308 K using the single test tube diffusion method. The gel was prepared in the test tube (length of 200 mm and diameter of 30 mm) using aqueous solutions of Na_2SiO_3 (33.7 g of 10.9 wt%), L-tartaric acid $C_4H_6O_6$ (28.8 g of 13.0 wt%), and CH_3COOH (24.6 g of 10.7 wt%) and aged for several days. A solution of $Ba(NO_3)_2$ (23.7 g of 15.5 wt%) was heated to boiling point until complete dissolution of $Ba(NO_3)_2$ and, after cooling to room temperature, was gently poured on top of the gel. The crystals were harvested after five months. Samples used for measurements were cut from transparent areas on the tips of crystal.

2.2 X-ray Crystal Structure Determination

The X-ray diffraction measurements were carried out using a Rigaku Saturn CCD X-ray diffractometer with graphite-monochromated Mo K_α radiation ($\lambda = 0.71073$ Å). The diffraction data were collected at 298 K using an ω scan mode with a crystal-to-detector distance of 40 mm, and processed using the CrystalClear software package. The intensity data were corrected for Lorentz polarization and absorption effects. The structure was solved by direct methods using the SIR2011 program and refined on F^2 by full-matrix least-squares methods using the SHELXL-2013 program in the WinGX package (Burla et al., 2012; Farrugia, 2012; Sheldrick, 2015).

Table 1. Crystal data, intensity collection, and structure refinement for $BaC_4H_4O_6$

Compound, M_r	$BaO_6C_4H_4$, 285.41
Crystal color	Colorless
Measurement temperature	298 K
Crystal system, space group	Orthorhombic, $P2_12_12_1$
Lattice constants	$a = 8.3986(2)$ Å, $b = 9.0370(3)$ Å, $c = 8.1924(3)$ Å
V, Z	621.79(3) Å^3, 4
D(cal.), μ(Mo K_α), F(000)	3.049 Mg m^{-3}, 6.365 mm^{-1}, 528
Sample shape, size in diameter	Sphere, $2r = 0.30$ mm
θ range for data collection	3.31–37.87 °
Index ranges	$-14 \leq h \leq 14$, $-15 \leq k \leq 15$, $-13 \leq l \leq 13$
Reflections collected, unique	17681, 3258 [R(int) = 0.0306]
Completeness to θ_{max}	97.9 %
Absorption correction type	Spherical
Transmission factor T_{min}–T_{max}	0.2428–0.2746
Date [$I > 2\sigma(I)$], parameter	3110, 117
Final R indices	$R_1 = 0.0190$, $wR_2 = 0.0360$
R indices (all data)	$R_1 = 0.0224$, $wR_2 = 0.0376$
Weighting scheme	$w = 1/[\sigma^2(F_o^2) + (0.0174P)^2 + 0.1072P]$, $P = (F_o^2 + 2F_c^2)/3$
Goodness-of-fit on F^2	1.057
Flack parameter	-0.032(7)
Extinction coefficient	0.0141(5)
Largest diff. peak and hole	1.721 and -1.198 eÅ^{-3}

2.3 Thermal Measurements

DSC and TG-DTA measurements were respectively carried out in the temperature ranges of 102–370 K and 300–1530 K, using DSC7020 and TG-DTA7300 systems from Seiko Instruments Inc. Aluminium (for DSC) and platinum (for TG-DTA) open pans without a cover were used as the measuring vessel and reference pan. Fine powder samples prepared by grinding the some cutting crystals of colorless were used for the measurements. The sample amount varied between 4.60 and 4.98 mg, and the heating rates were either 5 or 10 K min^{-1} under the flow of nitrogen gas.

3. Results and Discussion

3.1 Crystal Structure

The crystal structure of $BaC_4H_4O_6$ was determined by the single-crystal X-ray diffraction at room temperature. The lattice parameters calculated from all observed reflections showed that the crystal belongs to orthorhombic system. The intensity statistics and systematic extinctions of the observed reflections strongly revealed that the crystal structure is an acentric space group of $P2_12_12_1$. Thus, the space group of $BaC_4H_4O_6$ was determined to be orthorhombic $P2_12_12_1$ with cell parameters $a = 8.3986(2)$ Å, $b = 9.0370(3)$ Å, and $c = 8.1924(3)$ Å. The atomic coordinates and thermal parameters for $BaC_4H_4O_6$, including the positions of all hydrogen atoms, were refined at room temperature. A final *R*-factor of 1.90% was calculated for 3110 unique observed reflections. The relevant crystal data, and a summary of the intensity data collection and structure refinement parameters are given in Table 1. Figure 1 shows a projected view of the $BaC_4H_4O_6$ crystal structure in the *bc*-plane. The positional parameters in fractions of the unit cell, and the thermal parameters are listed in Table 2. Selected bond lengths (in Å) and angles (in degrees) are given in Table 3. The hydrogen-bond geometry (in Å and degrees) is presented in Table 4.

Table 2. Atomic coordinates and thermal parameters ($\times 10^4$ Å^2) with standard deviations in brackets. The anisotropic thermal parameters are defined as $\exp[-2\pi^2(U_{11}a^{*2}h^2+U_{22}b^{*2}k^2+U_{33}c^{*2}l^2+2U_{23}b^*c^*kl+2U_{13}a^*c^*hl+2U_{12}a^*b^*hk)]$. The isotropic thermal parameters (Å^2) for H atoms are listed under U_{11}.

Atom	*x*	*y*	*z*	U_{11}	U_{22}	U_{33}	U_{23}	U_{13}	U_{12}
Ba	0.27986(2)	0.76534(2)	0.40953(2)	122.8(6)	144.8(5)	129.2(5)	-2.2(4)	7.2(5)	-10.8(4)
O(1)	0.2408(2)	0.0703(2)	0.3543(2)	150(9)	205(8)	184(8)	-1(6)	-15(7)	-28(6)
O(2)	0.4188(2)	0.1569(2)	0.1757(2)	185(9)	284(10)	120(8)	32(7)	-12(7)	-26(8)
O(3)	0.6070(2)	0.2646(2)	0.4085(2)	161(7)	206(7)	149(7)	-13(9)	10(6)	-72(6)
O(4)	0.5586(2)	-0.0593(2)	0.4695(2)	173(9)	138(7)	131(8)	-10(6)	20(7)	8(6)
O(5)	0.7266(3)	0.0118(2)	0.7356(3)	268(10)	295(10)	288(11)	-93(8)	-160(10)	111(10)
O(6)	0.5465(3)	0.1783(3)	0.8193(2)	221(9)	262(10)	150(8)	-61(7)	-3(7)	-8(8)
C(1)	0.3674(3)	0.1385(3)	0.3169(3)	121(10)	140(9)	133(10)	5(7)	-28(8)	14(7)
C(2)	0.4655(3)	0.1949(3)	0.4631(3)	131(10)	139(9)	125(9)	-6(8)	0(7)	-1(7)
C(3)	0.4995(3)	0.0579(3)	0.5683(3)	134(9)	140(9)	94(10)	-7(7)	-4(7)	-3(7)
C(4)	0.6010(3)	0.0859(3)	0.7193(3)	163(10)	170(10)	114(9)	6(8)	-18(8)	-19(8)
H(1)	0.635(5)	0.328(5)	0.490(5)	0.030(11)					
H(2)	0.602(5)	-0.041(5)	0.398(5)	0.024(10)					
H(3)	0.408(4)	0.257(4)	0.515(4)	0.015(8)					
H(4)	0.397(5)	0.029(4)	0.617(4)	0.020(9)					

The obtained crystal structure of $BaC_4H_4O_6$ is very similar to that previously reported by Silgo et al. (1999), and very different from those of the $CaC_4H_4O_6{\cdot}4H_2O$ and $SrC_4H_4O_6{\cdot}4H_2O$ crystals because it does not contain any water molecules (Boese & Heinemann, 1993; Fukami et al., 2016c; Hawthorne et al., 1982). The fundamental features of the obtained structure are as follows. Each Ba atom in the unit cell is surrounded by nine O atoms at distances from 2.771(2) to 2.884(2) Å, forming a BaO_9 polyhedron, as listed in Table 3. The average Ba–O distance is 2.818 Å. The volume of the BaO_9 polyhedron (at the distance of 2.818 Å) is larger than those of the CaO_8 and SrO_8 dodecahedra (at average Ca–O and Sr–O distances of 2.455 and 2.594 Å, respectively) in the calcium and strontium salts (Boese & Heinemann, 1993; Fukami et al., 2016c; Hawthorne et al., 1982). This is probably because of the difference in ionic radius between Ba^{2+} and Ca^{2+} (or Sr^{2+}) ions. Moreover, the presence or absence of internal water molecules in the structure may be related to the difference in volume between the BaO_9 polyhedron and the CaO_8 (or SrO_8) dodecahedron. The lengths of six O–C bonds in the $C_4H_4O_6$ molecule are in the range of 1.256(3)–1.422(3) Å, as listed in Table 3. The variation in O–C distances is probably related to differences in bond type. The lengths of single and double O–C bonds in organic molecules are around 1.43 and 1.22 Å, respectively. The O(3)–C(2) and O(4)–C(3) bond lengths (hydroxyl group) are about 1.42 Å. Thus, these two O–C bonds have single-bond character. Because the lengths of the remaining four O–C bonds are about 1.26 Å, these bonds have double-bond character. Additionally, the lengths of three C–C bonds are in the range of 1.523(3)–1.540(3) Å, similar to that (around 1.54 Å) of single C–C bonds in organic molecules, indicating these bonds have single-bond character. Thus, the characteristic features of the $C_4H_4O_6$ molecule in the $BaC_4H_4O_6$ crystal are very similar to that in the $SrC_4H_4O_6{\cdot}4H_2O$ crystal (Fukami et al., 2016c).

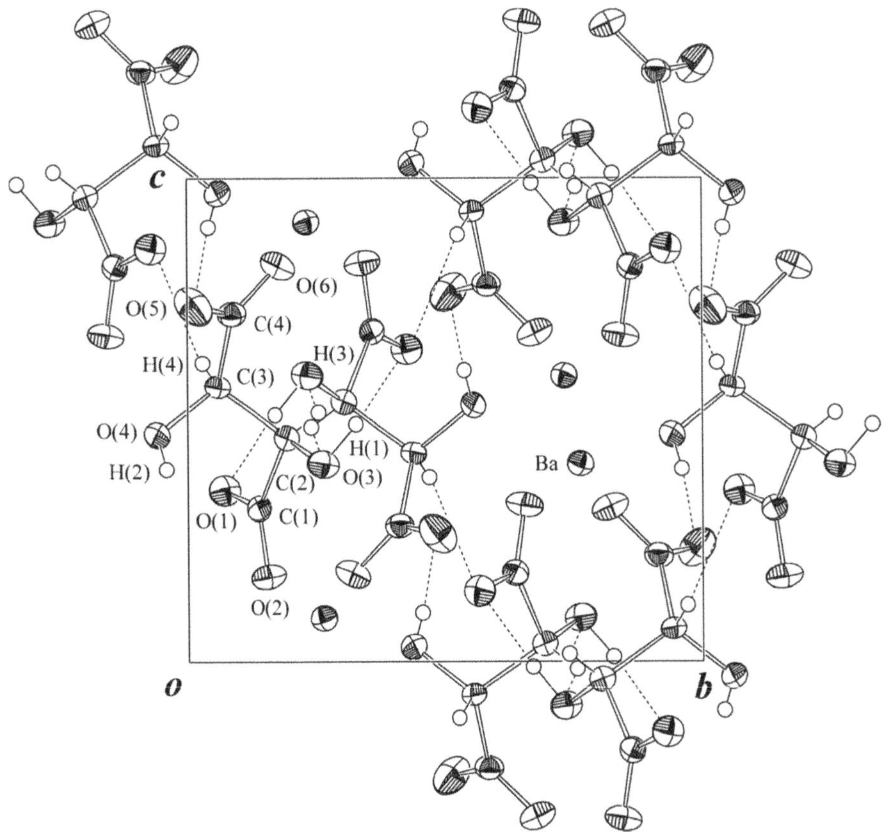

Figure 1. ORTEP projection in the *bc*-plane of $BaC_4H_4O_6$ structure with 70% probability-displacement thermal ellipsoids

The solid and short dashed lines show the O–H···O and C–H···O hydrogen bonds in Table 4.

Table 3. Selected interatomic distances (in Å) and angles (in degrees)

Ba–O(1)$^{(a)}$	2.812(2)	Ba–O(2)$^{(b)}$	2.802(2)
Ba–O(2)$^{(c)}$	2.834(2)	Ba–O(3)$^{(b)}$	2.773(2)
Ba–O(4)$^{(d)}$	2.811(2)	Ba–O(4)$^{(a)}$	2.870(2)
Ba–O(5)$^{(d)}$	2.808(2)	Ba–O(6)$^{(e)}$	2.771(2)
Ba–O(6)$^{(f)}$	2.884(2)	O(1)–C(1)	1.267(3)
O(2)–C(1)	1.245(3)	O(3)–C(2)	1.418(3)
O(4)–C(3)	1.422(3)	O(5)–C(4)	1.256(3)
O(6)–C(4)	1.256(3)	C(1)–C(2)	1.540(3)
C(2)–C(3)	1.536(3)	C(3)–C(4)	1.523(3)
O(1)–C(1)–O(2)	125.5(2)	O(1)–C(1)–C(2)	114.9(2)
O(2)–C(1)–C(2)	119.5(2)	O(3)–C(2)–C(1)	110.5(2)
O(3)–C(2)–C(3)	112.3(2)	O(4)–C(3)–C(2)	110.3(2)
O(4)–C(3)–C(4)	113.0(2)	O(5)–C(4)–O(6)	126.2(3)
O(5)–C(4)–C(3)	117.8(2)	O(6)–C(4)–C(3)	115.9(2)
C(3)–C(2)–C(1)	105.6(2)	C(4)–C(3)–C(2)	115.2(2)

Symmetry codes: (a) *x*,*y*+1,*z*; (b) –*x*+1,*y*+1/2,–*z*+1/2; (c) –*x*+1/2,–*y*+1,*z*+1/2; (d) *x*–1/2,–*y*+1/2,–*z*+1; (e) –*x*+1,*y*+1/2, –*z*+3/2; (f) –*x*+1/2,–*y*+1,*z*–1/2.

The observed crystal structure contains O–H···O and C–H···O hydrogen bonds formed between adjacent $C_4H_4O_6$ molecules, as shown in Fig. 1 and Table 4. The lengths of the hydrogen bonds are in the range of 2.667(3) –3.303(3) Å, where the C–H···O bonds are much longer than the O–H···O bonds. Thus, the bond strength of the C–H···O bonds is significantly lower than that of the O–H···O bonds. The structure based on the presence of the C–H···O hydrogen bonds is substantially different from that in the $SrC_4H_4O_6{\cdot}4H_2O$ crystal, which does not form a similar hydrogen bond network

(Fukami et al., 2016c). As shown in Fig. 1, there are zig-zag hydrogen-bonded chains along the *c*-axis, which consist of O(4)–H(2)···O(5) and C(3)–H(4)···O(1) bonds between adjacent $C_4H_4O_6$ molecules. Moreover, there are zig-zag hydrogen-bonded chains along the *a*-axis, which consist of O(3)–H(1)···O(1) and C(2)–H(3)···O(3) bonds between molecules.

The calculated angle between the two least-squares planes of atoms, O(1)O(2)O(3)C(1)C(2) and O(4)O(5)O(6)C(3)C(4), in the $C_4H_4O_6$ molecule is 67.17(8)°. The angles in the strontium tartaric acid, monohydrate racemic (MDL-) tartaric acid, L- tartaric acid, and D-tartaric acid crystals are 88.7(1)°, 72.79(4)°, 56.33(5)°, and 56.34(5)°, respectively (Fukami et al., 2016b, 2016c). Thus, the torsional strain in the $C_4H_4O_6$ unit of $BaC_4H_4O_6$ is smaller than that in the strontium salt and most similar to that in the MDL-tartaric acid crystal.

Table 4. Hydrogen bond distances (in Å) and angles (in degrees)

D–H···A	D–H	H···A	D···A	<D–H···A
O(3)–H(1)···O(1)[(a)]	0.91(4)	1.81(4)	2.696(3)	165(4)
O(4)–H(2)···O(5)[(b)]	0.71(4)	1.98(4)	2.667(3)	164(5)
C(2)–H(3)···O(3)[(c)]	0.85(4)	2.61(4)	3.210(3)	128(3)
C(3)–H(4)···O(1)[(d)]	0.98(4)	2.44(4)	3.303(3)	147(3)

Symmetry codes: (a) $x+1/2,-y+1/2,-z+1$; (b) $-x+3/2,-y,z-1/2$; (c) $x-1/2,-y+1/2,-z+1$; (d) $-x+1/2,-y,z+1/2$.

3.2 Thermal Analyses

DSC measurements on the powder sample of $BaC_4H_4O_6$ were performed in the temperature range from 102 to 370 K at a heating rate of 5 K min^{-1}. No endothermic or exothermic peak in the DSC curve was observed in the temperature range. In general, it is accepted that a clear peak in DSC curve is attributed to a change in exchange energy at a phase transition. Thus, the obtained result indicates that there is no phase transition in the temperature range of 102–370 K for the $BaC_4H_4O_6$ crystal.

Figure 2 shows the TG, differential TG (DTG), and DTA curves of $BaC_4H_4O_6$ in the temperature range of 300–1530 K. The sample weight (powder) used for the measurements was 4.98 mg, and the heating rate was 10 K min^{-1} under a nitrogen gas flow of 300 ml min^{-1}. The DTA curve shows eleven apparent small endothermic peaks, and the DTG curve has nine peaks, including some small peaks. The temperatures describing the DTA and DTG peaks are shown in Table 5 and indicated by arrows in Fig. 2. At temperatures below 1400 K, the DTG peaks correspond to the respective DTA peaks, except for very small peaks. The DTG curve, which is the first derivative of the TG curve, reveals the temperature dependence of the rate of weight loss. Thus, the DTA peaks are associated with maximum rates of weight loss in the TG curve due to thermal decomposition of the sample.

Table 5. Peak temperatures (K) obtained from DTA and DTG curves of $BaC_4H_4O_6$

DTA	537		647	733	999	1051	1078	1272		1323	1441	1456	1496
DTG	538	601	647	728	1002	1018		1272	1309	1322			

Table 6. TG results of thermal decomposition of $BaC_4H_4O_6$

Temp. range [K]	Weight loss (obs.) [%]	Weight loss (cal.) [%]	Eliminated molecules
450–545	1.9	1.4	$2H_2$
545–655	16.6	19.6	2CO
655–840	9.7	9.8	CO
840–1060	7.5	5.6	$(1/2)O_2$
1060–1530	12.3	11.2	O_2
Total	47.9	47.7	

The TG curve shows the temperature dependence of the weight loss of the $BaC_4H_4O_6$ crystal. The weight loss around 537 K in the TG curve was determined to be 1.9% in the temperature range from 450 to 545 K. We assume that the weight loss is caused by the elimination of two moles of H_2 gas. The theoretical weight loss is calculated to be 1.4% [=4×1.01/285.41], and is very close to the experimental weight loss of 1.9% at around 537 K. The weight losses around 647 and 733 K were determined to be 16.6% and 9.7% in the temperature ranges of 545–655 K and 655–840 K, respectively. We assume that the weight losses are caused by the elimination of CO gas (two moles and one mole). The theoretical weight losses are calculated to be 19.6% [=2×(12.01+16.00)/285.41] and 9.8% [=(12.01+16.00)/285.41] and again, are close to the experimental weight losses of 16.6% and 9.7% at around 647 and 733 K, respectively. The different evaporation temperatures of CO gas likely result from the different routes of CO generation, from the hydroxyl (H–C–O–H) or carbonyl (O–C=O) groups of the $C_4H_4O_6$ molecule. The weight losses around 999 and 1051 K were determined to be 7.5% and 12.3% in the temperature ranges of 840–1060 K and 1060–1530 K, respectively. We assume that the weight losses are caused by the evolution of O_2 gas (half mole and one mole). The theoretical weight losses are calculated to be 5.6% [=16.00/285.41] and 11.2% [=2×16.00/285.41], and are close to the experimental weight losses of 7.5% and 12.3% at around 999 and 1051 K, respectively. It is implied that the half mole of O_2 gas is generated from one O atom in the carbonyl group remained by the evolution of 3CO gases in the temperature range of 545–840 K mentioned above. Table 6 summarizes the experimental and theoretical weight losses in each temperature range.

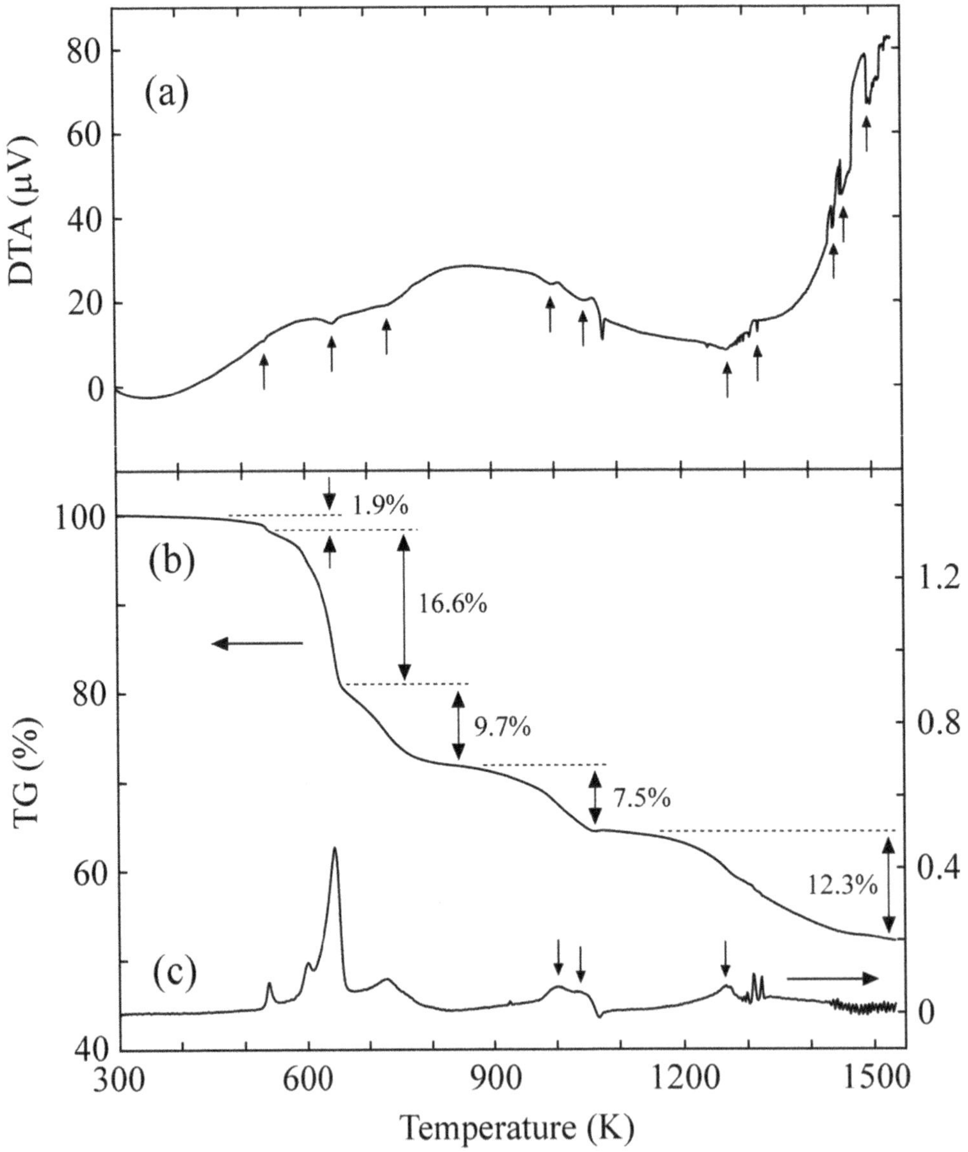

Figure 2. (a) DTA, (b) TG, and (c) DTG curves of $BaC_4H_4O_6$ crystal on heating. The positions of small peaks are indicated by arrows

Sample weight (powder) was 4.98 mg, and the heating rate was 10 K min^{-1} under N_2 gas flow of 300 ml min^{-1}.

The total theoretical weight loss from $BaC_4H_4O_6$ is 47.7%. This value is very close to the total experimental weight loss of 47.9%. Thus, the slight differences between the experimental and theoretical weight loss values at each temperature range in Table 6 are probably due to the overlapping of the temperature ranges corresponding to the decomposition reactions. Summarizing the considerations mentioned above, the chemical reactions involved in the thermal decomposition of $BaC_4H_4O_6$ can be described by the following chemical equations:

$BaC_4H_4O_6 \rightarrow BaC_4O_6 + 2H_2$
$BaC_4O_6 \rightarrow BaC_2O_4 + 2CO$
$BaC_2O_4 \rightarrow BaCO_3 + CO$
$BaCO_3 \rightarrow BaCO_2 + (1/2)O_2$

$BaCO_2 \rightarrow BaC + O_2$.

After all gases had evolved, barium monocarbide (BaC) remains in the vessel, as suggested by the chemical equations listed above. In fact after the measurements (heated up to 1530 K), the inside surface of the open vessel was found to be turned black in color. The black substance is considered to be barium monocarbide. Similarly, a chalky white substance was confirmed to remain in the vessel after the $SrC_4H_4O_6{\cdot}4H_2O$ crystal was heated up to 1250 K, and was suggested to be strontium oxide SrO (Fukami et al., 2016c). The electron configurations for Ba and Sr atoms are [Kr] $5s^2$ $5p^6$ $6s^2$ and [Kr] $5s^2$, respectively. Thus, both the atoms convert into a divalent positive ion when two electrons are removed from the outermost shell. Moreover, the valence electron configuration [Kr] $5s^2$ $5p^3$ $6s^1$ for barium is a stable electron structure, and the Ba atom can convert into a tetravalent positive ion. This is also deduced from the presence of BaO_2 produced by the reaction of BaO and $(1/2)O_2$. It is concluded that tetravalent Ba^{4+} ion is necessary to produce BaC at high temperatures. The observed process during the thermal decomposition and the remaining substance in the vessel are very different from those in the case of the $SrC_4H_4O_6{\cdot}4H_2O$ crystal (Fukami et al., 2016c).

4. Conclusion

Single crystals of barium L-tartrate, $BaC_4H_4O_6$, were grown in silica gel medium at 308 K by the diffusion method. The thermal properties and crystal structure of the single crystals are studied by DSC, TG-DTA, and X-ray diffraction. The room temperature crystal structure, including the positions of the hydrogen atoms, is determined to be orthorhombic with space group $P2_12_12_1$ by means of single-crystal X-ray diffraction. It is confirmed that the structure consists of BaO_9 polyhedra, $C_4H_4O_6$ molecules, and zig-zag hydrogen-bonded chains along the *a*- and *c*-axes linked by O–H···O and C–H···O hydrogen bonds between adjacent $C_4H_4O_6$ molecules. No phase transition is observed in the temperature range of 102–450 K. Weight losses during the thermal decomposition of $BaC_4H_4O_6$ occurs in the temperature range of 450–1530 K. The chemical equations illustrating the decomposition reactions of $BaC_4H_4O_6$ are presented with the corresponding temperature ranges. It is suggested that the weight losses are caused by the evolution of $2H_2$, 2CO, CO, $(1/2)O_2$, and O_2 gases, and that the residual black substance in the vessel after decomposition is barium monocarbide (BaC), which comprises tetravalent barium (Ba^{4+}) ion.

References

Boese, R., & Heinemann, O. (1993). Crystal structure of calcium tartrate tetrahydrate, $C_4H_4O_6Ca(H_2O)_4$. *Z. Kristallogr.*, *205*(1–2), 348–349. http://dx.doi.org/10.1524/zkri.1993.205.12.348

Bootsma, G. A., & Schoone, J. C. (1967). Crystal structures of mesotartaric acid. *Acta Crystallogr*, *22*(4), 522–532. http://dx.doi.org/10.1107/S0365110X67001070

Burla, M. C., Caliandro, R., Camalli, M., Carrozzini, B., Cascarano, G. L., Giacovazzo, C., ... & Spagna, R. (2012). SIR2011: a new package for crystal structure determination and refinement. *J. Appl. Crystallogr.*, *45*(2), 357–361. http://dx.doi.org/10.1107/S0021889812001124

Buschmann, J., & Luger, P. (1985). Structure of potassium hydrogen (+)-tartrate at 100 K, $K^+.C_4H_5O_6^-$. *Acta Crystallogr.*, *C*41(2), 206–208. http://dx.doi.org/10.1107/S0108270185003341

Derewenda, Z. S. (2008). On wine, chirality and crystallography. *Acta Crystallogr.*, *A64*(1), 246–258. http://dx.doi.org/10.1107/S0108767307054293

Farrugia, L. J. (2012). WinGX and ORTEP for Windows: an update. *J. Appl. Crystallogr.*, *45*(4), 849–854. http://dx.doi.org/10.1107/S0021889812029111

Firdous, A., Quasim, I., Ahmad, M. M., & Kotru, P. N. (2010). Dielectric and thermal studies on gel grown strontium tartrate pentahydrate crystals. *Mull. Mater. Sci.*, *33*(4), 377–382. http://www.ias.ac.in/matersci/bmsaug2010/377.pdf

Fukami, T., Hiyajyo, S., Tahara, S., & Yasuda, C. (2016a). Thermal properties, crystal structure, and phase transition of racemic $CaC_4H_4O_6{\cdot}4H_2O$ single crystals. *Am. Chem. Sci. J.*, *16*(3), 28258 1–11. http://dx.doi.org/10.9734/ACSJ/2016/28258

Fukami, T., Tahara, S., Yasuda, C., & Nakasone, K. (2016b). Structural refinements and thermal properties of L(+)-tartaric, D(–)-tartaric, and monohydrate racemic tartaric acid. *Inter. J. Chem.*, *8*(2), 9–21. http://dx.doi.org/10.5539/ijc.v8n2p9

Fukami, T., Tahara, S., Yasuda, C., & Nakasone, K. (2016c). Crystal structure and thermal properties of $SrC_4H_4O_6{\cdot}4H_2O$ single crystals. *Int. Res. J. Pure Appl. Chem.*, *11*(1), 23674 1–10. http://dx.doi.org/10.9734/IRJPAC/2016/23674

Gal, J. (2013). Citation for chemical breakthrough awards: Choosing pasteur's award-winning publication. *Bull. Hist. Chem.*, *38*(1), 7–12. http://www.scs.illinois.edu/~mainzv/HIST/awards/Citations/v38-1%20p7-12.pdf

Hawthorne, F. C., Borys, I., & Ferguson, R. B. (1982). Structure of calcium tartrate tetrahydrate. *Acta Crystallogr.*, *B38*(9), 2461–2463. http://dx.doi.org/10.1107/S0567740882009042

Kader, M. M. A., El-Kabbany, F., Taha, S., Abosehly, A. M., Tahoon, K. K., & El-Sharkawy, A. A. (1991). Thermal and electrical properties of ammonium tartrate. *J. Phys. Chem. Solids*, *52*(5), 655–658. http://www.sciencedirect.com/science/journal/00223697/52/5

Labutina, M. L., Marychev, M. O., Portnov, V. N., Somov, N. V., & Chuprunov, E. V. (2011). Second-order nonlinear susceptibilities of the crystals of some metal tartrates. *Crystallogr. Rep.*, *56*(1), 77–79. http://dx.doi.org/10.1134/S1063774510061082

Lowry, T. M. (1923). Pasteur as chemist. *Proc. Roy. Soc. Med.*, *16*, 16–20. http://www.ncbi.nlm.nih.gov/pmc/articles/PMC2104002/

Pasteur, M. L. (1848). Sur les relations qui peuvent existre la forme cristalline, la composition chimique et Le sens de la polarisation rotatoire. *Ann. Chim. Phys.*, *24*, 442–463. http://gallica.bnf.fr/ark:/12148/bpt6k65691733/f446.image.r=Annales%20de%20chimie%20et%20de%20physique.langEN

Shah, A., & Patel, I. B. (2010). FTIR and XRD study of barium tartrate ($BaC_4H_4O_6$) crystals grown by gel method. *AIP conf. Proc.*, *1249*, 192–193. http://dx.doi.org/10.1063/1.3466554

Sheldrick, G. M. (2015). Crystal structure refinement with SHELXL. *Acta Crystallogr.*, *C71*(1), 3–8. http://dx.doi.org/10.1107/S2053229614024218

Silgo, C. G., Platas, J. G., Pérez, C. R., López, T., & Torres, M. (1999). Barium L-tartrate. *Acta Crystallogr.*, *C55*(5), 740–742. http://dx.doi.org/10.1107/S0108270198016709

Song, Q. B., Teng, M. Y., Dong, Y., Ma, C. A., & Sun, J. (2006). (2S,3S)-2,3-Dihydroxy-succinic acid monohydrate. *Acta Crystallogr.*, *E62*(8), o3378–o3379. http://dx.doi.org/10.1107/S1600536806021738

Torres, M. E., Peraza, J., Yanes, A. C., López, T., Stockel, J., López, D. M., ... & Silgo, C. G. (2002). Electrical conductivity of doped and undoped calcium tartrate. *J. Phys. Chem. Solids*, *63*(4), 695–698. http://www.sciencedirect.com/science/journal/00223697/63/4

Weil, M. (2015). Crystal structure of lead(II) tartrate: a redetermination. *Acta Crystallogr.*, *E71*(1), 82–84. http://dx.doi.org/10.1107/S2056989014027376

18

The Fascinating Symmetries of Golden Clusters Containing Two Skeletal Elements [M-2] in the Cluster Nucleus

Enos Masheija Rwantale Kiremire

Correspondence: Enos Masheija Rwantale Kiremire, Department of Chemistry and Biochemistry University of Namibia, Private Bag 13301 Windhoek, Namibia, E-mail: kiremire15@yahoo.com.

Abstract

The golden clusters containing two skeletal atoms in the cluster nucleus have been analyzed more carefully with the help of the 4n series method. Their shapes can reasonably and readily be predicted by mapping the [M-2] capped clusters of 3 to 13 nuclearity onto Pivoriunas geometrical model. Some of these clusters were historically categorized as being spherically centered.

Keywords: series, spherical, capped, geometrical-model, nuclearity

1. Introduction

The shapes of golden clusters have fascinated scientists for some time (Pauling, 1977; Cotton and Wilkinson, 1980;Mingos,1984,1987; Steigelmann, et al, 1993;Kappen, et al, 1995a, 1995b;Greenwood and Earnshaw, 1998; Raithby,1998; Vincente, et al, 1998;Kilmartin, 2010; Kwok-Ming,2011;Pei, et al, 2011; Yuan, et al, 2013;Sarip, 2013;Puls,2014; Konishi, 2014). Their geometries were classified as centered, non-centered and exo-attached polyhedral clusters (Konishi, 2014). It has been found that the series method is generally very precise in categorizing and predicting the shapes of clusters such as boranes and their relatives, carbonyl and Zintl clusters (Kiremire, 2016a). The categorization of some golden clusters has been done using skeletal numbers and some of the work published (Kiremire, 2016b, 2016c, 2017). In this paper the emphasis is to closely focus mainly on selected samples of capped golden clusters whose nuclei are comprised of two skeletal elements. The major significance of this paper apart from using the series method to identify the composition of the cluster nucleus is to assign a possible geometrical framework of how other skeletal elements are arranged around the two nuclear skeletal elements to generate the overall skeletal geometry.

2. Results and Discussion

2.1 Background

The series method has evolved to a level where skeletal numbers were assigned to main group and transition metal elements and a few ligands (Kiremire, 2016a, 2016b, 2016c, 2017). The use of skeletal numbers makes it exceedingly easy to categorize and even predict the shapes of some clusters. In many cases, the series method agrees with the literature results and in some gives a more precise categorization and structural prediction. In this work, the skeletal numbers (Kiremire, 2016a, 2016b, 2016c, 2017) will be applied to categorize and characterize a selected sample of golden clusters. It has also been found that a given k(n) parameter defines the shape of the skeletal elements in most cases(k = skeletal linkages of a cluster and n = number of skeletal elements). For instance, $Os_6(CO)_{18}^{2-}$, k = 6(5)-18-1 = 11; $Rh_6(CO)_{16}$,k = 6(4.5)-16 = 11;$B_6H_6^{2-}$, k = 6(2.5)-3-1 = 11; $SFe_2Ru_3(CO)_{14}^{2-}$, k = 1(1)+2(5)+3(5)-14-1 = 11; $Se_2Mn_4(CO)_{12}^{2-}$, k = 2(1)+4(5.5)-12-1 = 11; $Co_4P_2(CO)_{10}$; k = 4(4.5)+2(1.5)-10 = 11; $Ru_6I(CO)_{17}$, k = 6(5)-2-17 = 11;and $Re_6(C\ (CO)_{19}^{2-}$, k = 6(5.5)-2-19-1 = 11. All these clusters are interrelated by the parameter k(n) = 11(6). That is, 11 cluster linkages and 6 skeletal atoms. The characteristic feature of each one of the complexes is the presence of an ideal octahedral skeletal geometry, O_h.

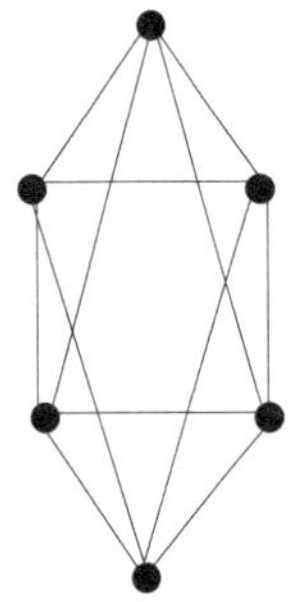

O_h SYMMETRY

2.2 Categorization of Golden Clusters Using Skeletal Numbers

A golden cluster, like any other cluster can be categorized by using skeletal numbers (Kiremire, 2016, 2017). Let us take a few examples for illustrations.

Ex-1. $(AuL)_3V(CO)_5$

According to series method, the skeletal elements are the providers of the skeletal linkages while the ligands are the consumers of some or all those skeletal linkages. When a cluster comprising of original 'naked' skeletal elements and ligands is formed, the ligands utilize some or all of the skeletal linkages. Then the skeletal linkages remaining [k(n)] are the ones that dictate the symmetry of the cluster as pointed out in the case of the O_h geometry discussed earlier.

The skeletal number of Au is 3.5 and the ligand L, k = -1(Kiremire, 2016, 2017), hence [AuL] fragment has a k value given by k = 3.5-1 = 2.5, V = 6.5, CO, k = -1(Kiremire, 2016, 2017). It is interesting to note that the fragment [AuL] is isolobal to boron[B]. That is, [AuL]⇔[B]. The k(n) value for the $(AuL)_3V(CO)_5$ cluster is given by k = 3[2.5]+1(6.5)-5(1) = 9; hence, k(n) = 9(4). For a given series S = 4n+q, k = 2n-q/2, and q = 2[2n-k] = 2[2(4)-9] = 2[-1] = -2 (Kiremire, 2016c). This gives us the cluster series S = 4n-2, and capping series Cp = $C^1+C^1 = C^2C$[M-2]. The number of valence electrons is given by V = 14n-2 = 14(4)-2 = 54. We can verify this value from the formula of the complex, $(AuL)_3V(CO)_5$; V = 3(11+2)+1(5)+5(2) =54. The symbol [M-2] means that there are 2 skeletal elements which are members of the CLOSO family S = 4n+2 and with k =2n-1 = 2(2)-1 =3. These 2 nuclear skeletal elements will be linked by a triple bond upon which the two capping skeletal elements will be superimposed. This is graphically sketched in F-1.

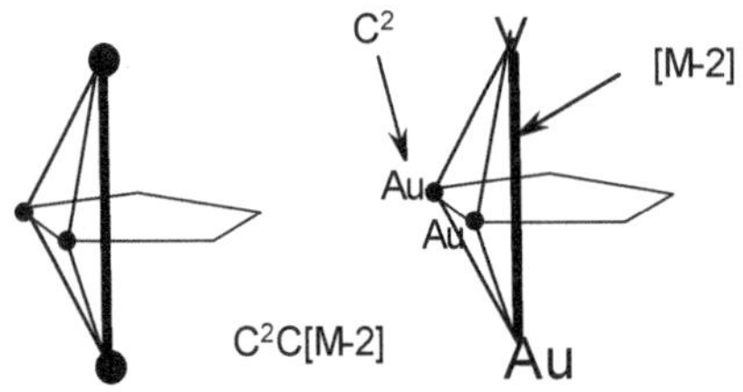

F-1. Graphical representation of the skeletal shape of $(AuL)_3V(CO)_5$

Thus, in this geometrical framework, the nuclear skeletal atoms [M-2] form the vertical axis and the capping skeletal atoms successively occupy the 5 'equatorial' positions around the [M-2] axis. This geometrical framework is adopted from (Pivoriũnas,2005).

Ex-2. $(AuL)_4Mn(CO)_4^+$; k = 4[2.5]+1[5.5]-4(1)+0.5 = 12, n = 4+1 = 5, k(n) = 12(5).The q value, is calculated in the same way as in Ex-1, q = 2[2n-k] = 2[2(5)-12] = 2[-2] = -4. This gives us S = 4n-4, and Cp = $C^1+C^2 = C^3C$[M-2]. The number of valence electrons can readily be verified V = 14n-4 = 14(5)-4 = 66. This agrees with the calculation from the cluster formula $(AuL)_4Mn(CO)_4^+$, V = (11+2)(4)+7+8-1 = 66. Based on the capping symbol Cp = C^3C[M-2], there will be three skeletal capping atoms around the two nuclear skeletal atoms. This is sketched in F-2 and agrees with the observation (Pivoriũnas, 2005).

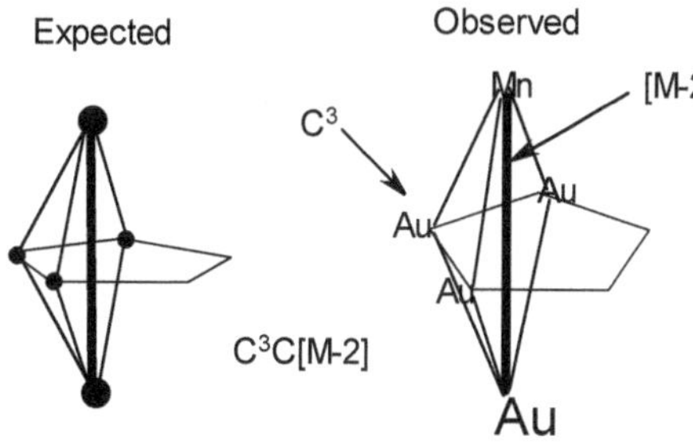

F-2. Graphical representation of the skeletal shape of $(AuL)_4Mn(CO)_4^+$

Ex-3. $(AuL)_5Fe(CO)_3^+$; k = 5[2.5]+1(5)-3+0.5 = 15; n =5+1 = 6, k(n) = 15(6). This means we have six skeletal elements. The q value = 2[2n-k] = 2[2(6)-15] = 2[-3] = -6, S = 4n-6, Cp = C^1+C^3 = C^4C[M-2]. The number of valence electrons V = 14n-6 = 14(6)-6 = 78. Verification from formula, V = 5(13)+1(8)+3(2)-1 = 78. Thus, according to the series method and the Pivoriūnas geometrical framework, the [M-2] skeletal elements will lie along the axis while the remaining 4 skeletal atoms will be distributed around the equatorial positions. This is sketched in F-3 in agreement with the observation (Pivoriūnas,2005).

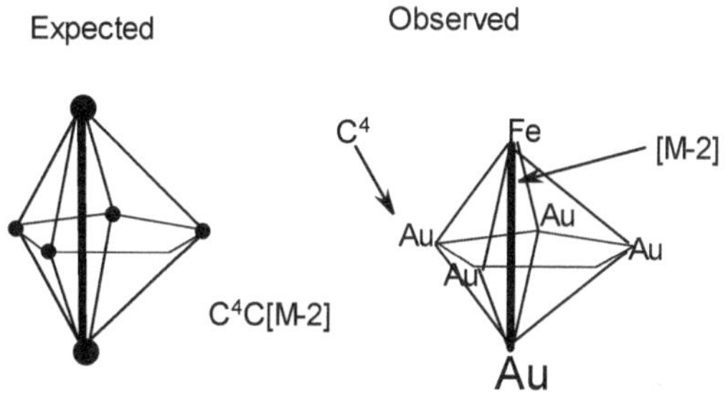

F-3. Graphical representation of the skeletal shape of $(AuL)_5Fe(CO)_3^+$

Ex-4. $Au_7L_7^+$, L = PPh_3; k = 7(3.5)-7+0.5 = 18, k(n) = 18(7), q = 2[2(7)-18] = 2[-4] = -8; S = 4n-8, Cp = C^5C[M-2]. Valence electrons V = 14n-8 = 14(7)-8 = 90. Verification from the formula, V = 7(11)+7(2)-1 = 90. Clearly, this case will have all the 'equatorial' positions filled up as sketched in F-4. It is not a surprise to that the cluster of that shape is referred to as pentagonal bipyramid(Mingos,1984;Konishi,2014) as in $B_7H_7^{2-}$(Greenwood, et al, 1998) with k(n) = 13(7). However, unlike the $B_7H_7^{2-}$ cluster, the $Au_7L_7^+$ complex has a k(n) parameter value of 18(7). This implies that we will expect a different type of cluster bonding.

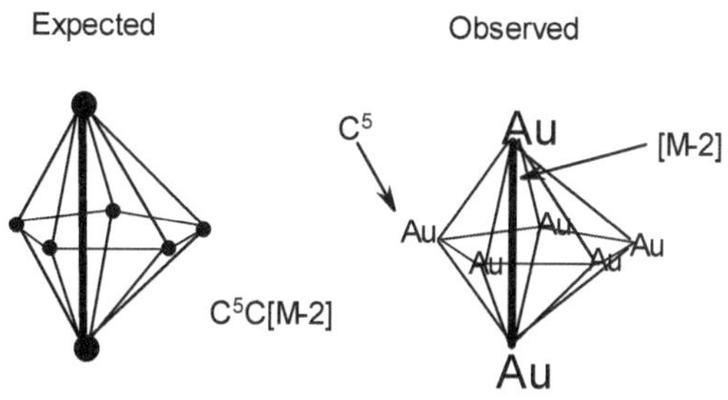

F-4. Graphical representation of the skeletal shape of $(AuL)_7^+$

Ex-5. $(AuL)_7Mo(CO)_3^+$; following the same procedure as in the above examples, in this case we get, k(n) = 21(8), S = 4n-10,Cp = C^6C[M-2]. Valence electrons V = 14n-10 = 14(8)-10 = 102. Verification from the formula, V = 7(13)+1(6)+3(2)-1 = 102. Here we have 6 capping skeletal elements. The 5 of the capping skeletal elements will complete the first ring of the equatorial arrangement, and the sixth will commence the second ring just above the [M-2] axis. This sketch is shown in F-5 and agrees with what is observed (Pivoriūnas, 2005).

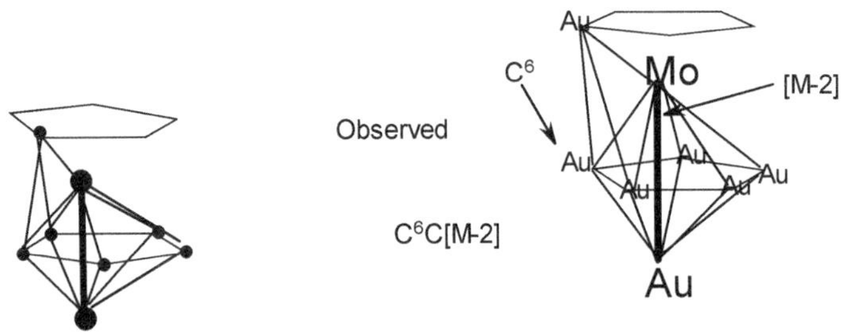

F5. Expected F-5. Graphical representation of the skeletal shape of $(AuL)_7Mo(CO)_3{}^+$

Ex-6. $(AuL)_8Pt(CO)^{2+}$.Following the same procedure as in the previous examples, k(n) = 24(9), S = 4n-12, Cp = C^7C[M-2]. The valence electrons V = 14n-12 = 14(9)-12 = 114. Verification from formula, V = 8(13)+1(10)+1(2)-2 = 114. From the capping symbol, the 7 skeletal elements will be capping around the [M-2] axis. The graphical skeletal shape for the predicted and observed (Gimeno,2008) structures are sketched as shown in F-6.

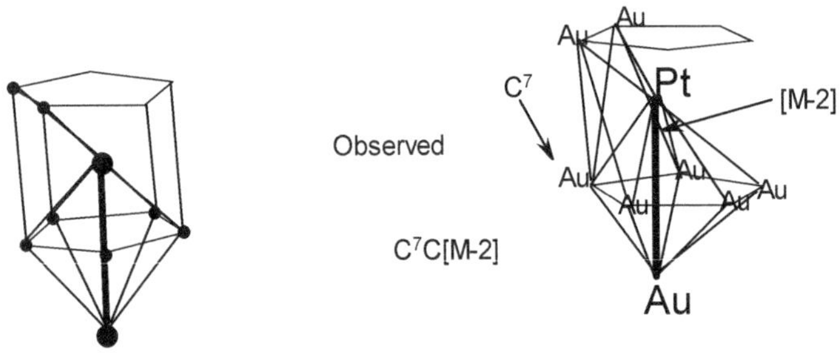

F-6. Expected F-6. Graphical representation of the skeletal shape of $(AuL)_8Pt(CO)^{2+}$

Ex-7. $(AuL)_6(AuCl)_3Pt(CO)$; k(n) = 27(10), S = 4n-14, Cp = C^8C[M-2]. The valence electrons V = 14n-14 = 14(10)-14 = 126. Verification from the formula, V = 6(13)+3(12)+1(10)+1(2) = 126. The proposed predicted and observed (Pivoriūnas, 2005) shapes are sketched in F-7.

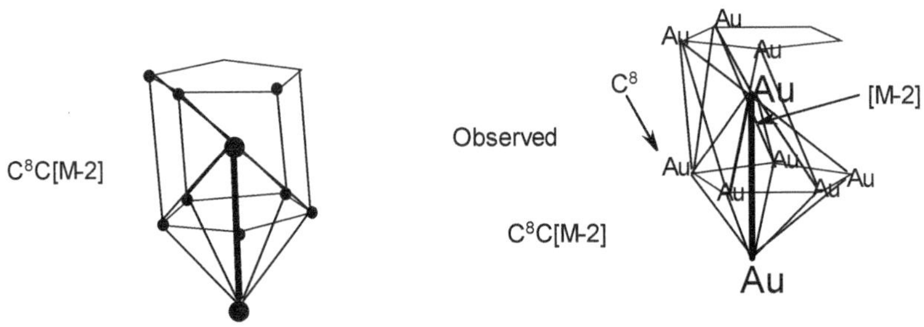

F-7. Expected F-7. Graphical representation of the skeletal shape of $(Au)_{10}(L)_8{}^+$

Ex-8. Au13Cl2L103+; k(n) = 36(13), S = 4n-20, Cp = C11C[M-2]. The valence electrons V = 14n-20 = 14(13)-20 = 162. Verification from formula, V = 13(11)+2(1)+10(2)-3 = 162. The proposed skeletal sketch is shown in F-8 and is in agreement with the observation (Konishi, 2014).

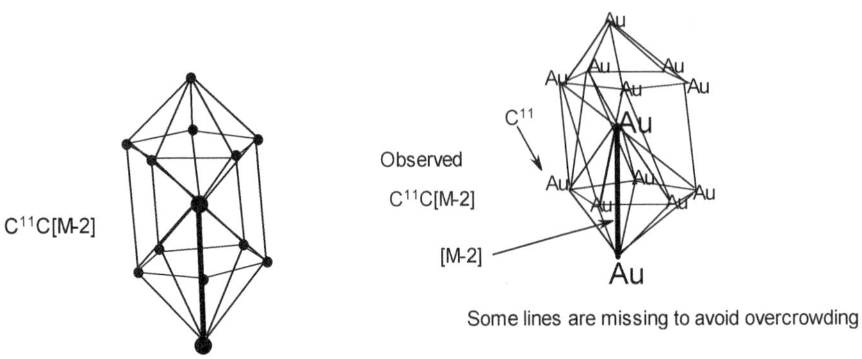

F-8. Expected F-8. Graphical representation of the skeletal shape of $(Au)_{13}Cl(_2L)_{10}{}^{3+}$

Ex-9. [Pt(H)(AgNO3)(AuL)8+

K = 1[4]+1[3.5]+8[2.5]-0.5-0.5+0.5 = 27→k(n) =27(8), 4n-14, Cp = C8C[M-2]. The valence electrons V = 14n-14 = 14(10)-14 = 126. Verification from formula, V = 1[10]+1(1)+1(11)+1(1)+8(13)-1 = 126. The shape will be based on a two-centered skeletal atom nucleus. The other 8 atoms will be capping around [M-2]. This is sketched in F-9 in agreement with the literature (Kappen, 1995b).

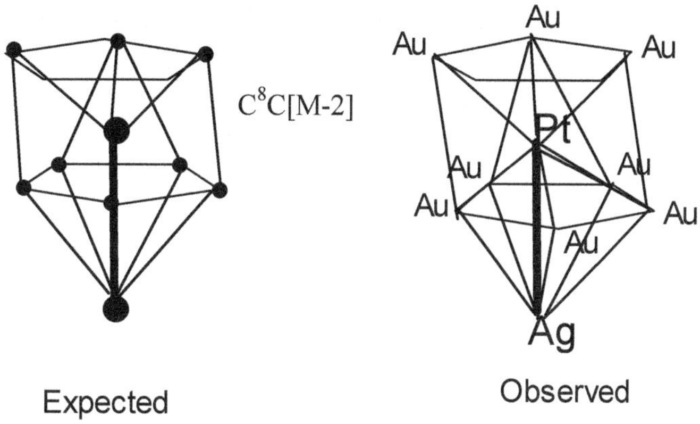

F-9. Skeletal sketch of $[Pt(H)(Ag)(AuL)_8^+$

Ex-10. Pt(H)(AgNO3)2(AuL)8+

k = 1[4]+2[3.5]+8[2.5]-0.5+2(0.5)+0.5 = 30; k(n) = 30(11), S = 4n-16, Cp = C1+C8 = C9C[M-2]. The valence electrons V = 14n-16 = 14(11)-16 = 138. Verification, V = 1(10)+1(1)+2(11+1)+8(13)-1 = 138. When the nine capped and two centered [M-2] nucleus cluster are mapped onto the geometrical framework, we get F-10 as has been observed (Kappen, 1995b).

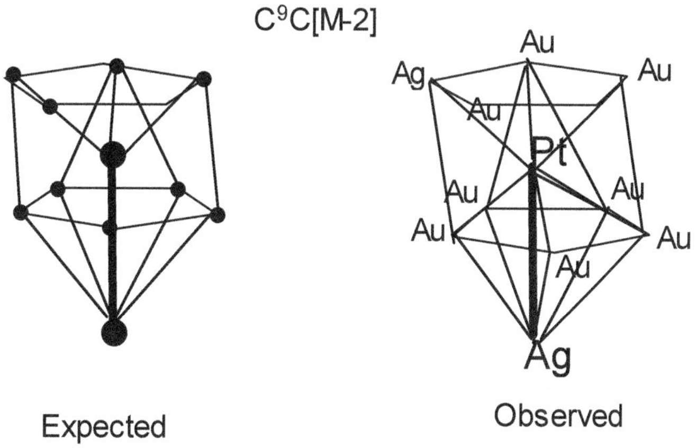

F-10. Skeletal sketch of $[Pt(H)(Ag)_2(AuL)_8^{2+}$

Ex-11. Au12L10Cl3+

k = 12[3.5]-10(1)-1(0.5)+1.5 = 33; k(n) = 33(12), S = 4n-18, Cp = C10C[M-2]. The valence electrons V = 14n-18 = 14(12)-18 = 150. Verification from formula, V = 12(11)+10(2)+1(1)-3 = 150. This means that there are 10 capping skeletal atoms while the remaining 2 will be the nucleus, [M-2]. The expected structure is sketched in F-11.

$C^{10}C[M\text{-}2]$

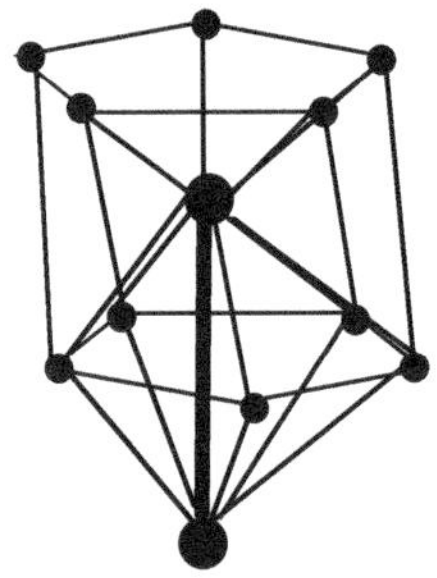

F-11

A selected sample of [M-2] based golden clusters are given in Table 1. Golden clusters have been divided into broad groups such SPHERICAL CLUSTERS and TOROIDAL CLUSTERS (Mingos, 1984). On close analysis with series, the spherical type are found to be those clusters which have a BINUCLEAR INDEX [M-2](Kiremire, 2016b). Examples include, Au9L8+, C7C[M-2], Au11(I)3L7, C9C[M-2], and Au13(Cl)2L103+, C11C[M-2] while the toroidal type are MONONUCLEAR INDEX [M-1] based clusters (Kiremire, 2016b). These include, Au8L72+, C7C[M-1]; Au9L83+, C8C[M-1] and Au10 L6 (Cl)3+, C9C[M-1].

Table 1. Selected [M-2] based Gold Clusters

CLUSTER	k VALUE	k(n)	q VALUE	SERIES	CATEGORY
$Mo_2(Cp)_2(CO)_4$	3	3(2)	2	4n+2	CLOSO[M-2]
$Au_3L_3X_3$	6	6(3)	0	4n+0	C^1C[M-2]
$(AuL)_2Fe(CO)_4$	6	6(3)	0	4n+0	C^1C[M-2]
$(AuL)_3O^+$	9	9(4)	-2	4n-2	C^2C[M-2]
$(AuL)_3V(CO)_5$	9	9(4)	-2	4n-2	C^2C[M-2]
$(AuL)_3Co(CO)_3$	9	9(4)	-2	4n-2	C^2C[M-2]
$(AuL)_3Fe(CO)_4^+$	9	9(4)	-2	4n-2	C^2C[M-2]
$Au_4L_4I_2$	9	9(4)	-2	4n-2	C^2C[M-2]
$Fe_2Ir_2(CO)_{12}(AuL)^+$	12	12(5)	-4	4n-4	C^3C[M-2]
$(AuL)_4Re(CO)_4^+$	12	12(5)	-4	4n-4	C^3C[M-2]
$(AuL)_4Co(CO)_3^+$	12	12(5)	-4	4n-4	C^3C[M-2]
$(AuL)_4Mn(CO)_4^+$	12	12(5)	-4	4n-4	C^3C[M-2]
$(AuL)_4Re(CO)_4^+$	12	12(5)	-4	4n-4	C^3C[M-2]
$(AuL)_5Fe(CO)_3^+$	15	15(6)	-6	4n-6	C^4C[M-2]
$(AuL)_5Mo(CO)_4^+$	15	15(6)	-6	4n-6	C^4C[M-2]
$(AuL)_6Re(CO)_3^+$	18	18(7)	-8	4n-8	C^5C[M-2]
$(AuL)_6V(CO)_4^+$	18	18(7)	-8	4n-8	C^5C[M-2]
$Au_7L_7^+$	18	18(7)	-8	4n-8	C^5C[M-2]
$(AuL)_6Co(CO)_2^+$	18	18(7)	-8	4n-8	C^5C[M-2]
$(AuL)_6Mn(CO)_3^+$	18	18(7)	-8	4n-8	C^5C[M-2]
$Au_8L_8^{2+}$	21	21(8)	-10	4n-10	C^6C[M-2]
$(AuL)_7Mo(CO)_3^+$	21	21(8)	-10	4n-10	C^6C[M-2]
$(AuL)_7Co(CO)_2^{2+}$	21	21(8)	-10	4n-10	C^6C[M-2]
$(AuL)_8Pt(CO)^{2+}$	24	24(9)	-12	4n-12	C^7C[M-2]
$Au_9L_8^+$	24	24(9)	-12	4n-12	C^7C[M-2]
$(AuL)_6(AuCl)_3Pt(CO)$	27	27(10)	-14	4n-14	C^8C[M-2]
$Au_{11}L_7Cl_3$	30	30(11)	-16	4n-16	C^9C[M-2]
$Au_{11}L_7I_3$	30	30(11)	-16	4n-16	C^9C[M-2]
$Au_{11}L_{10}^{3+}$	30	30(11)	-16	4n-16	C^9C[M-2]
$Au_{13}L_7Cl_5$	36	36(13)	-20	4n-20	$C^{11}C$[M-2]
$Au_{13}Cl_2L_{10}^{3+}$	36	36(13)	-20	4n-20	$C^{11}C$[M-2]
$Au_{13}L_{12}^{5+}$	36	36(13)	-20	4n-20	$C^{11}C$[M-2]

2.3 Hypothetical Model for Predicting the Symmetries of [M-2] Based Golden Clusters

When the clusters in Table 1 are carefully scrutinized, they are found to form a set of capping series based on a closo nucleus comprising of two skeletal elements linked by a triple bond according to the series method. The k(n) parameter varies as follows: 3(2)→6(3) →9(4) →12(5) →15(6) →18(7) →21(8) →24(9) →27(10) →30(11) →33(12) →36(13). These correspond to the capping series: 4n+2{C^0C[M-2]} →C^1C[M-2]→C^2C[M-2] →C^3C[M-2] →C^4C[M-2] →C^5C[M-2] →C^6C[M-2] →C^7C[M-2] →C^8C[M-2] →C^9C[M-2] →$C^{10}C$[M-2] →$C^{11}C$[M-2]. According to the series method, the capping series vary by $\Delta k = 3$ and $\Delta n = 1$. When the [M-2] capping series are superimposed onto the geometrical framework proposed by Pivoriūnas (Pivoriūnas, 2005) sketched in D-2, then we get the cluster symmetries as sketched in D-3. What is interesting is that the derived symmetries agree very well with what is observed in the [M-2] based golden clusters. Some of these shapes have been described as being SPHERICALLY CENTERED (Mingos, 1984) as visualized in D-1. We very much know that simple transition metal complexes have a tendency take up specific geometries. For instance, ML_2(linear), ML_3(trigonal planar), ML_4(tetrahedral/square planar), ML_5 (trigonal bipyramid/square pyramid), ML_6(octahedral) and so on. Similarly, we can apply series method in combination with the Pivoriūnas geometrical model to readily provide a tentative prediction of shapes for golden clusters C^1C[M-2] to $C^{11}C$[M-2].

D-1

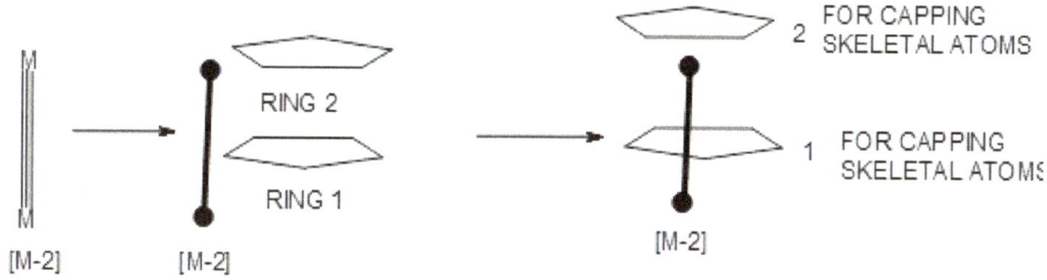

D-2. Geometric framework for generating observed sh pes of [M-2] based golden c lusters

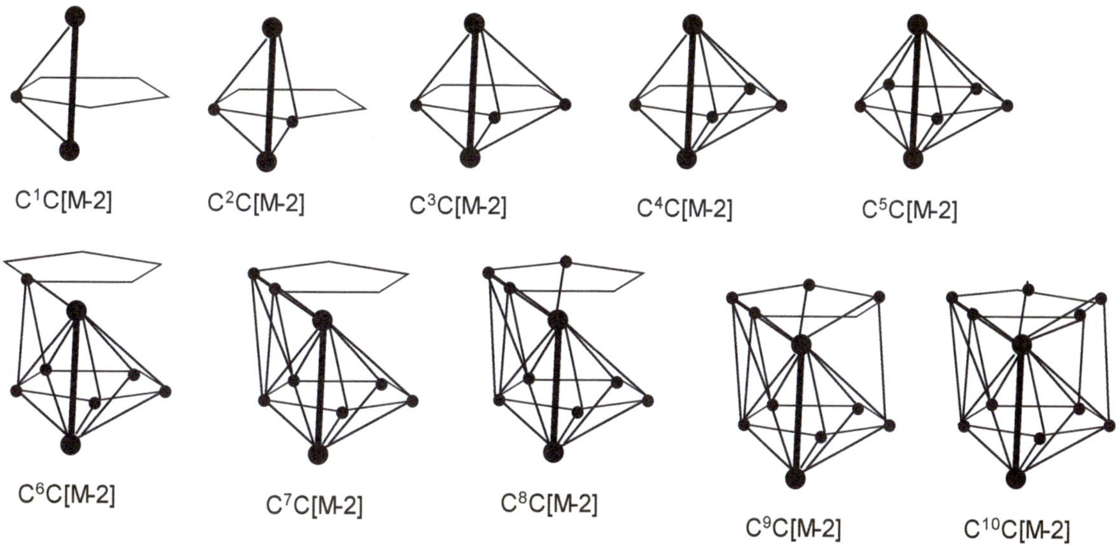

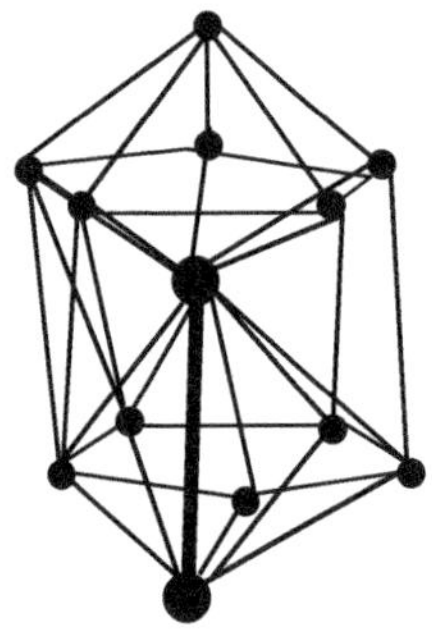

D-3. Sketch Illustrating The Formation of [M-2] Based Golden Capped Clusters

3. Conclusion

The series method has successfully been found to categorize the golden clusters and assist in predicting some of their ideal symmetries. This is reflected in the cases of a large number of [M-2] based golden clusters. In this regard, a combination of the [M-2] capping series with the Pivoriūnas geometrical framework has been extremely useful.

Acknowledgement

The author wishes to acknowledge the University of Namibia for financial support and providing the enabling environment, NAMSOV, Namibia for financial assistance and Merab Kambamu Kiremire for proof-reading the draft.

References

Cotton, F. A., & Wilkinson, G. (1980). *Advanced Inorganic Chemistry, 4th Ed.*, John Wiley and Sons, New York, 1980

Gimeno, M. C. (2008). Modern Supramolecular Gold-Metal Interactive and Applications. Edited A. Laguna, 2008. Wiley-VCH, Weinheim.

Greenwood, N. N., & Earnshaw, A. (1998). *Chemistry of the Elements, 2nd Ed.* Butterworth, Oxford.

Kappen, T. G. M. M., Schlebos, P. P. J., Bour, J. J., Bosman, W. P., Smits, J. M. M., Beurskens, P.,T., & Steggerda, J. J. (1995a). Synthesis and Characterization of Platinum-Gold Clusters Fused to Tetrahedral Copper Units. *J. Am. Chem. Soc., 117*, 8327-8334. https://doi.org/10.1021/ja00137a005

Kappen, T. G. M. M., Schlebos, P. P. J., Bour, J. J., Bosman, W. P., Smits, J. M. M., Beurskens, P. T., & Steggerda, J. J. (1995b). Cluster Growth: Some Representative Reactions. *Inorg. Chem., 34,* 2121-2132. https://doi.org/10.1021/ic00112a027

Kilmartin, J. (2010). Molecular Gold Clusters as Precursors to Heterogeneous Catalysis. PhD Thesis.

Kiremire, E. M. R. (2016a). The Application of the 4n Series Method to Categorize Metalloboranes. *Int. J. Chem., 8*(3), 62-73. https://doi.org/10.5539/ijc.v8n3p62

Kiremire, E. M. R. (2016b). Clusters of Gold Containing P-Block Elements. *Am. J. Chem., 6*(5), 126-144.

Kiremire, E. M. R. (2016c). A Hypothetical Model for the Formation of Transition metal Carbonyl Clusters Based Upon 4n Series Skeletal Numbers. *Int. J. Chem., 8*(4), 78-110. https://doi.org/10.5539/ijc.v8n4p78

Kiremire, E. M. R. (2017). The Golden Series and Clusters of Gold-unique Shapes and Bonding. *Int. J. Chem., 9*(1), 38-57. https://doi.org/10.5539/ijc.v9n1p38

Konishi, K. (2014). Phosphine-Coordinated Pure-Gold Clusters and Unique Optical Properties /Responses. *Structure and Bonding, 161*, 49-86. https://doi.org/10.1007/430_2014_143

Kwok-Ming, L. (2011). Synthesis, characterization and photophysical properties of Chalcogenido, phosphinidene and alkyl Complexes of Gold(I) and its Congener and their Supramolecular assembly arising from metal-metal interactions. PhD Thesis.

Mingos, D. M. P. (1984). Gold Cluster Compounds: Are they materials in miniature? *Gold Bull., 17*(1), 5-12. https://doi.org/10.1007/bf03214670

Mingos, D. M. P. (1987). Complementary Spherical Electron Density model for Coordination Compounds. *Pure and Appl. Chem., 59*(2), 145-154. https://doi.org/10.1351/pac198759020145

Pauling, L. (1977). Structure of Transition Metal Cluster Compounds: Use of an additional orbital resulting from f, g character of spd bond orbitals. *Proc. Natl. Acad. Sci., USA, 74*(12), 5235-5238. https://doi.org/10.1073/pnas.74.12.5235

Pei, Y., Shao, N., Jian, D., & Zeng, X. C. (2011). Hollow Polyhedral Structures in Small Gold-Sulfide Clusters. *J. Am. Chem. Soc.,* 1441-1449. https://doi.org/10.1021/nn103217z

Pivoriũnas, G. (2005). Neuer Goldcluster. PhD Thesis.

Puls, A. B. (2014). A New Synthesis Concept for Metal Doped Gold Clusters. PhD Thesis.

Raithby, P. R. (1998). The Build-Up of Bimetallic Transition Metal Clusters. *Platinum Metal Rev., 42*(4), 146-157.

Sarip, R. (2013). Gold Molecular Clusters to Nanoparticles: A Bottom-up Approach to Supported Nanoparticles for Heterogeneous Catalysis. PhD Thesis.

Steigelmann, O., Bissing, P., & Schmidbaur, H. (1993). 1,1,1,1-Tetrakis(trioorganylphosphineaurio(I) ethanium(+) Tetrafluoroborates-Hypercoordinated Species Containing $H_3C\text{-}C(AuL)_4^+$ cations. *Z. Naturforsch, 48b,* 72-78.

Van Der Velden, J. W. A. (1983). Preparation and Properties of Gold Cluster Compounds. PhD Thesis.

Vincente, Chicote, M. T., Abrisqueta, M. D., Gonnzález-Hererro, & Guerrero, R. (1998). Recent Advances in the Chemistry of Gold(I) Complexes with C-,N-,and S-Donor Ligands. *Gold Bulletin, 31*(3), 83-87. https://doi.org/10.1007/BF03214767

Yuan, Y., Cheng, L., & Yang, J. (2013). Electronic Stability of Phosphine-protected Au_{20} Nanocluster: Superatomic Bonding. *J. Phy. Chem., 117*, 13276-13282.

19

The Golden Series and Clusters of Gold-unique Shapes and Bonding

Enos Masheija Rwantale Kiremire[1]

[1]Enos Masheija Rwantale Kiremire, Department of Chemistry and Biochemistry, University of Namibia, Private Bag 13301, Windhoek, Namibia

Correspondence: Enos Masheija Rwantale Kiremire, Department of Chemistry and Biochemistry, University of Namibia, Private Bag 13301, Windhoek, Namibia. E-mail: kiremire15@yahoo.com

Abstract

Golden clusters have been analyzed and categorized using the series method. They have been found to be intensely capped. The complexity of capping is likely responsible for the unique shapes of clusters. The predicted structures of the golden clusters with one or two nuclear indices have been found to be in agreement with the observations from structural determinations. According to the series, the golden clusters have tendency of occurring in an overlapping area which is normally dominated by naked metal clusters. The series approach method is easy and very reliable in analyzing and characterizing clusters. A brief background to the series method was included to ease readability of the paper.

Keywords: gold-clusters, cluster-code, nuclear-index, cluster-map, 4n-series, Skeletal-numbers, conservation, principle

1. Introduction

The polyhedral skeletal electron pair theory (PSEPT) or simply known as Wade-Mingos rules has been extremely useful in analyzing and categorizing clusters for a long time (Wade, 1971; Mingos, 1972; Welch, 2013). Relatively recently, the 4n series method was developed and has further been refined for rapid application to analyze and characterize clusters with great success (Kiremire, 2016a, 2016b). Golden clusters portray a wide range of fascinating shapes and have attracted increased intensity in research (Cotton and Wilkinson, 1980; Mingos, 1984, 1987; Greenwood and Earnshaw, 1998; Pauling, 1977; Pivoriūnas, 2005; Gimeno, 2008; Konishi, 2014). In addition, gold nanoparticles have shown enormous potential for applications in medicine, catalysis and electronics (Tiwari, et al, 2011; Daniel, et al, 2004). It became tempting to try and analyze the golden clusters using the skeletal numbers derived from 4n series method. This paper presents an analysis and categorization of selected number of golden clusters using the newly introduced skeletal numbers and the conservation of cluster skeletal linkages principle (Kiremire, 2016b).

2. Results and Discussion

2.1 Background

In order to analyze and categorize the gold clusters, the data for skeletal numbers has been reproduced (Kiremire, 2016b) as this is new information for many readers. The skeletal numbers are given in Tables 1 and 2. Their derivation from the series is explained in the same reference (Kiremire, 2016b) and it is not necessary to repeat it in this work. The skeletal numbers have been utilized to calculate the cluster linkage value (K) for a skeletal square pyramid geometry. Selected examples are given in Table 3.

Table 1. Skeletal numbers of main group elements

7.5	7	6.5	6	5.5	5	4.5	4	3.5	3
Sc	Ti	V	Cr	Mn	Fe	Co	Ni	Cu	Zn
Y	Zr	Nb	Mo	Tc	Ru	Rh	Pd	Ag	Cd
La	Hf	Ta	W	Re	Os	Ir	Pt	Au	Hg

Table 2. Skeletal values of transition metals

7.5	7	6.5	6	5.5	5	4.5	4	3.5	3
Sc	Ti	V	Cr	Mn	Fe	Co	Ni	Cu	Zn
Y	Zr	Nb	Mo	Tc	Ru	Rh	Pd	Ag	Cd
La	Hf	Ta	W	Re	Os	Ir	Pt	Au	Hg

2.1.1 Conservation of Cluster Skeletal Linkages Principle

The concept of conservation of cluster skeletal linkages principle is simple. This principle was discovered by careful studies of the 4n series generated when a given cluster of skeletal elements is combined with ligands (Kiremire, 2016b).

The principle simply states that a given cluster comprising of naked skeletal elements contains an inherent fixed number of skeletal linkages. When such a cluster combines with ligands, some or all the skeletal linkages are used up (K_L). Then the sum of the used up linkages and those remaining in the cluster (K_C) are equal to the number of the original linkages (K_O) contained by the naked or uncombined (ligand free) cluster. Using this concept, the skeletal linkages of clusters (K_C) can readily be calculated. A selected sample is given in Table 3. The clusters in Table 3 have been numbered. To illustrate the concept of conservation of cluster skeletal linkages principle, take cluster 1, $K_L + K_C = 17+8 = 25 = K_O$; 6→ $K_L + K_C = 8.5+8 = 16.5 = K_O$; 8→ $K_L + K_C = 4.5+8 = 12.5 = K_O$;9 → $K_L + K_C = 3.5+8 = 11.5 = K_O$, and 11→ $K_L + K_C = 6.5+8 = 14.5 = K_O$. The beauty and simplicity of the Conservation of Cluster Skeletal Linkages Principle is that for each electron donated to a skeletal fragment by a ligand or simply a naked electron from a charge consumes or removes 0.5 linkage value of the skeletal fragment. Therefore, according to the series, 1H ≡1e (1^-) charge ≡$R^\bullet$(alkyl fragment) ≡ $X^\bullet$ = Cl, Br, I; K = -0.5 and 1 :CO ≡$2H^\bullet$ ≡2e (2^-) charge≡ η^2-C_2H_4 ≡ L = PPh_3, K = 2(-0.5) = -1;η^4-C_4H_4 ≡4e , K = 4(-0.5) = -2; (and η^5-C_5H_5 ≡ 5e, K = 5(-0.5) = -2.5. The negative sign is used for ligands to denote the removal of skeletal values from the naked cluster 'basket'. For example, a boron skeletal element (B) is assigned a K = 2.5(see Table 1). Therefore a B_6 naked skeletal cluster possesses a total of K =6x2.5 = 15 skeletal linkages in its basket. If this skeletal fragment combines with 6H ligands and two negative charges (2e) as ligands, they will consume or remove from the naked B_6 cluster ≡ 6(0.5)+2(0.5) = 4 skeletal linkages. What is left over will be = 15-4 = 11. On the other hand, the osmium cluster, $Os_6(CO)_{18}^{2-}$, the Os skeletal element is assigned K=5(see Table 2). Hence the naked osmium skeletal cluster Os_6 has K =6x5 =30 linkages in its basket. When the naked cluster is combined with 18 CO ligands and 2 negative charges, the linkages removed from its basket will be = 18x1+2(0.5) = 19. What will be left in the Os_6 basket = 30-19 = 11. In this regard, viewing the clusters in this way, it is not a surprise that the skeletal shapes of $B_6H_6^{2-}$ and $Os_6(CO)_{18}^{2-}$ are similar, that is, O_h symmetry as sketched in F-2.

Table 3. Application of skeletal linkage numbers to generate a square pyramid geometry

	CLUSTER	Skeletal Elements	Linkage Contribution By Skeletal Elements, K_o	Ligands	Ligand Linkage Consumption, K_l	Net Cluster Linkages, K_c	n	K(n) code
1	$Ru_5(C)(CO)_{15}$	5Ru	5x5 = 25	1C+15CO	1(2)+15(1) =17	25-17 = 8	5	8(5)
2	$Fe_5(C)(CO)_{14}^{2-}$	5Fe	5x5 =25	1C+14CO+2e	1(2)+14(1)+2(0.5) = 17	25-17 =8	5	8(5)
3	$Fe_5(N)(CO)_{14}^{-}$	5Fe	5x5 =25	1N+14CO+1e	1(2.5)+14(1)+1(0.5)= 17	25-17 =8	5	8(5)
4	$Fe_5(C)(CO)_{12}(L)_3$	5Fe	25	1C+12CO+3L	2+12+3 = 17	25-17=8	5	8(5)
5	L = PMe_2Ph							
6	$(Rh)_2(Cp)_2B_3H_7$	2Rh+3B	2(4.5)+3(2.5) =16.5	2Cp+7H	2(2.5)+7(0.5) = 8.5	16.5-8.5 =8	2+3 = 5	8(5)
7	$Ru_2B_3(Cp)_2(H)_9$	2Ru+3B	2(5)+3(2.5) =17.5	2Cp+9H	2(2.5)+9(0.5) = 9.5	17.5-9.5 =8	2+3 =5	8(5)
8	B_5H_9	5B	5(2.5) =12.5	9H	9(0.5) = 4.5	12.5-4.5 =8	5	8(5)
9	$C_2B_3H_7$	2C+3B	2(2)+3(2.5) =11.5	7H	3.5	11.5-3.5 =8	5	8(5)
10	$C_5H_5^+$	5C	5(2) = 10	5H-1e	2.5-0.5 =2	10-2 = 8	5	8(5)
11	$B_4H_8Co(Cp)$	4B+1Co	4(2.5)+1(4.5) = 14.5	8H+Cp	4+2.5 =6.5	14.5-6.5 =8	5	8(5)
12	$B_4H_8[Fe(CO)_3]$	4B+1Fe	4(2.5)+1(5) =15	8H+3CO	4+3 =7	15-7 =8	5	8(5)
13	$S_2Fe_3(CO)_9$	2S+3Fe	2(1)+3(5) = 17	9CO	9	17-9 =8	5	8(5)
14	$Te_2Fe_3(CO)_9$	2Te+3Fe	17	9CO	9	17-9=8	5	8(5)
15	$Se_2Fe_2Mn(CO)_9^{-}$	2Se+2Fe+1 Mn	2(1)+2(5)+1(5. 5)=17.5	9CO+1e	9.5	17.5-9.5=8	5	8(5)
16	$Re_5(C)(CO)_{16}(H)^{2-}$	5Re	5(5.5)=27.5	1C+16CO+1H+2e	2+16+0.5+2(0.5) =19.5	27.5-19.5 = 8	5	8(5)
17	$Fe_2Mo(CO)_{10}(S)(Te)$	2Fe+1Mo+ 1S+1Te	2(5)+1(6)+1(1) +1(1) =18	10CO	10	18-10 =8	5	8(5)

2.1.2 Derivation of Cluster Series for the Code K(N) and Its Geometry

Given the values of n and the corresponding K value, we can derive the corresponding series. Take the above example of K(n) = 8(5). We know from the series that K = 2n-q/2. Hence, 8 = 2(5)-q/2, q/2 = 10-8 = 2 and q =4. Hence the cluster series is given by S = 4n+q = 4n+4. The 4n+4 series is known as the NIDO family. Therefore all the clusters from 1 to 17 belong to the NIDO family. The K(n) = 8(5) means that the 5 skeletal elements are bound by 8 linkages. The ideal shape of such cluster is normally a square pyramid shown in F-1.

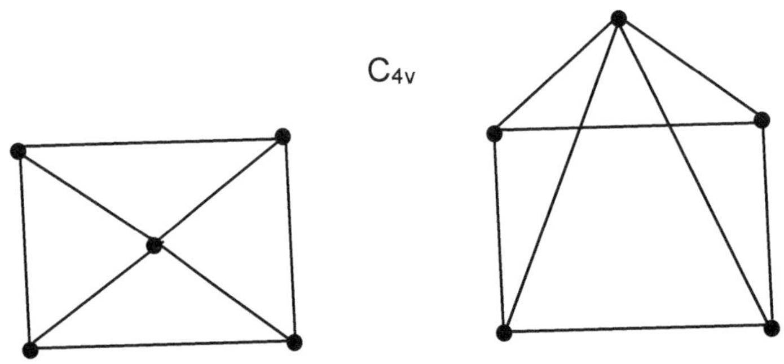

F-1 A sketch of the ideal skeletal shape of K(n) = 8(5)

2.1.3 Valence Electrons of a Cluster

The number of the valence electrons of a cluster is simply derived from the series formula. If the cluster is comprised of main group skeletal elements only, then the series formula is applied as it is. In the case of S = 4n+4, the valence formula becomes V = 4n+4. In the case of K(n) = 8(5), the valence electrons value is, V = 4(5)+4 = 24. In the case of transition elements, V = 14n+4 = 14(5)+4 = 74. If the cluster has mixed skeletal elements, such as cluster number 6 in Table 3, then an adjustment is made to account for this. Using cluster number 6 as an illustration, there 2 transition metal atoms and 3 main group skeletal elements. Therefore, in this case, we use V= 4n+4 = 4(5)+4 = 24. But then since 4n+4 +10n = 14n+4, we have to add 10n = 10(2) = 20 to make allowance for the presence of 2 metal skeletal atoms. Hence, the adjusted number of valence electrons will be given by V = 24+20 = 44. This value can be compared with that calculated from the cluster formula, $Rh_2(Cp)_2B_3H_7$; V = 9x2+5x2+3x3+7 = 44.

Let us consider the transformation of the cluster code from K(n) = 8(5) to 11(6) by adding ΔK(n) = 3(1). The code K(n) = 11(6), gives us K =2(6)-q/2 = 11, q/2 = 1 and q =2.This produces S = 4n+2. The series S = 4n+2 comprises of CLOSO family members. The clusters with a code K(n) = 11(6), normally adopt an ideal octahedral skeletal shape as sketched in F-2. If all the skeletal atoms are transition elements, then the number of valence electrons will be given by V = 14n+2 = 14(6)+2 = 86. A sample of selected clusters with the code K(n) = 11(6) are shown in Scheme 1.

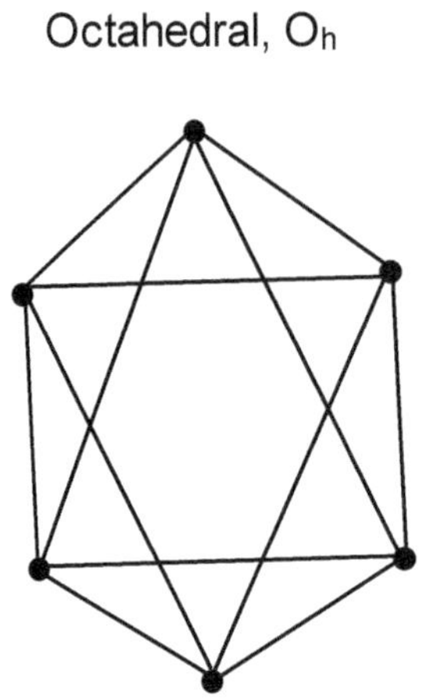

F-2 A sketch of the ideal skeletal shape of K(n) = 11(6)

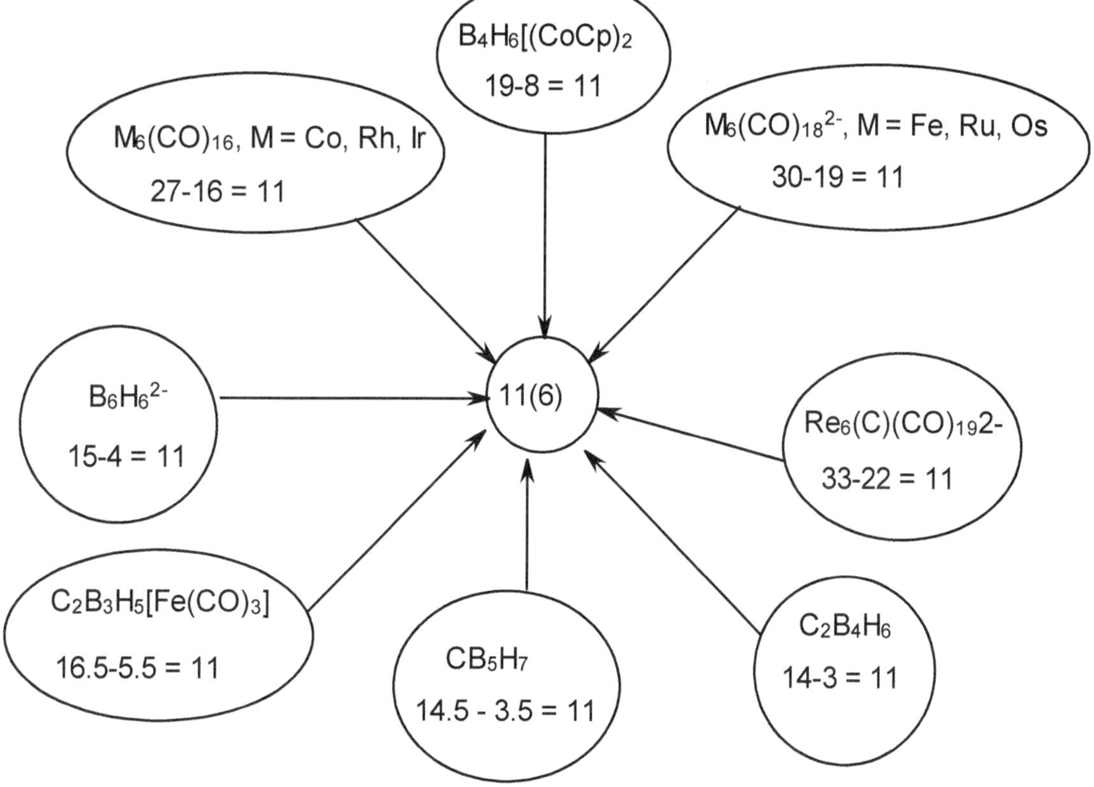

Scheme 1. Some of the cluster systems that generate the ideal O_h symmetry

2.2 Categorization of Golden Clusters

The application of skeletal linkage numbers and the conservation of Cluster Skeletal Linkages Principle, the cluster codes K(n) for golden clusters were determined and hence their respective series. A good number of worked out examples of categorized golden clusters are given in G-0 to G-8 and the rest of the examples are given in a summary form. As can be seen from examples G-1 to G-17, all these golden clusters are intensely capped. The major capping series included C^nC[M-0],n =8-20; C^nC[M-1],n =3-20; C^nC[M-2],n =2-11; C^nC[M-0],n =8-20 and C^nC[M-3],n =3-10. Apart from the 17 golden clusters that have been categorized for illustration, more others were added giving a total of more than 50 clusters and are given in Table 4. It has also been found that those golden clusters centered on one nuclear skeletal element [M-1] are normally referred to as CENTERED TORROIDAL, while those centered on two nuclear skeletal elements, [M-2] are usually referred to as CENTERED SPHERICAL(see Table 8). The capping phenomenon was extensively discussed in previous work (Kiremire, 2016c). In this paper, only brief highlights will be covered by examples. We know that the complex, $Os_6(CO)_{18}^{2-}$ has an octahedral geometry, K = 6(5)-18-1 =11, S = 4n+2(CLOSO family), cluster code =K(n) =11(6)→[M-6] nuclear symbol. A mono-capped cluster of an octahedral osmium complex is $Os_7(CO)_{21}$, K =7(5)-21 =14, S =4n+0, K(n) =14(7), C^1C[M-6]; $Os_8(CO)_{22}^{2-}$,K = 8(5)-22-1 =40-23 = 17, K(n) = 17(8), S =4n-2, C^2C[M-6]→bi-capped octahedral; $Os_9(H)(CO)_{24}^{-}$, K = 9(5)-0.5-24-0.5 = 45-25 =20, K(n) = 20(9), S =4n-4, C^3C[M-6]→tri-capped octahedron, $Os_{10}(CO)_{26}^{2-}$,K =10(5)-26-1 = 50-27 =23, K(n) = 23(10), S =4n-6, C^4C[M-6] tetra-capped octahedron. These examples demonstrate that all these capping clusters have closo nuclei of 6 octahedral skeletal atoms. What is interesting about the golden clusters, is that they do portray clusters whose capping is centered around one skeletal atom (1), [M-1], two skeletal atoms(2), [M-2]or three skeletal atoms(3), [M-3] at the nucleus and in some cases there is no atom in the closo nucleus, [M-0]. This is radically different from the clusters such those of osmium mentioned above whose nuclei are composed of 6 octahedral atoms. There are even giant clusters whose capping is centered around [M-6]. These include $Ni_{38}Pt_6(CO)_{48}(H)^{5-}$(Rossi, et al, 2012); K =38(4)+6(4)-48-0.5-2.5 = 176-51=125,K(n) = 125(44), S =4n-74, $C^{38}C$[M-6], V = 14n-74 = 14(44)-74 = 542 and $Pt_{38}(CO)_{44}^{2-}$,K = 38(4)-44-1 = 152-45 = 107, K(n) = 107(38), S = 4n-62, $C^{32}C$[M-6],V= 14n-62 = 14(38)-62 = 470;$Rh_{23}(CO)_{38}(N)_4^{-}$ (Belyakova, 2003); K =23(4.5)-38-10-0.5 = 103.5-48.5 =55, K(n) =55(23), S =4n-18, $C^{10}C$[M-13].These giant clusters have large "nuclear index". On the other hand, the golden clusters have a tendency of having very small "nuclear indices" regardless as to whether the cluster is of low, medium or high nuclearity. Let us consider the golden cluster (Gimeno,

2008) $Pt(AuL)_8^{2+}$, L =PPh_3. Its K value = 25 has been calculated as shown in G-0 example below. How the 25 linkages are distributed as guided by series is also explained.G-0 $Pt(AuL)_8^{2+}$, L = PPh_3

K
1(4)
+8(3.5)
-8
+1

→ 25 → K(n) = 25(9),q/2 = 2(9)-25 = -7, q = -14

S = 4n-14, C^8C[M-1]

V= 14n-14 = 14(9)-14 =112

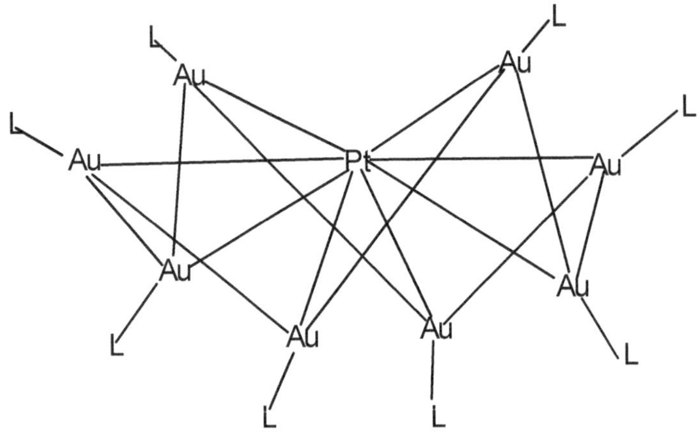

C^8C[M-1]

K = 25

F-3 A skeletal sketch of $Pt(AuL)_8^{2+}$

The series method indicates that one skeletal element will be a CLOSO NUCLEUS and the other 8 skeletal elements will be capped around it. Since the capping linkage of a single skeletal element is derived from the series formula S = 4n-2 and K = 2n+1 = 2(1)+1 =3, this agrees with the 3 connections attached to each of the 8 capping elements. The total linkages associated with 8 skeletal atoms = 3x8 = 24. The [M-1] CLOSO symbol means that the skeletal nucleus atom is a family member of the CLOSO series S = 4n+2, n = 1 and K =2n-1 = 2(1)-1 = 1. This result is quite important because the closo nucleus contains within itself 1 skeletal linkage value K = 1. Thus, C^8C[M-1] is loaded with K = 8x3+1 = 25. In this way, the series method could be regarded as a "miniature microscope" that is utilized to view molecular fragments and clusters however some chemical sense must accompany it for the interpretation of what is observed. This result of K= 25 is in total agreement with the literature report (Gimeno, 2008). A graphical skeletal representation is given in F-3. Suppose the cluster is reacted with a carbonyl ligand to form $Pt(CO)(AuL)_8^{2+}$ complex ,then transformation of thegolden the golden cluster $Pt(AuL)_8^{2+}$ will take place.

$$Pt(AuL)_8^{2+} \xrightarrow{CO} Pt(CO)(AuL)_8^{2+}$$

We know that $Pt(AuL)_8^{2+}$, has K =25. Since a CO ligand is a 2 electron donor, it will remove a value of one (K =1) from the linkage basket of $Pt(AuL)_8^{2+}$. This means that the new K value will have one K value lower, that is, K =25-1 = 24 for $Pt(CO)(AuL)_8^{2+}$ cluster. We can confirm this by the series method of using skeletal linkages, K = 1(4)+8(3.5)-1-8+1= 24, K(n) = 24(9), q/2 =2(9)-24 =-6, q =-12, S = 4n-12, C^7C[M-2], V= 14n-12 =14(9)-12 = 114. The symbol, C^7C[M-2],means that there will be 2 skeletal nuclear atoms surrounded by 7 capping skeletal elements. What has happened is that by removing 1 skeletal linkage, one of the capping skeletal element is forced to uncap and go to the nucleus. This requires skeletal structure re-arrangement. This prediction is observed in the skeletal structure of $Pt(CO)(AuL)_8^{2+}$ reported in literature(Gimeno,2008;Pivoriũnas,2005) and the golden skeletal structure is sketched in F-0(Gimeno, 2008; Pivoriũnas, 2005).

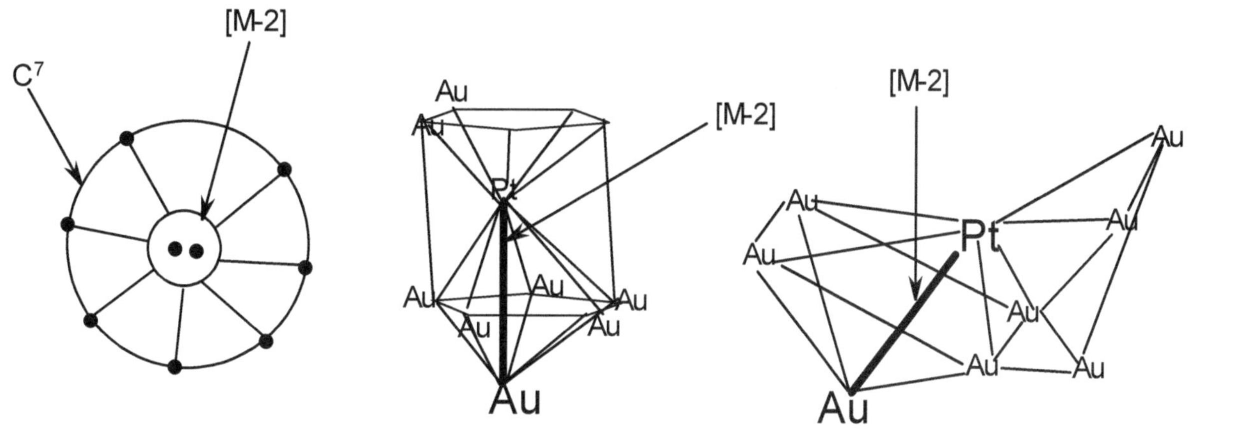

Series prediction   Pivoriunas model prediction   Observed sketch of $Pt(CO)(AuL)_8^{2+}$

[Pivoriunas,2005]   [Gimeno, 2008]

F-0. Graphical sketches of $Pt(CO)(AuL)_8^{2+}$

According to the series, the symbol [M-2] means that the 2 nuclear atoms belong to the CLOSO family, S = 4n+2 and K =2n-1 =2(2)-1 =3. That is, the two nuclear atoms are supposed to be linked by a triple bond, M≡M. It will be interesting to determine which 2 skeletal atoms of the golden cluster are triply bonded. Arising from the complexity of capping and unique geometrical architecture of golden clusters, a desire to construct K(n) cluster map was prompted. The map will facilitate comparing and contrasting the golden clusters with clusters of other transition metals. This resulted in the production of Table 5.

G-1 $Pd(AuL)_8^{2+}$

1(4)
+8(3.5-1) ⟶ k = 25 , n =1+8 = 9
+2(0.5)

k = 2n-q/2, q/2 = 2n-k =2(9)-25 = -7, q = -14, S = 4n-14, Cp = $C^1+C^7 = C^8C[M-1]$

G-2 $Pt(AuL)_8^{2+}$ Cp = $C^8C[M-1]$

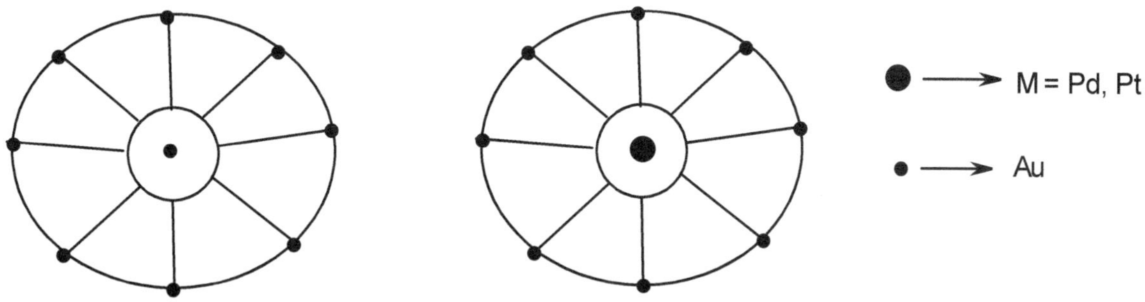

Predicted   Observed

F-4 Skeletal sketch of $M(AuL)_8^{2+}$

G-3 $Pd(CO)(AuL)_8^{2+}$

1(4) +8(3.5-1) -1 +2(0.5)	⟶	k = 24	, n =1+8 = 9

$k = 2n\text{-}q/2$, $q/2 = 2n\text{-}k = 2(9)\text{-}24 = -6$, $q = -12$, $S = 4n\text{-}12$, $Cp = C^1\text{+}C^6 = C^7C[M\text{-}2]$

G-4 $Pt(CO)(AuL)_8^{2+}$ $Cp = C^7C[M\text{-}2]$

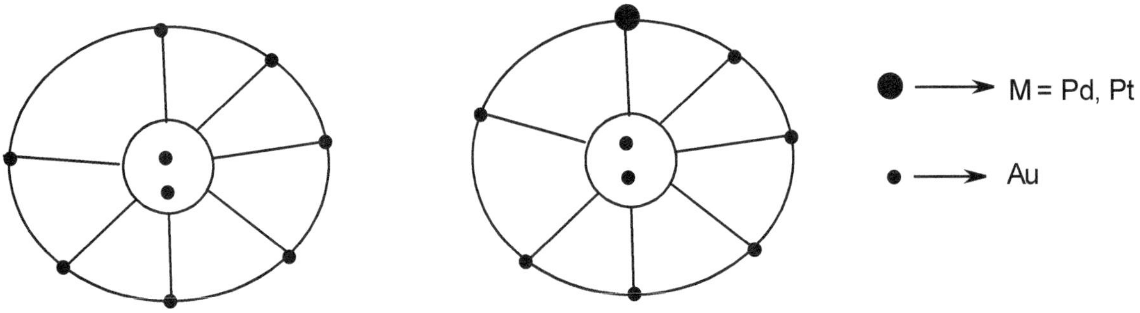

Predicted Observed

F-5 Skeletal sketch of $M(AuL)_8^{2+}$

G-5 $Au_{10}L_6Cl_3^+$

10(3.5) -6 -1.5 +1(0.5)	⟶	k = 28	, n = 10

$k = 2n\text{-}q/2$, $q/2 = 2n\text{-}k = 2(10)\text{-}28 = -8$, $q = -16$, $S = 4n\text{-}16$, $Cp = C^1\text{+}C^8 = C^9C[M\text{-}1]$

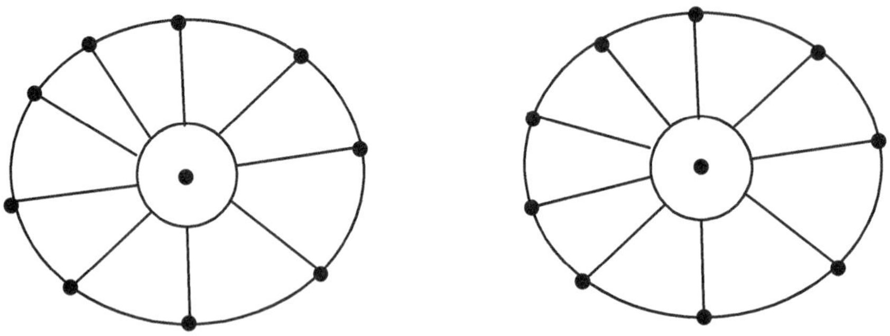

Predicted Observed

F-6 Skeletal sketch of $Au_{10}L_6Cl_3^+$

G-6 $Au_{10}L_8Cl^+$

$$\left.\begin{array}{l} 10(3.5) \\ -8 \\ -0.5 \\ +1(0.5) \end{array}\right| \longrightarrow \quad k = 27 \quad , n = 10$$

$k = 2n-q/2$, $q/2 = 2n-k = 2(10)-27 = -7$, $q = -14$, $S = 4n-14$, $Cp = C^1+C^7 = C^8C[M-2]$

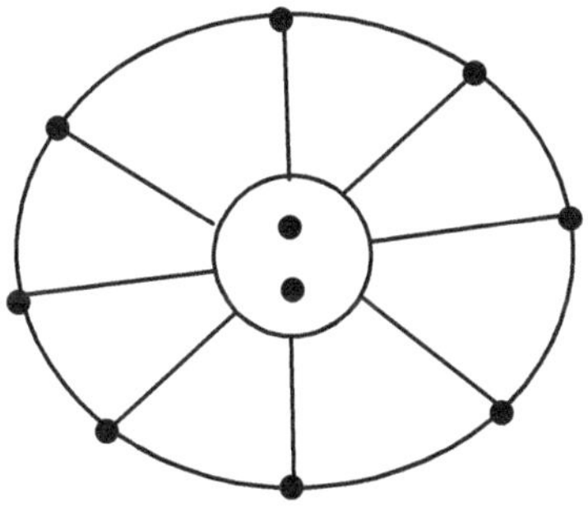

Predicted

F-7 Skeletal sketch of $Au_{10}L_8Cl^+$

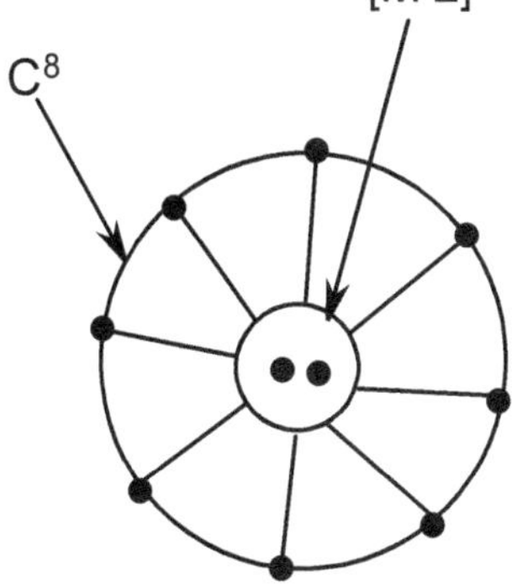

F-7 Observed graph of $Au_{10}L_8Cl^+$ [Konishi, 2014]

Series method

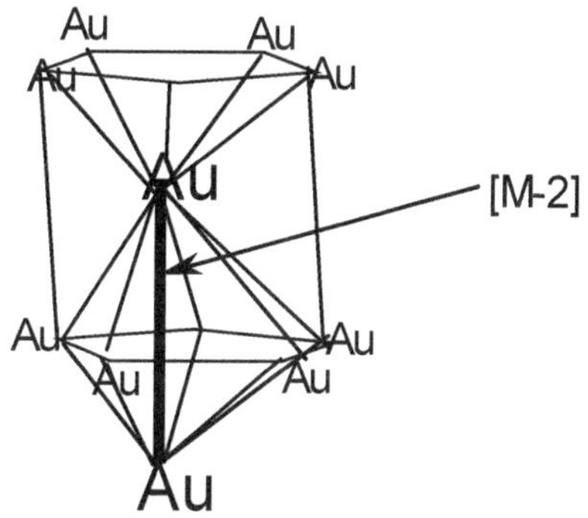

F-7 Skeletal sketch of actual shape of cluster $Au_{10}L_8Cl^+$ [Konishi, 2014]

Pivoriunas model, 2005

G-7 $Au_8L_6^{2+}$

8(3.5) -6 +1	$\longrightarrow$	k = 23	, n = 8

k = 2n-q/2, q/2 = 2n-k =2(8)-23 = -7, q = -14, S = 4n-14, Cp = $C^1+C^7 = C^8C$[M-0]

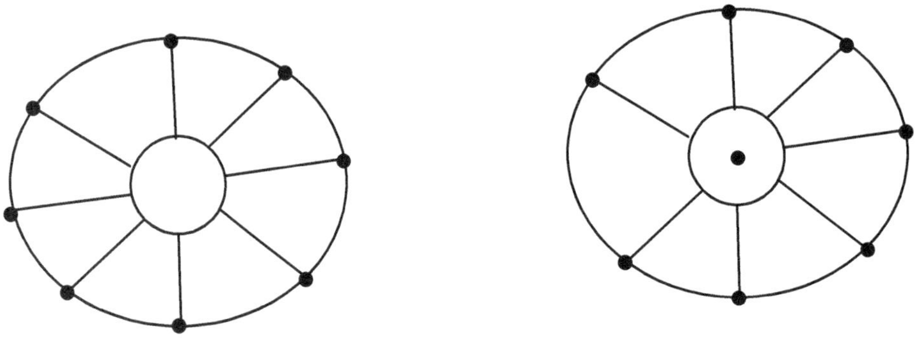

Predicted Observed

F-8 Skeletal sketch of $Au_8L_6^{2+}$

G-8 $Au_8L_8Cl_2^{2+}$ L_2 = dpppe

8(3.5) -8 -1 +1	$\longrightarrow$	k = 20	, n = 8	k(n) = 20(8)

k = 2n-q/2, q/2 = 2n-k =2(8)-20 = -4, q = -8, S = 4n-8, Cp = $C^1+C^4 = C^5C$[M-3]

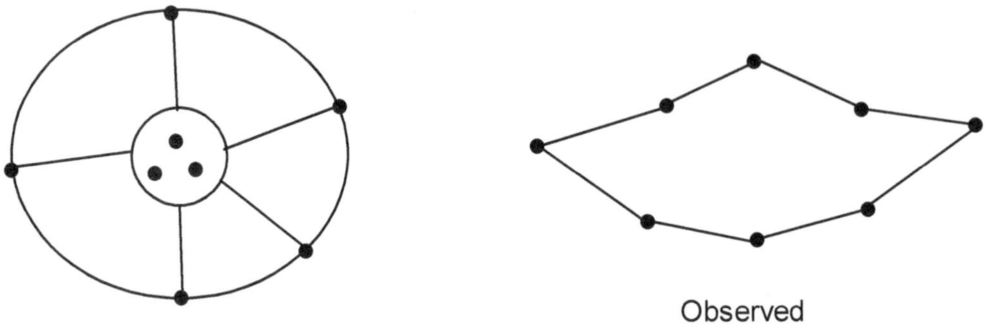

Predicted Observed

F-9 Skeletal sketch of $Au_8L_6Cl_2^{2+}$

G-9 $Au_4(I)_2L_4$, L = PPh_3

k

4(3.5)
-1
-4
→ = 9

k(n) = 9(4), S = 4n-2, Cp = $C^2C[M-2]$

$C^2C[M-2]$

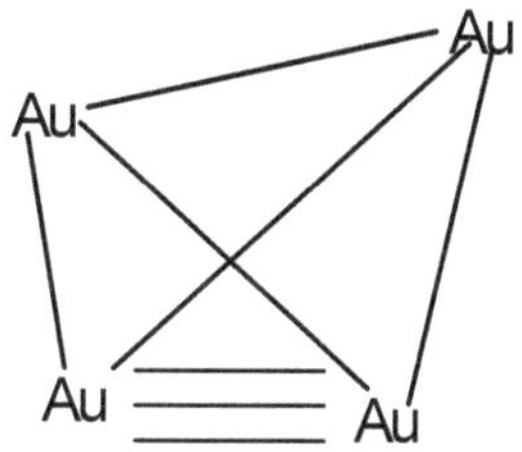

k = 9

F-10

G-10 $Au_6L_6^{2+}$

k = 6(3.5)-6+1 = 16, k(n) = 16(6), S = 4n-8, Cp = $C^5C[M-1]$

V = 14n-8 = 14(6)-8 = 76

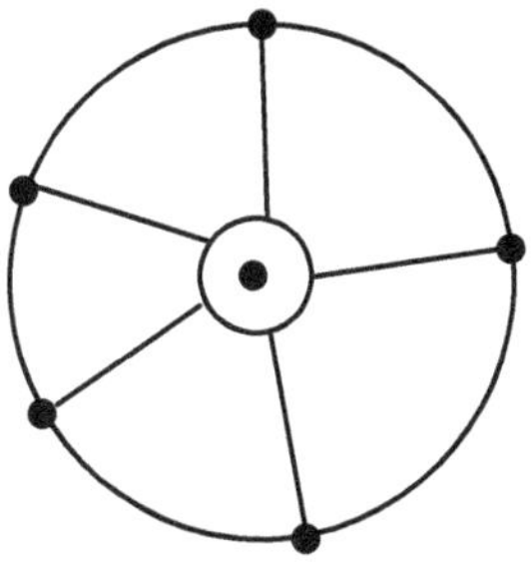

$C^5C[M-1]$

F-11 Rough sketch of predicted structure of G-10

G-11 $Au_7L_7^+$ $V = 7x11+7x2-1 = 90$

$k = 7(3.5)-7+0.5 = 18$, $k(n) = 18(7)$, $S = 4n-8$, $C^5C[M-2]$ $V = 14n-8 = 14(7)-8 = 90$

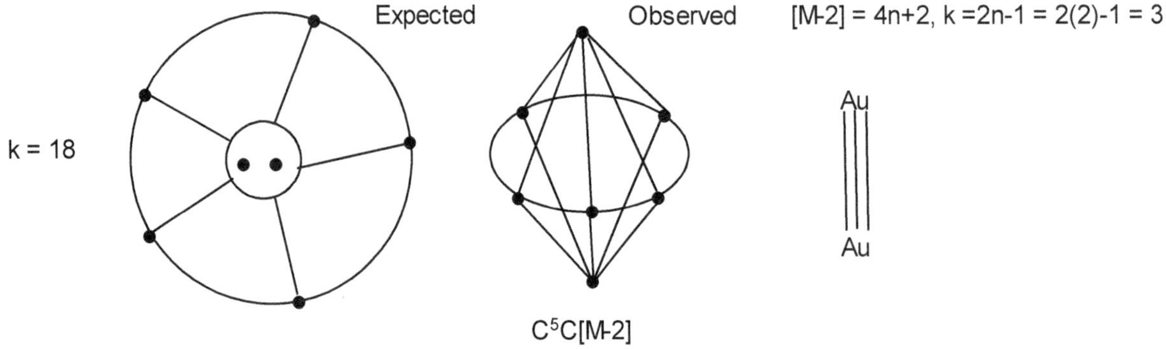

F-12 Sketch of G-11

G-12 $Au_8L_7^{2+}$

$L = PPh_3$

$k = 8(3.5)-7+1 = 22$, $k(n) = 22(8)$, $S = 4n-12$, $C^7C[M-1]$

$V = 14n-12 = 14(8)-12 = 100$

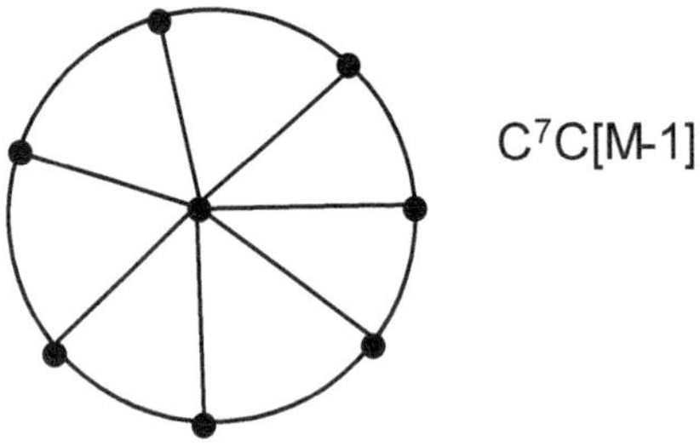

F-13 Graphical skeletal sketch representing G-12

G-13 $Au_9L_8^{3+}$

$k(n) = 25(9)$, $S = 4n-14$, $C^8C[M-1]$

$V = 14n-14 = 14(9)-14 = 112$

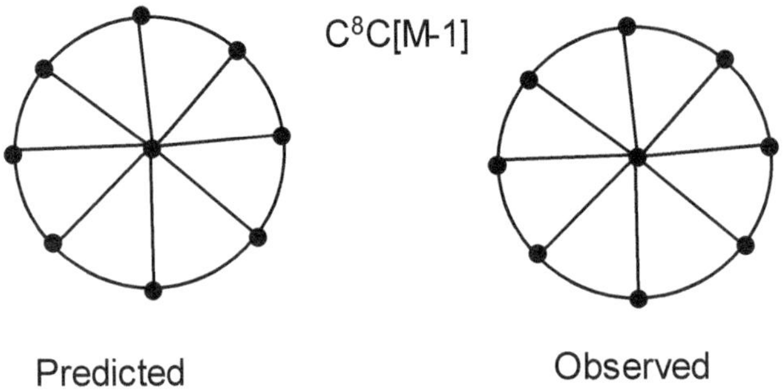

F-14 Sketch of graphical skeletal representation of G-13

G-14 $Au_{10}L_6 Cl_3^+$

k(n) = 28(10), S = 4n-16, $C^9C[M-1]$

V = 14n-16 = 14(10)-16 = 124

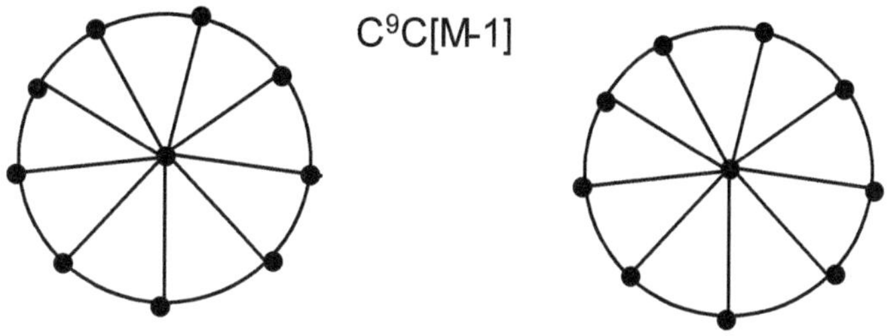

Predicted Observed

F-15 Sketch of graphical skeletal representation of G-14

G-15 $Au_8L_8^{2+}$

k(n) = 21(8), S = 4n-10, $C^6C[M-2]$

V = 14n-10 = 14(8)-10 = 102

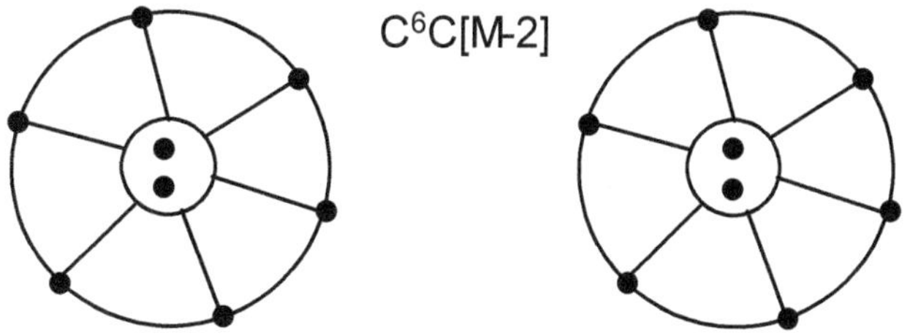

Predicted Observed

F-16 Sketch of graphical skeletal representation of G-15

G-16 $Au_{11}L_7I_3$

$k(n) = 30(11)$, $S = 4n-16$, $C^9C[M-2]$

$V = 14n-16 = 14(11)-16 = 138$

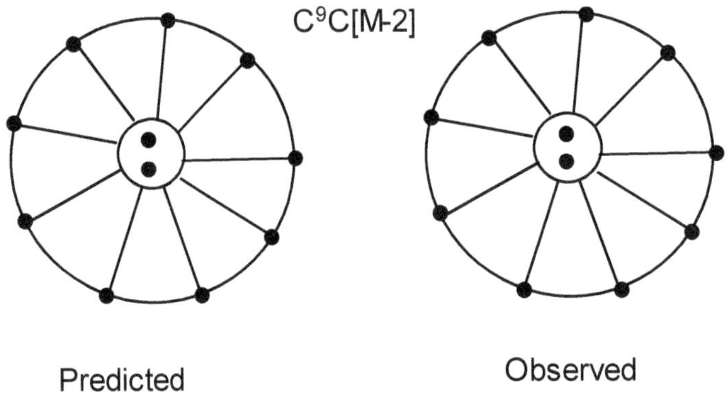

F-17 Sketch of graphical skeletal representation of G-16

G-17 $Au_{13}L_{10}Cl_2^{3+}$

$k(n) = 36(13)$, $S = 4n-20$, $C^{11}C[M-2]$

$V = 14n-20 = 14(13)-20 = 162$

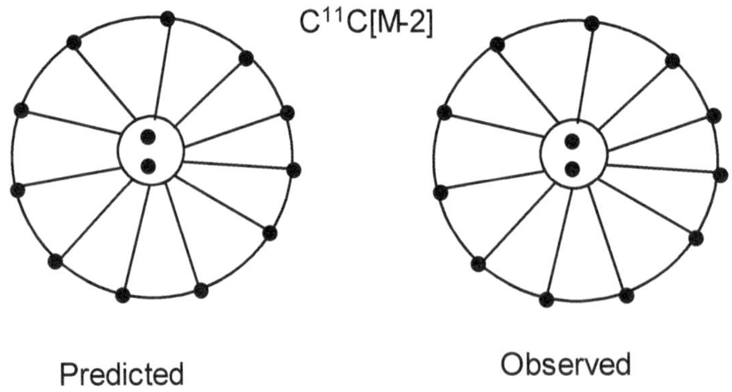

F-18 Sketch of graphical skeletal representation of G-17

Table 4. Selected categorized golden clusters

Cluster	k value	Series	n	V	Category	Code, k(n)
$(AuR)_3(SH)^-$	4	4n+4	3	46	Arachno	4(3)
$Ru_3Cl(CO)_{10}(AuL)$	7	4n+2	4	58	Closo	7(4)
$IrRu_3(CO)_{13}(AuL)$	9	4n+2	5	72	Closo	9(5)
$FeCo_3(CO)_{12}(AuL)$	9	4n+2	5	72	Closo	9(5)
$Au_8L_6^{2+}$	23	4n-14	8	98	C^8C[M-0]	23(8)
$Au_9L_8^{3+}$	26	4n-16	9	110	C^9C[M-0]	26(9)
$Au_{11}L_{10}Cl_2^{3+}$	32	4n-20	11	134	$C^{11}C$[M-0]	32(11)
$Au_{20}L_{10}Cl_4^{2+}$	59	4n-38	20	242	$C^{20}C$[M-0]	59(20)
Au_4L_4	10	4n-4	4	52	C^3C[M-1]	10(4)
$Au_6L_6^{2+}$	16	4n-8	6	76	C^5C[M-1]	16(6)
$Au_6L_4Cl_2$	16	4n-8	6	76	C^5C[M-1]	16(6)
$Au_8L_7^{2+}$	22	4n-12	8	100	C^7C[M-1]	22(8)
$Au_8L_6(I)^+$	22	4n-12	8	100	C^7C[M-1]	22(8)
$Au_9L_8^{3+}$	25	4n-14	9	112	C^8C[M-1]	25(9)
$Au_9L_5R_3$	25	4n-14	9	112	C^8C[M-1]	25(9)
$Pd(AuL)_8^{2+}$	25	4n-14	9	112	C^8C[M-1]	25(9)
$Pt(AuL)_8^{2+}$	25	4n-14	9	112	C^8C[M-1]	25(9)
$Au_{10}L_6Cl_3^+$	28	4n-16	10	124	C^9C[M-1]	28(10)
$Ag_{10}L_8^{2+}$	28	4n-16	10	124	C^9C[M-1]	28(10)
$Au_{11}L_{10}^{5+}$	31	4n-18	11	136	$C^{10}C$[M-1]	31(11)
$Au_{11}L_{10}^{5+}$	31	4n-18	11	136	$C^{10}C$[M-1]	31(11)

Table 4 continued

Cluster	k value	Series	N	V	Category	Code, k(n)
$Au_4L_4I_2$	9	4n-2	4	54	C^2C[M-2]	9(4)
$Au_7L_7^+$	18	4n-8	7	90	C^5C[M-2]	18(7)
$Au_8L_8^{2+}$	21	4n-10	8	102	C^6C[M-2]	21(8)
$Pd(CO)(AuL)_8^{2+}$	24	4n-12	9	114	C^7C[M-2]	24(9)
$Pt(CO)(AuL)_8^{2+}$	24	4n-12	9	114	C^7C[M-2]	24(9)
$Au_9L_8^+$	24	4n-12	9	114	C^7C[M-2]	24(9)
$Au_{10}L_8Cl^+$	27	4n-14	10	126	C^8C[M-2]	27(10)
$Au_{11}L_7Cl_3$	30	4n-16	11	138	C^9C[M-2]	30(11)
$Au_{11}L_8Cl_2^+$	30	4n-16	11	138	C^9C[M-2]	30(11)
$Au_{11}L_{10}^{3+}$	30	4n-16	11	138	C^9C[M-2]	30(11)
$Au_{11}L_8Cl_2^+$	30	4n-16	11	138	C^9C[M-2]	30(11)
$Au_{11}L_{10}^{3+}$	30	4n-16	11	138	C^9C[M-2]	30(11)
$Au_{11}L_7X_3$	30	4n-16	11	138	C^9C[M-2]	30(11)
$Au_{13}L_{10}Cl_2^{3+}$	36	4n-20	13	162	$C^{11}C$[M-2]	36(13)
$Au_{13}L_8Cl_4^+$	36	4n-20	13	162	$C^{11}C$[M-2]	36(13)
$Au_6L_8^{2+}$	14	4n-4	6	80	C^3C[M-3]	14(6)
$Au_8L_8Cl_2^{2+}$	20	4n-8	8	104	C^5C[M-3]	20(8)
$Pd(AuL)_8^{2+}$	23	4n-10	9	116	C^6C[M-3]	23(9)
$Pt(AuL)_8^{2+}$	23	4n-10	9	116	C^6C[M-3]	23(9)
$Au_{13}L_{10}Cl_2^+$	35	4n-18	13	164	$C^{10}C$[M-3]	35(13)
$Au_{11}L_{12}^{3+}$	28	4n-12	11	142	C^7C[M-4]	28(11)
$Pd(AuL)_8(AuCl)_4$	34	4n-16	13	166	C^9C[M-4]	34(13)
$Au_{38}(SR)_{24}$	109	4n-66	38	466	$C^{34}C$[M-4]	109(38)

Table 4 continued

Cluster	k value	Series	n	V	Category	Code
$Au_6L_8^{2+}$	12	4n+0	6	84	C^1C[M-5]	12(6)
$HIrRu_3(CO)_{12}(AuL)_2$	12	4n+0	6	84	C^1C[M-5]	12(6)
$IrRu_3(CO)_{12}(AuL)_3$	15	4n-2	7	96	C^2C[M-5]	15(7)
$Au_{25}(SR)_{18}^-$	69	4n-38	25	312	$C^{20}C$[M-5]	69(25)
$Ru_6(C)(CO)_{16}(AuL)_2$	17	4n-2	8	110	C^2C[M-6]	17(8)
$Os_{10}(C)(CO)_{24}(AuL)^-$	26	4n-8	11	146	C^5C[M-6]	26(11)
$Au_{20}L_{10}Cl_4^{3+}$	49	4n-18	20	262	$C^{10}C$[M-10]	49(20)
$Au_{20}L_9Cl_4^{2+}$	49	4n-18	20	262	$C^{10}C$[M-10]	49(20)
$Au_{22}L_{12}$	55	4n-22	22	286	$C^{12}C$[M-10]	55(22)

Table 5. The K(n) Map of Cluster Series

4n-44	**68(23)**	**70(24)**	**72(25)**	74(26)	76(27)	78(28)	80(29)	82(30)	84(31)	86(32)	C^{23}
4n-42	**65(22)**	**67(23)**	**69(24)**	71(25)	73(26)	75(27)	77(28)	79(29)	81(30)	83(31)	C^{22}
4n-40	**62(21)**	**64(22)**	**66(23)**	68(24)	70(25)	72(26)	74(27)	76(28)	78(29)	80(30)	C^{21}
4n-38	**59(20)**	**61(21)**	**63(22)**	65(23)	67(24)	69(25)	71(26)	73(27)	75(28)	77(29)	C^{20}
4n-36	**56(19)**	**58(20)**	**60(21)**	62(22)	64(23)	66(24)	68(25)	70(26)	72(27)	74(28)	C^{19}
4n-34	**53(18)**	**55(19)**	**57(20)**	59(21)	61(22)	63(23)	65(24)	67(25)	69(26)	71(27)	C^{18}
4n-32	**50(17)**	**52(18)**	**54(19)**	56(20)	58(21)	60(22)	62(23)	64(24)	66(25)	68(26)	C^{17}
4n-30	**47(16)**	**49(17)**	**51(18)**	53(19)	55(20)	57(21)	59(22)	61(23)	63(24)	65(25)	C^{16}
4n-28	**44(15)**	**46(16)**	**48(17)**	50(18)	52(19)	54(20)	56(21)	58(22)	60(23)	62(24)	C^{15}
4n-26	**41(14)**	**43(15)**	**45(16)**	47(17)	49(18)	51(19)	53(20)	55(21)	57(22)	59(23)	C^{14}
4n-24	**38(13)**	**40(14)**	**42(15)**	44(16)	46(17)	48(18)	50(19)	52(20)	54(21)	56(22)	C^{13}
4n-22	**35(12)**	**37(13)**	**39(14)**	41(15)	43(16)	45(17)	47(18)	49(19)	51(20)	53(21)	C^{12}
4n-20	**32(11)**	**34(12)**	**36(13)**	38(14)	40(15)	42(16)	44(17)	46(18)	48(19)	50(20)	C^{11}
4n-18	**29(10)**	**31(11)**	**33(12)**	35(13)	37(14)	39(15)	41(16)	43(17)	45(18)	47(19)	C^{10}
4n-16	**26(9)**	**28(10)**	**30(11)**	32(12)	34(13)	36(14)	38(15)	40(16)	42(17)	44(18)	C^{9}
4n-14	**23(8)**	**25(9)**	**27(10)**	29(11)	31(12)	33(13)	35(14)	37(15)	39(16)	41(17)	C^{8}
4n-12	**20(7)**	**22(8)**	**24(9)**	26(10)	28(11)	30(12)	32(13)	34(14)	36(15)	38(16)	C^{7}
4n-10	**17(6)**	**19(7)**	**21(8)**	23(9)	25(10)	27(11)	29(12)	31(13)	33(14)	35(15)	C^{6}
4n-8	**14(5)**	**16(6)**	**18(7)**	20(8)	22(9)	24(10)	26(11)	28(12)	30(13)	32(14)	C^{5}
4n-6	**11(4)**	**13(5)**	**15(6)**	17(7)	19(8)	21(9)	23(10)	25(11)	27(12)	29(13)	C^{4}
4n-4	**8(3)**	**10(4)**	**12(5)**	14(6)	16(7)	18(8)	20(9)	22(10)	24(11)	26(12)	C^{3}
4n-2	**5(2)**	**7(3)**	**9(4)**	11(5)	13(6)	15(7)	17(8)	19(9)	21(10)	23(11)	C^{2}
4n+0	**2(1)**	**4(2)**	**6(3)**	8(4)	10(5)	12(6)	14(7)	16(8)	18(9)	20(10)	C^{1}
4n+2	[-1](0)	1(1)	3(2)	5(3)	7(4)	**9(5)**	**11(6)**	13(7)	15(8)	17(9)	closo
[M-x]	**M-0**	**M-1**	**M-2**	**M-3**	**M-4**	**M-5**	**M-6**	**M-7**	**M-8**	**M-9**	4n+2
4n+4			0(1)	2(2)	4(3)	**6(4)**	**8(5)**	10(6)	12(7)	14(8)	Nido
4n+6	*	*	*		1(2)	**3(3)**	**5(4)**	7(5)	9(6)	11(7)	Arac

2.3 Construction of Table 5

In Scheme 1, the code k(n) =11(6) was used as a unifying point for all clusters with the ideal octahedral symmetry. Taking this code as an example, k = 11 refers to the skeletal linkages the cluster has and n = 6 refers to the number of skeletal atoms. Numerous cluster complexes with the ideal octahedral could be cited which fulfill the requirement k(n) = 11(6). Thus this code unites all the family members which possess an octahedral shape. However, some cases with k(n) = 11(6) may not have an ideal octahedral shape. Such shapes may be considered to portray some form of "k-isomerism. From the series we know that a capping step involves a change $\Delta n = 1$ and $\Delta k = 2n+1 = 2(1)+1 = 3$, then we can easily construct a column of capping series in terms of the cluster code k(n) for an ideal octahedral system as follows: 11(6)→14(7) →17(8) →20(9) →23(10) →26(11) →29(12), and so on. Starting from 11(6), we can extend the column downwards as follows: 11(6) →8(5) →5(4) →2(3) →-1(2). The k(n) = -1(2) makes little chemical sense and therefore it is not necessary to go beyond this point. In earlier work (Kiremire, 2016c), the capping osmium carbonyl fragment was found to be $Os(CO)_3$, S = 14n-2, k = 2n+1=3, V = 14 and it involves changing the type of series. However the horizontal movement along the row the type of series remain the same and $\Delta n = \pm 1$ and $k = \pm 2$. Taking the k(n) = 11(6) as an illustration, and moving horizontally along the row, from left to right, we get the series 11(6)→13(7) →15(8) →17(9) →19(10) and so on. Starting from 11(6) and moving from right to left we get the CLOSO series 11(6)→9(5) →7(4) →5(3) →3(2) →1(1) →-1(0). In principle, starting from one k(n) code, an infinite array of k(n) values can be constructed. Table 5 is extremely important and may be regarded as a "map" of clusters. Let us consider the following cluster codes taken from Table -5**;** 10(4), 16(6), 22(8), 25(9), 28(10) and 31(11). All these code lie on [M-1] axis while the codes 9(4), 18(7), 21(8), 24(9), 27(10), 30(11) and 36(13) lie on [M-2] axis. Clearly, all these golden cluster codes are located on the far left of the cluster code map(Table 5). The K(n) parameter in table 5 means that (k)stands for cluster linkages while the(n) in parenthesis stands for the number of the skeletal elements corresponding to that k value. The middle region of the cluster map and the extreme right of the map is the area occupied mainly by the transition metal clusters. The clusters such as boranes, hydrocarbons, and main group elements clusters lie bellow the CLOSO AXIS and are on the right portion of the cluster code map. This is summarized in Table-6. Thus, Table 6 is a derivative of Table 5. The significance of the table is that it shows that the clusters of the 'lower' series such as boranes and hydrocarbons (nido, arachno, klapo) are rare or non-existent on the extreme left side of the table. Only one possible member, K(n) = 2(2) appears at [M-3] column(see Table 5). The closo nuclear baseline [M-x] is shown in red while its corresponding K(n) values along S = 4n+2 are shown in blue color above it. Also shown in blue below the [M-x] axis are the K(n) values corresponding to 4n+4 and 4n+6 series. The number of lower series clusters increases as the nuclear [M-x] index (x) increases from 4 to values of x>4. In addition, a capping code, C^nC[M-x] may be expressed as point on the vertical axis as illustrated in Table 7. The degree of cluster capping increases as you move from right to the left in Table 5. As can be seen from Table 2, the 'periodic law' of the skeletal numbers decreases from left to right starting with k value of 7.5 to k value of 3. Gold belongs to a period with a k value of 3.5. It appears that the lower the k value of the skeletal element, the greater the tendency of the element form capping clusters with low nuclear indices. This is probably the reason why Au forms capping series with [M-1] and [M-2] nuclear indices. If in the cluster code K(n), if we keep n constant and move along a diagonal from the right to the left the K value increases stepwise by one. This means that capping becomes more and more intense as we move along a diagonal from left to right in Table 5. Take the example K(n) = 9(6) in Table 5. This code represents clusters such as C_6H_6, which has isomers such as benzene and prismane or $Co_6(P)(CO)_{16}^-$. The diagonal movement left-wise, we get the sequences of cluster codes as follows; 9(6)→10(6) →11(6)[octahedral clusters) →12(6)[mono-capped trigonal bipyramid→13(6) →14(6) →15(6) →16(6). When we reach at K(n) = 16(6), we encounter clusters such as $Au_6L_6^{2+}$, L = PPh_3. At K(n)= 9(6), q/2 = 2(6)-9 = 3, q = 6, S =4n+6 (ARACHNO family)[Trigonal Prism skeletal shape] while at K(n) = 16(6), q/2 = 2(6)-16 = -4, q = -8, S = 4n-8, C^5C[M-1][Penta-capped mono-nuclear cluster]. This is a major transformation in symmetry. That is why the skeletal shape of $Co_6(P)(CO)_{16}^-$ (K = 9) is radically different from that of $Au_6(PPh_3)_6^{2+}$(K =16).

2.4 Geometries and Bonding in Golden Clusters

Arising from the complexity of the capping nature of gold clusters, their geometries and nature of bonding are expected to be complex. This will be illustrated by considering a few examples. Consider the cluster $Au_4(I)_2(L)_4$, L =PPh_3 is described as having a tetrahedral geometry(Mingos, 1984). The cluster has a k value of 9. This means it embodies 9 skeletal linkages within 4 skeletal elements, with a code k(n) = 9(4). The cluster belongs to S = 4n-2 series with the capping Cp = C^2C [M-2]. According to the 4n series, this means the cluster has two skeletal elements which are capping on other 2 atoms which are linked by a triple bond as indicated in F-10. On the other hand a transition metal carbonyl cluster such as $Ir_4(CO)_{12}$, k = 6, code k(n) = 6(4), S = 4n+4. This means it belongs to the nido series. This cluster has 4 skeletal elements bound by 6 linkages as shown in F-2. The skeletal shape is expected to have a genuine tetrahedral symmetry. Clearly, there is a fundamental difference between k(n) = 9(4) and k(n) =6(4). The difference can be traced on the k(n) diagonal:6(4)→7(4) →8(4) →9(4). The 6(4) lies in the nido region while 9(4) lies in the bi-capped region close to naked metal cluster region. Another good example is $Au_6L_6^{2+}$(L = PPh_3). This cluster is described as having an octahedral skeletal geometry (Cotton, et al, 1980; Mingos, 1984). But the cluster has a k value of 16, k(n) = 16(6) and belongs to S = 4n-8 series with Cp = C^5C[M-1]. The k =16 can readily be calculated from the cluster formula as follows: k =6(3.5)-6+1 = 16 and the series formula derived as explained in examples G-1 to G-8. As we have seen, an ordinary octahedral shape has k(n) = 11(6) code, and S = 4n+2, closo. Let us trace the changes in series from 11(6) to 16(6). The series are 11(6) →4n+2, 12(6) →4n+0, 13(6) →4n-2, 14(6) →4n-4, 15(6) →4n-6, and 16(6) →4n-8, Cp = C^5C[M-1]. This means the cluster is expected to have one nuclear element with a cluster content k = 1 surrounded by 5 capped skeletal elements with 15 linkages. These changes can be traced diagonally in Table 4A. In short, the gold cluster k(n) = 16(6) of 6 skeletal elements and 16 linkages is expected to have a much more complex structure than, k(n) = 11(6) with 6 skeletal atoms and 11 linkages. Following the procedure adopted in sketching the sketch F-3 adopted from the literature (Gimeno, 2008), we can produce a possible tentative shape of the yellow cluster (Cotton, et al, 1980) $Au_6L_6^{2+}$ as indicated in F-z guided by k values from the series. The skeletal atom around which other 5 elements are capped has been enlarged for emphasis. In three dimensional set up, the shape gives a semblance of an octahedral geometry.

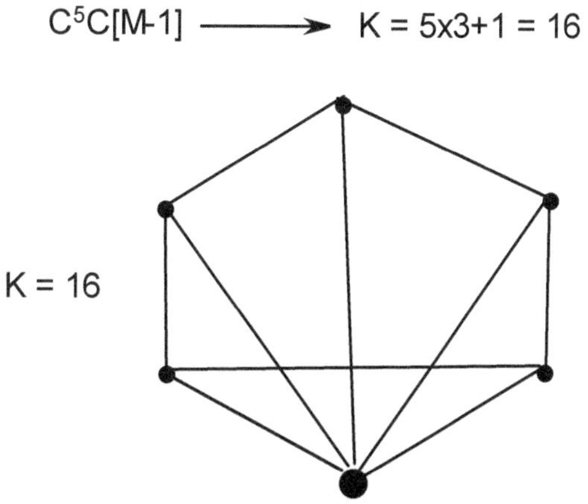

F-z Tentative possible sketch of $Au_6L_6^{2+}$

Consider another good example, $Au_7L_7^+$; k(n) = 18(7), S =4n-8, Cp = C^5C[M-2]. If we assume that the 5 skeletal atoms are capped around two others joined by a triple bond, then we can represent this by a proposed possible skeletal sketch shown in F-19.

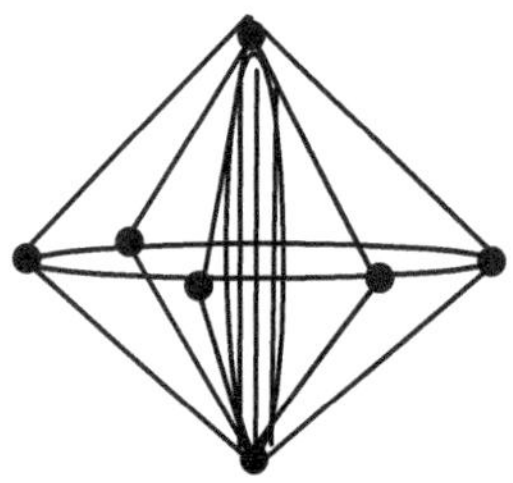

F-19

Surprisingly, this geometrical model was found to be in agreement with that of Pivoriūnas (Pivoriūnas, 2005). The cluster has been described as pentagonal bipyramid (Mingos, 1984). According to the 4n series, it has two atoms linked by a triple bond around which 5 other skeletal elements are joined by 15 linkages. An ordinary pentagonal bipyramid cluster such as $B_7H_7^{2-}$, k(n) = 13(7), has seven skeletal atoms with a content of 13 linkages as opposed to 18 linkages for the 7 golden skeletal atoms.

Let us take one more example $Au_{10}Cl_3L_6^+$. This cluster has a cluster code k(n) = 28(10) with S = 4n-16, V = 14n-16 = 124 and Cp = C^9C[M-1]. According to 4n series, this cluster has one skeletal atom at the nucleus around which 9 other skeletal elements are capped with a total of 28 skeletal linkages. Indeed this is what is observed (Mingos, 1984; Konishi, 2014). A rough skeletal sketch of this is shown in F-20. Mingos describes clusters of this nature as follows "The higher nuclearity clusters are characterized by the presence of an additional central gold atom and the presence of this atom makes a fundamental contribution to the bonding because the radial metal-metal bonding to the central metal atom is stronger than the peripheral metal-metal bonding on the surface of the cluster" (Mingos, 1984). This powerful statement underpins the power of series method.

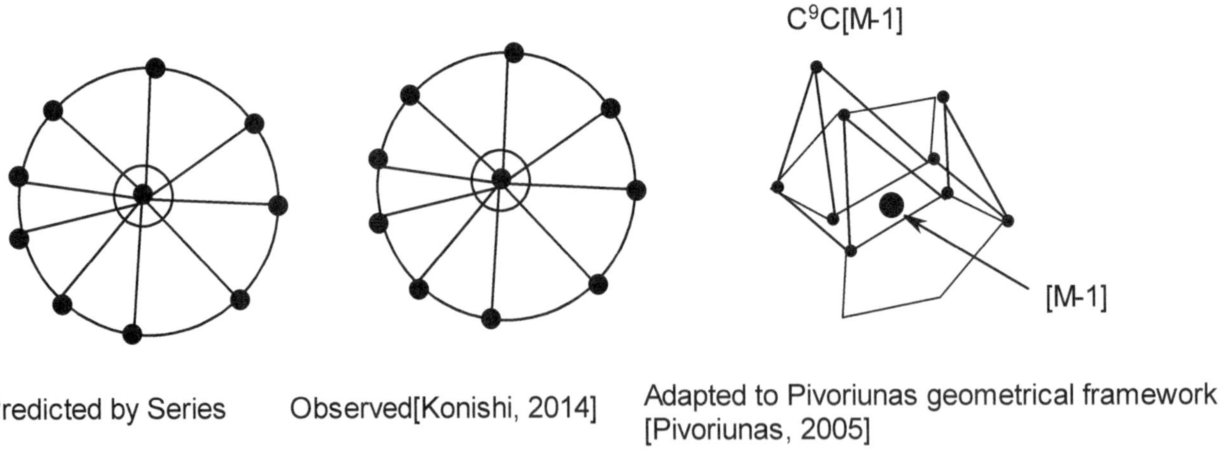

F-20. Graphical sketches of $Au_{10}Cl_3L_6^+$

2.5 Observing the Nuclear Centers of Clusters.

The capping symbol that was introduced in earlier work is C^nC[M-x]. The [M-x] is the cluster nucleus. The value of x = 1, 2, 3, 4, 5, 6, 7, 8, 9, 10 and so on represent the number of skeletal elements present in the nucleus. What has been witnessed in the gold clusters are some of the clusters of the type C^nC[M-1] given the name single atom-centered toroidal clusters. For giant clusters, nuclei such as C^nC[M-x](x>6) have been encountered. It will be interesting if the atoms which constitute [M-x] nuclei could easily be identified by x-ray structural analysis as predicted by the series approach.

Table 6. Sketch to show the relative positions of clusters on the K(n) map

	10	
	9	
NAKED METAL	8	
	7	
GOLD CLUSTERS(x = -4→0)	6	GOLD CLUSTERS(x = 0→4)
	5	
	4	
	3	
	2	TRANSITION METAL CARBONYLS(x = 0→ ABOVE)
	1	METALLOBORANES(x = 0 →ABOVE)
[M-x] -5 -4 -3 -2 -1	0	1 2 3 4 5 6 7 8 9 10 11 12
CLOSO, NIDO,ARACHNO, KLAPO		CLOSO BORANES(x = 1→ABOVE)
NON-EXISTENT		NIDO CARBORANES(x = 1→ABOVE)
		ARACHNO METALLOBORANES(x = 1→ABOVE)
		HYPHO METALLACARBORANES(x = 1→ABOVE)
		KLAPO ZINTL IONS(x = 1→ABOVE)
		HYDROCARBONS(x = 1→ABOVE)
		TRANSITION METAL CARBONYLS(x = 1→ABOVE)
		MAIN GROUP ELEMENT CLUSTERS(x = 1→ABOVE)
		NON-CAPPING REGION

Table 7. Geometrical representation of capping cluster on a given [M-x] column

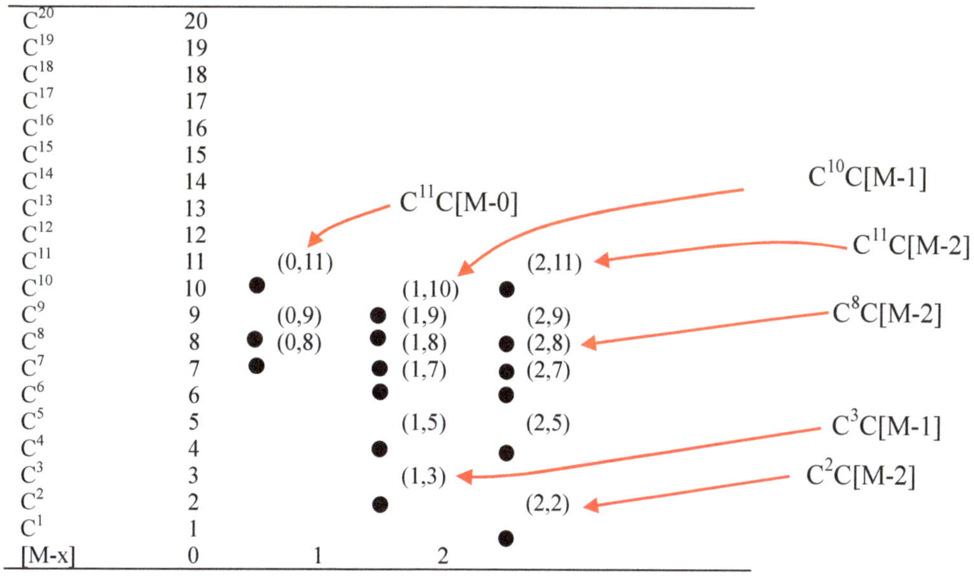

Table 8. Comparing series results of a selected sample of golden clusters with literature information

T-8	COMPARE									
Cluster	k value	Series	n	V series	Category Series	Code, k(n)	Lit.	Lit.	V Lit.	Ref.
$Au_4L_4I_2$	9	4n-4	4	54	$C^2C[M-2]$	9(4)	tetrahedral		54	1
$Au_6L_6^{2+}$	16	4n-8	6	76	$C^5C[M-1]$	16(6)	octahedral		76	1,2,3
$Au_7L_7^{+}$	18	4n-8	7	90	$C^5C[M-2]$	18(7)	Pentagonal bipyramid		88	1
$Au_6L_8^{2+}$	14	4n-4	6	80	$C^3C[M-3]$	14(6)	Di-edge-bridged T_d		80	1
$Au_8L_7^{2+}$	22	4n-12	8	100	$C^7C[M-1]$	22(8)	Icosahedron	Centered TOR	100	1
$Au_9R_3L_5$	25	4n-14	9	112	$C^8C[M-1]$	25(9)		Centered TOR		
$Au_9L_8^{3+}$	25	4n-14	9	112	$C^9C[M-1]$	25(9)	Centered crown	Centered TOR	112	1,2,3
$Au_8L_8^{2+}$	21	4n-10	8	102	$C^6C[M-2]$	21(8)	Capped centered chair	Centered SP	102	1
$Au_9L_8^{+}$	24	4n-12	9	114	$C^7C[M-2]$	24(9)	Bi-capped centered chair		114	1
$Au_{10}L_6Cl_3^{+}$	28	4n-16	10	124	$C^9C[M-1]$	28(10)		Centered TOR		1
$Au_{11}L_7X_3$	30	4n-16	11	138	$C^9C[M-2]$	30(11)	Dodecahedron	Centered SP	128	1,2,3
$Au_{13}L_{10}Cl_2^{3+}$	36	4n-20	13	162	$C^{11}C[M-2]$	36(13)	Icosahedron	Centered SP	152	1

TOR = TORROIDAL, SP = SPHERICAL, 1 = Mingos, 1984; 2 =Pauling,1977; 3 =Cotton, et al, 1980

3. Conclusion

Elemental gold forms clusters which are intensely capped. According to series approach, they tend to occur in a region overlapping with the area where naked metallic clusters are normally found. Although a good range of clusters have nuclei of the type [M-x], x = 1 or 2 (centered toroidal or centered spherical), x can take up values, 3, 4, 5, 6 and so on. But the majority of clusters are congested around x = 0-4 region. The series are quite reliable in analyzing and categorizing a wide range of clusters. A lot more needs to be done to fully understand the unique shapes and bonding in the golden clusters which portray complex capping even for a small number of skeletal atoms. Indeed, gold cluster chemistry is a rich mine of knowledge, yet much more waiting to be mined.

Acknowledgement

The financial assistance by the University of Namibia and NAMSOV, Namibia is greatly appreciated. Mrs Merab Kambamu Kiremire for proof-reading the draft is hereby acknowledged.

Dedication

This work is dedicated to two of the great historical scientists, namely Irving Langmuir and Gilbert Newton Lewis for their brilliant insights and contributions towards the 18 and 8 electron rules more than 90 years ago.

References

Belyakova, O. A., & Slovokhotova, Y. L. (2003). Structures of large transition metal clusters. Russian Chemical Bulletin. Inter. Ed., *52*(11), 1-29. https://doi.org/10.1023/B:RUCB.0000012351.07223.d4

Cotton, F. A., & Wilkinson, G. (1980). *Advanced Inorganic Chemistry, 4th Ed.*, John Wiley and Sons, New York, 1980.

Daniel, M. C., & Astruc, D. (2004). Gold Nanoparticles, Assembly, Supramolecular Chemistry, Quantum-Size-Related Properties, and Applications toward biology, Catalysis and Nanotechnology. *Chem. Rev., 104*, 293-346. https://doi.org/10.1021/cr030698+

Gimeno, M. C. (2008). Modern Supramolecular Gold-Metal Interactive and Applications. Edited A. Laguna, 2008. Wiley-VCH, Weinheim.

Greenwood, N. N., & Earnshaw, A. (1998). *Chemistry of the Elements, 2nd Ed.* Butterworth, Oxford.

Kiremire, E. M. R. (2016a). The Application of the 4n Series Method to Categorize Metalloboranes. *Int. J. Chem., 8*(3), 62-73. https://doi.org/10.5539/ijc.v8n3p62

Kiremire, E. M. R. (2016b). A Hypothetical Model for the Formation of Transition metal Carbonyl Clusters Based Upon 4n Series Skeletal Numbers. *Int. J. Chem., 8*(4), 78-110. https://doi.org/10.5539/ijc.v8n4p78

Kiremire, E. M. R. (2016c). Unusual underground capping Carbonyl Clusters of Palladium. *Int. J. Chem., 8*(1), 145-158. https://doi.org/10.5539/ijc.v8n1p145

Konishi, K. (2014). Phosphine-Coordinated Pure-Gold Clusters and Unique Optical Properties /Responses. Structure and Bonding, *161*, 49-86. https://doi.org/10.1007/430_2014_143

Mingos, D. M. P. (1972). A General Theory for Cluster and Ring Compounds of the Main Group and Transition Elements. *Nature (London), Phys. Sci., 236*, 99-102. https://doi.org/10.1038/physci236099a0

Mingos, D. M. P. (1984). Gold Cluster Compounds: Are they materials in miniature? *Gold Bull.*, *17*(1), 5-12. https://doi.org/10.1007/bf03214670

Mingos, D. M. P. (1987). Complementary spherical electron density model for coordination compounds. *Pure and Appl. Chem.*, *59*(2), 145-154. https://doi.org/10.1351/pac198759020145

Pauling, L. (1977). Structure of Transition Metal Cluster Compounds: Use of an additional orbital resulting from f, g character of SPD bond orbitals. *Proc. Natl. Acad. Sci., USA, 74*(12), 5235-5238. https://doi.org/10.1073/pnas.74.12.5235

Pivoriunas, G. (2005). Neuer Goldcluster. PhD Thesis

Rossi, F., & Zanello, P. (2011). Electron Reservoir Activity of High-Nuclearity Transition Metal Carbonyl Clusters. *Portugalia, Electrochimica Acta, 29*(1), 309-327. https://doi.org/10.4152/pea.201105309

Tiwari, P. M., Vig, K., Dennis, A., & Sing, S. R. (2011). Functionalized Gold Nanoparticles and Their Biomedical Applications. Nanomaterials, *1*, 31-63. https://doi.org/10.3390/nano1010031

Wade, K. (1971). The structural significance of the number of skeletal bonding electron-pairs in carboranes, the higher boranes and borane ions and various transition metal carbonyl cluster compounds. *Chem. Commun.,* 792-793. https://doi.org/10.1039/c29710000792

Welch, A. J. (2013). The significance of Wade's rules. *Chem. Commun., 49*, 3615-3616. https://doi.org/10.1039/c3cc00069a

Permissions

All chapters in this book were first published in IJC, by Canadian Center of Science and Education; hereby published with permission under the Creative Commons Attribution License or equivalent. Every chapter published in this book has been scrutinized by our experts. Their significance has been extensively debated. The topics covered herein carry significant findings which will fuel the growth of the discipline. They may even be implemented as practical applications or may be referred to as a beginning point for another development.

The contributors of this book come from diverse backgrounds, making this book a truly international effort. This book will bring forth new frontiers with its revolutionizing research information and detailed analysis of the nascent developments around the world.

We would like to thank all the contributing authors for lending their expertise to make the book truly unique. They have played a crucial role in the development of this book. Without their invaluable contributions this book wouldn't have been possible. They have made vital efforts to compile up to date information on the varied aspects of this subject to make this book a valuable addition to the collection of many professionals and students.

This book was conceptualized with the vision of imparting up-to-date information and advanced data in this field. To ensure the same, a matchless editorial board was set up. Every individual on the board went through rigorous rounds of assessment to prove their worth. After which they invested a large part of their time researching and compiling the most relevant data for our readers.

The editorial board has been involved in producing this book since its inception. They have spent rigorous hours researching and exploring the diverse topics which have resulted in the successful publishing of this book. They have passed on their knowledge of decades through this book. To expedite this challenging task, the publisher supported the team at every step. A small team of assistant editors was also appointed to further simplify the editing procedure and attain best results for the readers.

Apart from the editorial board, the designing team has also invested a significant amount of their time in understanding the subject and creating the most relevant covers. They scrutinized every image to scout for the most suitable representation of the subject and create an appropriate cover for the book.

The publishing team has been an ardent support to the editorial, designing and production team. Their endless efforts to recruit the best for this project, has resulted in the accomplishment of this book. They are a veteran in the field of academics and their pool of knowledge is as vast as their experience in printing. Their expertise and guidance has proved useful at every step. Their uncompromising quality standards have made this book an exceptional effort. Their encouragement from time to time has been an inspiration for everyone.

The publisher and the editorial board hope that this book will prove to be a valuable piece of knowledge for researchers, students, practitioners and scholars across the globe.

List of Contributors

Enos Masheija Rwantale Kiremire
Department of Chemistry and Biochemistry,University of Namibia, Private Bag 13301, Windhoek

Naofumi Naga and Yoshie Saito
Department of Applied Chemistry, Materials Science Course, College of Engineering, Shibaura Institute of Technology, 3-7-5 Toyosu, Koto-ku, Tokyo 135-8548, Japan

James O. Ojo and David A. Oyegoke
Department of Chemistry, Federal University of Technology, Akure, Ondo State, Nigeria

Udari Kodithuwakku, Chathura de Alwis, M.A.B Prashantha and Dilru R. Ratnaweera
Department of Chemistry, University of Sri Jayewardenepura, Nugegoda, Sri Lanka

Amir Sadeghi, Maximilian Sieber, Ingolf Scharf and Thomas Lampke
Chemnitz University of Technology, Institute of Materials Science, Chemnitz, Germany

Hosein Hasannejad
Arak University, Department of of Materials Engineering, Iran

Guan Wang, Dongsheng Chen, Yuanyuan Sun, Qing Cao, Bonan Li and Jianqi Li
Novel Technology Center of Pharmaceutical Chemistry, Shanghai Institute of Pharmaceutical Industry, Shanghai , P. R. China

Deming Wang and Zexi Wang
Department of Equipment Engineering, Chongqing Real Estate College, Chongqing401331, China

Ying Peng
Library, Chongqing Electric Power College, Chongqing400053, China

ugusto Rivera, Ingrid Miranda-Carvajal and Jaime Ríos-Motta
Universidad Nacional de Colombia, Sede Bogotá, Facultad de Ciencias, Departamento de Química, Carrera 30 No. 45-03, Bogotá, Código Postal 111321, Colombia

Doris F. Ogeleka, Harrison E. Avrukpere and Elias E. Elemike
Department of Chemistry, Federal University of Petroleum Resources, Effurun, Delta State, Nigeria

Oghenekohwiroro Edjere and Iwekumo E. Agbozu
Department of Environmental Management and Toxicology, Federal University of Petroleum Resources, Effurun, Delta State, Nigeria

Majaya Rachel Delfina, Sanele Mnkandla and Norah Basopo
Departments of Applied Biology and Biochemistry, National University of Science and Technology, P.O. Box AC 939, Ascot, Bulawayo, Zimbabwe

Donald Tapfuma
Departments of Environmental Science and Health, National University of Science and Technology, P.O. Box AC 939, Ascot, Bulawayo, Zimbabwe

Sergiy G. Krasutsky and Sheila H. Jacobo
Chemical Development Department, Albany Molecular Research Inc, United States

Mohamed Guitouni
X-ray fluorescence Laboratory, National Institute of Research and Physico-Chemical Analysis, Technological Pole Sidi Thabet 2020, Aryanah, Tunisia

kbal Bouatba, M'barek Benchanaa, Mustapha El Hammioui and Latifa Bilali
Laboratoire de Recherche, Reactivite des Materiaux et Optimisation des Procedes: REMATOP, Faculte des Sciences Semlalia, Universite Cadi Ayyad, Avenue Prince Moulay Abdellah, B.P. 2390, 40000 Marrakech, Maroc

Takanori Fukami, Seiya Hiyajyo, Shuta Tahara and Chitoshi Yasuda
Department of Physics and Earth Sciences, Faculty of Science, University of the Ryukyus, Japan

Index